Springer Theses

Recognizing Outstanding Ph.D. Research

Aims and Scope

The series "Springer Theses" brings together a selection of the very best Ph.D. theses from around the world and across the physical sciences. Nominated and endorsed by two recognized specialists, each published volume has been selected for its scientific excellence and the high impact of its contents for the pertinent field of research. For greater accessibility to non-specialists, the published versions include an extended introduction, as well as a foreword by the student's supervisor explaining the special relevance of the work for the field. As a whole, the series will provide a valuable resource both for newcomers to the research fields described, and for other scientists seeking detailed background information on special questions. Finally, it provides an accredited documentation of the valuable contributions made by today's younger generation of scientists.

Theses may be nominated for publication in this series by heads of department at internationally leading universities or institutes and should fulfill all of the following criteria

- They must be written in good English.
- The topic should fall within the confines of Chemistry, Physics, Earth Sciences, Engineering and related interdisciplinary fields such as Materials, Nanoscience, Chemical Engineering, Complex Systems and Biophysics.
- The work reported in the thesis must represent a significant scientific advance.
- If the thesis includes previously published material, permission to reproduce this must be gained from the respective copyright holder (a maximum 30% of the thesis should be a verbatim reproduction from the author's previous publications).
- They must have been examined and passed during the 12 months prior to nomination.
- Each thesis should include a foreword by the supervisor outlining the significance of its content.
- The theses should have a clearly defined structure including an introduction accessible to new PhD students and scientists not expert in the relevant field.

Indexed by zbMATH.

More information about this series at https://link.springer.com/bookseries/8790

Dhruv Khandelwal

Automating Data-Driven Modelling of Dynamical Systems

An Evolutionary Computation Approach

Doctoral Thesis accepted by
Eindhoven University of Technology, Eindhoven,
The Netherlands

Springer

Author
Dr. Dhruv Khandelwal
Eindhoven University of Technology
Eindhoven, The Netherlands

Supervisors
Dr. Roland Tóth
Eindhoven University of Technology
Eindhoven, The Netherlands

Dr. Maarten Schoukens
Eindhoven University of Technology
Eindhoven, The Netherlands

This work is supported by the Applied and Engineering Sciences (AES) domain of the Dutch Research Council (NWO) through project number 13852.

ISSN 2190-5053 ISSN 2190-5061 (electronic)
Springer Theses
ISBN 978-3-030-90342-8 ISBN 978-3-030-90343-5 (eBook)
https://doi.org/10.1007/978-3-030-90343-5

This Springer imprint is published by the registered company Springer Nature Switzerland AG
The registered company address is: Gewerbestrasse 11, 6330 Cham, Switzerland

Supervisors' Foreword

Mathematical models form an indispensable tool for science and technology in the modern world. Developing models of dynamical systems is a necessary preparatory step for many engineering applications, such as controlling the dynamics of an aircraft or predicting and managing the load on an electricity grid. To enable the industries to push the boundaries of engineering, we need tools and methodologies to build accurate dynamical models, many times directly from data. However, current data-driven modelling tools require a series of important decisions requiring serious expertise from the user.

Dr. Dhruv Khandelwal's thesis aims to automate the task of data-driven modelling by establishing a novel methodology that can tackle the challenges involved in (i) exploring various model structures and model complexities, (ii) optimizing models for multiple user-specified performance measures and (iii) automating model selection and validation. As a key contribution, a "grammar" for dynamic models using tree adjoining grammar (TAG) is developed, which generalizes the notion of model set, spanning over multiple classes of dynamic systems and stochastic noise models. By using this grammar, a computationally efficient evolutionary approach with consistency guarantees is developed. Based on Darwin's definition of natural selection in biology, the method "evolves" models by competing them in multiple, even conflicting, user-specified performance criteria, allowing to automatically discover the motion laws, friction dynamics, electromagnetic relations, stochastic effects, etc., that govern the system during its operation, but at the same time also optimizing the complexity, i.e. the details required to achieve the user set requirements of model performance. A key feature of the proposed method is that in the end, the user only needs to choose a model for the intended purpose a posteriori based on the estimated Pareto front. The developed approach represents a significant scientific step in automation of system identification, and its performance is demonstrated on several academic and real-world benchmark applications.

To develop his innovative results, Dr. Khandelwal joined to our Control Systems Research Group at the Eindhoven University of Technology in 2016. His research was conducted within the "control and data-driven modelling using symbolic methods" project funded by the Dutch Research Council (grant number: 13852), which represented a unique research endeavour between the high-tech companies ASML, Flanders Make, DataStories, National Instruments and the Dutch universities: the Eindhoven University of Technology and the Delft University of Technology. Based on a collaboration with scientist Dr. Cees Verdier, Dr. Manuel Mazo Espinosa, Dr. Robert Babuška and ourselves, Dr. Khandelwal has achieved serious development of theoretical results, software tools and their demonstration on applications as a joint work with the involved companies. The summary of his groundbreaking work forms the current thesis that he defended on 4 March 2020 and which has inspired further industrial collaboration and joint research work.

Eindhoven, The Netherlands
July 2021

Dr. Roland Tóth
Dr. Maarten Schoukens

Parts of this thesis have been published in the following articles:

Journal Papers (submitted or in preparation)

D. Khandelwal, M. Schoukens and R. Tóth (2020). A Tree Adjoining Grammar Representation for Models Of Stochastic Dynamical Systems. *Automatica, 119,* 109099

D. Khandelwal, M. Schoukens and R. Tóth. On Automated Multi-objective Identification Using Grammar-based Genetic Programming. *In preparation.*

Conference Proceedings

D. Khandelwal, M. Schoukens and R. Tóth (2019). Grammar-based Representation and Identification of Dynamical Systems. *Proceedings of the 18th IEEE European Control Conference*, Naples, Italy (June), pp. 1318–1323.

D. Khandelwal, M. Schoukens and R. Tóth (2019). Data-driven Modelling of Dynamical Systems Using Tree Adjoining Grammar and Genetic Programming. *Proceedings of the 2019 IEEE Conference on Evolutionary Computation*, Wellington, New Zealand (June), pp. 2673–2680.

D. Khandelwal, M. Schoukens and R. Tóth (2018). On the Simulation of Polynomial NARMAX Models. *Proceedings of the 2018 IEEE Conference on Decision and Control*, Miami Beach, Florida, USA (December), pp. 1445–1450.

D. Khandelwal, S. Weiland and A. Khalate (2017). Robust fault diagnosis by optimal input design for self-sensing systems. *Proceedings of the 20th IFAC World Congress*, Toulouse, France (July), pp. 1031–1036.

Workshop Abstracts

D. Khandelwal, M. Schoukens and R. Tóth (2019). Automating System Identification Using Grammar and Genetic Programming. *2019 Workshop on Nonlinear System Identification Benchmarks* , Eindhoven, The Netherlands (April).

D. Khandelwal, M. Schoukens and R. Tóth (2019). Automating System Identification: A Linguistic Approach. *38th Benelux Meeting on Systems and Control*, Lommel, Belgium (March).

D. Khandelwal, M. Schoukens and R. Tóth (2018). Grammar-based Representation and Identification of Dynamical Systems. *17th International Symposium on Intelligent Data Analysis*, 's-Hertogenbosch, The Netherlands (October).

D. Khandelwal, R. Tóth and P. M. J. Van den Hof (2018). Grammar-based Encoding of Well-posed Model Structures for Data-Driven Modelling. *37th Benelux Meeting on Systems and Control*, Soesterberg, The Netherlands (March).

D. Khandelwal, R. Tóth and P. M. J. Van den Hof (2017). Data-Driven Modelling of LTI systems using Symbolic Regression. *36th Benelux Meeting on Systems and Control*, Spa, Belgium (March).

D. Khandelwal and R. Tóth (2016). Data-Driven Modelling using Symbolic Regression. *35th Benelux Meeting on Systems and Control*, Soesterberg, The Netherlands (March).

Acknowledgements

Over the last four years, I have had the pleasure of working with and alongside a number of people who have made an indelible impression on me. I will cherish, for many years to come, the numerous memories I have shared with colleagues, friends and peers during the course of my doctoral research.

I owe a deep sense of gratitude to my supervisors Roland Tóth and Maarten Schoukens for their support and guidance. Roland's immaculate clarity of thought and attention to detail never ceased to amaze me. He inspired me to expect nothing short of the highest standards from my work. Maarten, in his own affable manner, motivated me to stay on course and maintain focus on the bigger picture, without being distracted by the many pitfalls of doctoral research. He kept me on my toes not only in my research activities but also on the squash courts.

The quality of this dissertation was certainly improved by the feedback provided by my Ph.D. committee members—Dr. Manuel Mazo Espinosa, Prof. Kieth Worden, Prof. Xavier Bombois, Prof. Mykola Pechenizkiy and Prof. Paul Van den Hof. I highly appreciate the time and effort they put into judging the quality of my thesis and am thankful for their valuable feedback.

The research presented in this dissertation was supported by NWO domain TTW, which is gratefully acknowledged. My research benefited greatly from discussions with the partners and user committee members of the CADUSY project—Cees Verdier, Manuel Mazo Espinosa, Robert Babuska, Koos van Berkel, Bruno Depraetere, Edward Kikken and Katya Vladislavleva. The frequent exchange of ideas with Cees, especially in the first half of the project, played an important role in determining the research directions I ended up pursuing. The case studies provided by the industrial partners allowed me to test the capabilities and identify the strengths and weaknesses of the ideas and methods developed in this work.

I would like to thank Siep Weiland whose keen insight and analytical thinking inspired me to consider the possibility of pursuing a Ph.D. track. I am also thankful to Paul Van den Hof and Johan Schoukens, who made a significant impact on not only my work but also on my general outlook on research in the field of system identification.

I would like to thank my friends and colleagues at the CS group for the *gezellig* environment over the last four years. I have shared with them many memories that are as important to me as the contents of this thesis, if not more. Veaceslav, Edwin and Tuan made sure I had an "unforgettable" beginning to my Ph.D. journey. In Ioannis, I found a reliable source of not just emotional and psychological support, but also valuable music recommendations. Over the years, many people have made an impression on me—Bahadir and Ruxandra with their wisdom, Ioannis with his humour, Daming with his cheerfulness, Ruben with his unbounded amounts of energy and Marcella with her kind ferocity. I have had the pleasure of enjoying many conferences, workshops, dinners, drinks, coffee-break conversations in the lovely company of Giuseppe, Harm, Ruben, Daming, Sofie, Giulio, Paul P., Zuan, Paul R., Carlos, Alejandro, Pepijn, David, Henrik, Tom S., Karthik, Shengling, Tom B., Clarissa and many more. Will, Diana, Hiltje, Lucia and (the late) Barbara have been very supportive during my stay in the group. To all the people of the CS group, I am thankful.

In the last four years, I have been lucky to find a "family" far from home. I look back to all the times I have spent with Rishi, Georgi and Nino in the last few years with a great deal of joy. Hopefully, there will be many more dinners, concerts, vacations and music nights to come. I am thankful for the companionship of my friends here in The Netherlands (and neighbouring countries)—Thomas, Annie, Edgar G., Stefanos, Sandra, Stefanie, Jordi, Alba, Joan, Antoine L., Joao and Claudia, many thanks for the lovely time we have spent together. My friends back in India have made every single trip back home fun and memorable; thank you Piyush, Rishika, Akshay, Hardik, Animesh and Utsab. I have also had the pleasure of playing and making music with the very talented folks of Music with Strangers and Particles. The last four years would have been a lot more monochromatic without the companionship of all these people.

I cannot sufficiently describe my gratitude towards my beloved family. My parents, sister and extended family members have been incredibly supportive of me over the years. Thank you Ma, Papa and Didi for your unconditional love, support and trust, and not to forget, the weekly series of entertaining Skype calls. When I was down for several weeks due to health-related problems, it was my immediate and extended family members that tirelessly supported me and helped me recover. I am forever indebted to them. Baba, you have always been a source of inspiration and will always be in my memories.

Finally, I owe a special note of thanks to Nino. She has been a constant source of joy, positivity, motivation and kindness. She has propelled me forward whenever I was down, and grounded me whenever my fantasies flew out of control. Madloba Nino!

Contents

Abbreviations

a.k.a	Also known as
AIC	Akaike information criterion
ARMA	Auto-regressive moving-average (model)
ARMAX	Auto-regressive moving-average (model) with exogenous inputs
ARX	Auto-regressive (model) with exogenous inputs
BFR	Best fit ratio
BIC	Bayesian information criterion
BJ	Box–Jenkins (model)
CMA-ES	Covariance matrix adaptation evolutionary strategy
CNN	Convolutional neural network
CT	Continuous time
CVA	Canonical variate analysis
DE	Differential evolution
DFT	Discrete-time Fourier transform
DGS	Data generating system
DT	Discrete time
EA	Evolutionary algorithm
EC	Evolutionary computing
ETFE	Empirical transfer function estimate
FIR	Finite impulse response (model)
FRF	Frequency response function
GA	Genetic algorithm
GFRF	Generalized frequency response function
GP	Genetic programming
i.i.d	Independent and identically distributed
ILS	Iterative least squares
IO	Input–output (representation)
IV	Instrumental variables
LFR	Linear fractional representation
LPV	Linear parameter-varying (representation)
LS	Least squares

LSTM	Long short-term memory
LTI	Linear time-invariant (system)
LTV	Linear time-varying (system)
MA	Memetic algorithm
MAP	Maximum *a posteriori*
MC	Monte Carlo (simulations)
MCMC	Markov chain Monte Carlo
MIMO	Multiple-input multiple-output (system or model)
ML	Maximum likelihood
MOEA	Multi-objective evolutionary algorithm
MOEA/D	MOEA based on decomposition
MOESP	Multivariable output-error state space (method)
MOO	Multi-objective optimization
MPC	Model predictive control
MSE	Mean squared error
MSP	Multi-step prediction
N4SID	Numerical algorithms for subspace state space system identification
NARMA	Nonlinear auto-regressive moving-average (model)
NARMAX	Nonlinear auto-regressive moving-average (model) with exogenous inputs
NARX	Nonlinear auto-regressive (model) with exogenous inputs
NBJ	Nonlinear Box–Jenkins (model)
NFIR	Nonlinear finite impulse response (model)
NIIR	Nonlinear infinite impulse response (model)
NL	Nonlinear (system or model)
NLTI	Nonlinear time-invariant (system)
NN	Neural network
NOE	Nonlinear output error (model)
NSGA-II	Non-dominated sorting genetic algorithm II
ODE	Ordinary differential equation
OE	Output error (model)
PAES	Pareto archived evolution strategy
PDE	Partial differential equation
PEM	Prediction error minimization
PID	Proportional-integral-derivative
P-NARMAX	Polynomial nonlinear auto-regressive moving-average (model) with exogenous inputs
P-NARX	Polynomial nonlinear auto-regressive (model) with exogenous inputs
PNLSS	Polynomial nonlinear state space (model)
POI	Point-of-interest
PPEM	Parametric prediction error minimization
PSO	Particle swarm optimization
PWA	Piece-wise affine
RBF	Radial basis function

RHS	Right-hand side
RKHS	Reproducing kernel Hilbert space
RMS	Root mean square
RNN	Recurrent neural network
SI	System identification
SISO	Single-input single-output (system or model)
SL	Switched-linear (system)
SNR	Signal-to-noise ratio
SPEA	Strength Pareto evolutionary algorithm
SPEA2	Strength Pareto evolutionary algorithm 2
SR	Symbolic regression
SS	State space (representation)
SVD	Singular value decomposition
TAG	Tree adjoining grammar
TLS	Total least squares
w.p.	With probability
w.r.t.	With respect to

Symbols

Constants, Coefficients and Parameters

$\theta_{\mathrm{p}}, \hat{\theta}_{\mathrm{p}}$	Predictor model parameter vector and its estimate
$\theta_{\mathrm{s},l}, \hat{\theta}_{\mathrm{s},l}$	l-approximate simulation model parameter vector and its estimate
n_{obj}	Number of identification criteria

Datasets

$\mathcal{D}_{\mathrm{est}}, N_{\mathrm{est}}$	Estimation dataset and its length
$\mathcal{D}_N, N$	Measurement dataset and its length N
$\mathcal{D}_{\mathrm{test}}, N_{\mathrm{test}}$	Test dataset and its length
$\mathcal{D}_{\mathrm{val}}, N_{\mathrm{val}}$	Validation dataset and its length
n_{p}	Number of periods in a periodic excitation signal

Evolutionary Algorithms

κ	Crossover operator
ρ_i^j	Derivation tree representation of individual i in generation j of an evolutionary algorithm (EA)
δ_i^j	Derived tree representation of individual i in generation j of an EA
μ	Mutation operator
$\mathcal{P}_J$	The Pareto front
$D^{(j)}$	Derived tree representation of population in iteration j of an EA
ρ_i^j, δ_i^j	Derivation tree and derived tree representation of individual i in iteration j
L	Number of iterations (generations) in an EA

$\mathcal{M}^{(j)}$	Population of models (phenotype) in iteration j of an EA
n_{S}	Population size in an EA
m_{a}	Maximum archive size
m_{d}	Maximum depth hyper-parameter
m_{id}	Maximum depth of a tree during initialization
P, P_{c}, P_{m}	Transition matrices of genetic programming (GP), crossover and mutation, respectively
$\pi(j)$	Markov chain probability distribution corresponding to the GP population in iteration j
p_c	Probability of crossover
p_m	Probability of mutation
p_s	Probability of selection
ς	Selection operator
$\tilde{X}^{(j)}$	Proposed population of models (genotype) in iteration j of an EA
$X^{(j)}$	Population of models (genotype) in iteration j of an EA
x_i^j	Individual i in generation j of an EA

Formal and Linguistic Grammars

δ	A derived tree
ρ	*A* derivation tree
$\mathcal{A}$	Set of auxiliary trees associated with a TAG
α	An auxiliary tree associated with a TAG
$\cdot[\![\cdot, \cdot]\!]$	Adjunction operation
$\Delta_G(\cdot)$	Mapping from derivation tree to derived tree based on grammar G
E	Set of edges of a finite tree
e	An edge in a finite tree
f	Foot node of an auxiliary tree
G	A grammar
$\mathcal{I}$	Set of initial trees associated with a TAG
η	An initial tree associated with a TAG
$l(\cdot), g(\cdot)$	Labelling function for vertices and edges of a tree, respectively
$\varepsilon(\cdot)$	Efficient set of a grammar-based model set
$\mathcal{M}(\cdot)$	Grammar-based model set specification
$L_{\mathrm{D}}(\cdot)$	Set of valid derivation trees of a grammar
G_{NBJ}	Polynomial NBJ grammar
G_{N}	Polynomial NARMAX grammar
G_{SS}	State space grammar
$L(\cdot)$	String language of a grammar
$L_{\mathrm{T}}(\cdot)$	Tree language of a grammar
$\mathfrak{N}$	Set of non-terminal node labels
$p_G(v, \psi)$	Gorn address of node v in tree ψ
$\prod_f$	Model expression (in string form) of a symbolic model f

r	Root node of a finite tree
$\mathfrak{S}$	Set of terminal and non-terminal node labels
S	Start node label associated with GP or a grammar
$\cdot[\cdot, \cdot]$	Substitution operation
$\mathfrak{T}$	Set of terminal node labels
ψ	A finite tree
V	Set of vertices of a finite tree
v	A vertex (node) in a finite tree

Functions and Operators

$\lfloor\cdot\rfloor$	Floor operator
$\Phi(\cdot)$	Genotype–phenotype mapping
$\phi_{\mathrm{P}}, \varphi_{\mathrm{P}}, k$	Regressor matrix for one step ahead and its kth row, respectively
$\phi_\tau, \varphi_\tau(k)$	Regressor matrix for τ-step-ahead outputs and its kth row, respectively
$d_t(\cdot)$	Depth of a tree
$E_\xi[\cdot]$	Expectation operator w.r.t. the distribution of ξ
J	Objective function
$\Lambda(\cdot)$	Operator that maps a tree to its leaves
$\mathrm{MSE}(\cdot)$	Mean squared error
$n_e(v, \psi)$	Number of edges between vertex v and root node of tree ψ
q^{-1}	Discrete-time backward shift
$r_c(\cdot)$	Crowding-distance-based rank of a model
$r_{\mathrm{ND}}(\cdot)$	Non-domination-based rank of a model
$\mathrm{rms}(\cdot)$	Root mean square

Sets, Sequences, Relations and Ordering

$\prec_{\mathrm{sel}}$	Selection ordering relations
$\preceq, \prec$	Partial and strict partial Pareto ordering in objective space, respectively
$U_a^b(k)$	Set of input terms ranging from $u(k-a)$ to $u(k-b)$
$\Xi_a^b(k)$	Set of noise terms ranging from $\xi(k-a)$ to $\xi(k-b)$
$Y_a^b(k), Y_{\mathrm{s},}l_a^b(k)$	Set of output y and simulated output y_s terms, respectively, ranging from time $k-a$ to $k-b$

Signals, Systems and Related Variables

γ_i	A subset of terms in the *i*th term of a polynomial NARMAX model
A, B, C, D, F	Polynomial functions associated with IO representations of systems
A_s, B_s, C_s, D_s	Matrices associated with SS representations of systems
a, b, c, d	Model coefficients associated with parametric input–output (IO) model structures
$b_{.,.}, a_{.,.}, d_{.,.}$	Exponents of the input, output and noise terms in a NARMAX model, respectively
$\bar{d}_i$	Exponents of the noise term $\xi(k-i)$
$\varepsilon_p, \varepsilon_s$	One-step-ahead prediction error and simulation error, respectively
$\tilde{f}(\cdot)$	Hermite polynomial-based representation of function f
$f_l(\cdot)$	l-step-ahead recursive formulation of f
$f_p(\cdot), f_s(\cdot)$	Nonlinear prediction function and nonlinear simulation function, respectively. Typically associated with IO structure
$f(\cdot), g(\cdot)$	Nonlinear functions, typically associated with IO structure
$\Psi_{\tau,k}$	Row k of the Jacobian of the τ-step-ahead prediction output
G, H	Rational functions associated with IO representations of systems
$\mathcal{M}$	Model set
$M, M(\theta)$	*A* dynamical model and its parameterized form, respectively
n_u	Maximum time lags in input terms
n_1	Maximum time lags in all terms of a discrete-time model
n_υ	Maximum time lags in (filtered) noise terms
n_ξ	Maximum time lags in noise terms
n_y	Maximum time lags in output terms
p	Number of terms in a NARMAX model
S	The true data generating system
$\sigma\xi$	Standard deviation of noise ξ
τ	Time steps for multi-step-ahead prediction
u	Input
υ	Noise or disturbance, typically assumed to be filtered
x, x_i	Latent variable (or state) vector and its *i*th component, respectively
ξ	Noise or disturbance
u_m, y_m	Measured input and output, respectively
y_0	Disturbance free output
$\hat{Y}, \hat{Y}_T$	Vector of one step ahead and τ-step-ahead predicted outputs, respectively
$\hat{y}, y_s$	One-step-ahead predicted output and simulation output, respectively
$y_{s,l}$	l-approximate simulation output
$\bar{y}_s$	Empirical simulation output
y	Output

Spaces and Fields

$\mathcal{B}_{\mathbb{R}}$	Borel σ-algebra on $\mathbb{R}$
$\mathscr{H}$	A Hilbert space
$\mathbb{R}, \mathbb{R}_{>0}, \mathbb{R}_{\geq 0}$	The field of reals, positive reals and non-negative reals, respectively
$\Theta(M)$	Parameter space determined by the structure of model M
$\mathbb{Z}, \mathbb{Z}_{>0}, \mathbb{Z}_{\geq 0}$	The set of integers, positive integers and non-negative integers, respectively

Chapter 1
Introduction

> *Science is a bit like the joke about the drunk who is looking under a lamppost for a key that he has lost on the other side of the street, because that's where the light is. It has no other choice.*
>
> *Noam Chomsky*

Mathematical models form an indispensable tool for science and technology in the modern world. The use of mathematical models has revolutionized virtually every industry in the 20th and 21st centuries, and plays an increasingly vital role in modern society. Accurate models of physical phenomena have been used to, for instance, transform the music industry by making available entire suites of virtual instruments in hand-held computers, save thousands of lives by predicting devastating storms days in advance and save millions of dollars by enabling accurately-controlled vertical landing of first stage systems of rockets for eventual re-use.

To enable scientists and industries to build such transformative technologies, we need tools and methodologies to build accurate models. To this extent, a wide variety of modelling techniques have been developed, ranging from *first-principles modelling* on one end of the spectrum, to *black-box modelling* techniques on the other. As engineering systems and corresponding performance specifications become more complex, several *data-driven modelling* approaches have been developed to cope with the increased modelling challenges in the context of *dynamical systems*. This has resulted in a complex landscape between the (i) nature of dynamics of systems being modelled, (ii) the mathematical representations used to describe the system, and (iii) the available modelling techniques that are suitable for the corresponding choices in items (i) and (ii). From a user's perspective, it can be challenging and cumbersome to navigate this complex landscape. This motivates the main research theme adopted in this thesis: How can we automate the task of data-driven modelling from the user's perspective?

D. Khandelwal, *Automating Data-Driven Modelling of Dynamical Systems*, Springer Theses, https://doi.org/10.1007/978-3-030-90343-5_1

1.1 Mathematical Models

1.1.1 Introduction

Since the very nascent stages of civilized society, human beings have attempted to understand the physical world around them. Long before the era of scientific inquiry, explanations of various real-life phenomena were considered to be the dominion of mythology and mysticism. In the subsequent centuries, the development of a number of organized human activities, including art, religion and scientific discovery, have been attributed, in part, to our desire to make sense of the world around us [11, 13, 34].

In the era of scientific inquiry, the quest to understand the intricate mechanisms of the physical world emerged as the search for objective truth, to supplement the realm of subjective truth, as in art or religion. In the age of the scientific method, the approach to scientific discovery was based on the development of rigorous theories which *had to be validated* (or invalidated) by experimentally establishing the truth of inferences obtained from the theory. Application of the scientific inquiry to various domains of the physical world around us lead to the development of various formal science and mathematics. In mathematics, we found a universal language to objectively and succinctly describe physical and abstract phenomena, enabling scientists to develop *mathematical models*. These mathematical models elucidate the relationship between quantities, physical or conceptual, that eventually give rise to the phenomena under study. For example, rigorous analysis of empirical observations led Isaac Newton to the formulation of the law of universal gravitation and Johannes Kepler to the laws of planetary motion.

1.1.2 Utility of Mathematical Models

In the words of celebrated physicist Richard Feynman, "mathematics is *not* just another language. Mathematics is a language plus reasoning; it is a language plus logic. Mathematics is a tool for reasoning" [7, p. 40]. Indeed, equipped with a suitable mathematical model, one can perform a number of logical deductions, predictions or actions. For instance, one can predict eventualities that are yet-to-be-observed, but nonetheless, captured by the mathematical relations of the model. Continuing with the previous example, while empirical evidence was used to develop the law of universal gravitation, the disagreement of the mathematical model with the observations of the orbit of Uranus led the astronomer Urbain Le Verrier to postulate the existence of the planet Neptune, only to be later confirmed by physical observations. More recently, the first empirical evidence of a black hole, a prediction derived from Einstein's theory of general relativity in early twentieth century, was obtained almost a century later [1, 2]. Besides prediction, one can also leverage the understanding of laws that govern the behaviour of real-world systems in order to expertly manipulate them.

For example, a keen understanding of aerodynamics and stability allowed Wright brothers to develop effective wing control mechanisms in order to achieve sustained engine-powered flight.

In modern times, a mathematical model may serve several purposes—predict data inferred from real-world systems, understand a given system to improve its design, understand consumer behaviour and emerging trends, predict the variations in stock markets and commodity prices, predict weather, control and optimize efficiency, and so on. As the variety in these examples suggests, mathematical models are relevant for multiple scientific domains and research communities—econometry [29], machine learning [16], systems and control [20], computational biology [18] and metrology [24], to name a few. In each of these scientific domains, mathematical models are used to solve a plethora of problems.

1.1.3 Reality Versus Abstraction

The behaviour of physical objects can be infinitely complex. The physical reality can be seen as a manifestation of interactions between countless objects and actions across scales ranging from microscopic to macroscopic. Capturing such complex behaviour would not just be impossible, but also undesirable. A mathematical model is typically built by simplifying reality up to a desired level of resolution and by isolating the object being modelled from its surrounding elements that are irrelevant for the intended study. This approach to modelling is referred to as the *systems* approach (see [32] for a detailed treatment).

In the systems approach, the object being modelled, also known as (a.k.a) the *system*, can be extracted from its environment, and can be seen as an entity that interacts with its environment. This procedure is known as *abstraction*. Furthermore, a complex system can also be further decomposed into sub-systems that interact with each other, with evident cause-and-effect relationships. At this level, it becomes easier to formulate the mathematical equations that describe the relationship between the variables (or the physical quantities) of each sub-system. Once the sub-systems have been described to a sufficient level of detail, they can be inter-connected via sharing of variables, thereby resulting in a model of the overall system. Willems [32] describes this modelling process as the *tearing, zooming and linking* approach to modelling systems.

To illustrate the concept of abstraction, consider the example of heat propagating through a metal bar given in Fig. 1.1. The temperature profile of a metal bar may depend on a number of factors such as ambient temperature, initial temperature profile, exposure to heating sources and electro-magnetic radiation, the shape and material properties of the object, and so on. Furthermore, to truly capture the causal forces resulting in heat propagation, inter-molecular and intra-molecular forces that give rise to conduction, convection and radiation must be taken into account. However, in order to study the propagation of heat in the system, the effects of various external and internal aspects can be either ignored due the insignificance of their influence, or simplified in order to describe their effects at the desired level of detail

(a) Heat propagation in a metal bar. (b) Rotational motion of a metal bar.

Fig. 1.1 An example of distributed and lumped dynamics in a system

and accuracy. This process of ignoring, abstracting and simplifying leads to Fourier's heat propagation equation

$$\frac{\partial T}{\partial t} = \alpha \left(\frac{\partial^2 T}{\partial x^2} + \frac{\partial^2 T}{\partial y^2} + \frac{\partial^2 T}{\partial z^2} \right), \tag{1.1}$$

where, $T(x, y, z, t)$ is the temperature at point (x, y, z) and time t, and α is the diffusivity constant of the material. Equation 1.1 reveals that the rate of change of temperature at a point, denoted by $\frac{\partial T}{\partial t}$, is a function of the temperature of the neighbouring points (described by the higher order spatial derivatives of the temperature). In order to derive (1.1), physical phenomena such as radiation losses are *ignored* since conduction is the predominant phenomenon of heat propagation in metals. Furthermore, the system is considered to be an *abstract* collection of infinitesimally small points, distributed in continuous space, with each point associated with its temperature. In doing so, the physical notion of molecules sharing heat energy through vibrations and collisions is *simplified* in order to highlight the physical phenomenon under study—heat propagation.

1.2 Models of Dynamical Systems

In this section, we introduce the concept of dynamical systems, illustrate some of the common utilization scenarios of models of dynamical systems and give an overview of some of the commonly used approaches for deriving such models.

1.2.1 Dynamical Systems

A *dynamical system* is a system in which the relationship between the variables depend not only on their values at a given point in time, but also on their values in the past, i.e., it is a system with *memory*. Conceptually, one may consider the variables

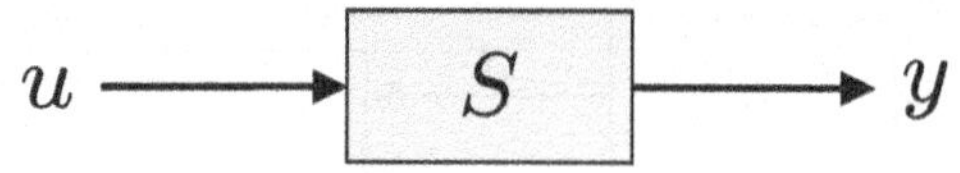

Fig. 1.2 Block-diagram representation of a system

to follow certain trajectories over a given time horizon, and consequently, the mathematical description of the dynamical system describes the relationship between the trajectories of the variables of the system and the external environment. The variables of a system can be influenced by the external environment via exogenous excitations called *inputs*, denoted as u. The dynamical system may also influence the external environment via *output* signals, denoted as y. This concept can be illustrated in a block-diagram representation as depicted in Fig. 1.2.

Dynamical systems may be categorized in several ways, as will be discussed in more details in Sect. 2.2 (also see [6, chap. 1]). For the purpose of introducing the research direction adopted in this dissertation, we present a characterization of dynamical systems based on spatial characteristics of the system. Such a characterization defines two types of dynamical systems—distributed or lumped dynamical systems.

A distributed system is a dynamical system in which the variable are functions of not only time, but also space. In other words, the variable of a distributed system can be conceptualized as trajectories over temporal and spatial dimensions. Typically, such dynamical systems are described using partial differential equations. The heat propagation model of a metal bar in (1.1) is an example of a distributed dynamical system as it describes the evolution of temperature over time and space (see Fig. 1.1a).

In contrast, a lumped dynamical system is a dynamical system in which the variables are functions of time alone. Realistically, all physical systems are distributed systems. However, in many cases, the spatial dynamics of a system can effectively be lumped at a single or a finite collection of points. For example, the metal bar in Fig. 1.1a can be abstracted as an interconnection of several smaller metal bar segments. For each bar segment, spatial dependency of the thermal dynamics can be lumped to a single variable by integrating the temperature profile over the length of the segment. As the bar segments get smaller, the approximation error decreases and the abstraction becomes closer to reality. As an alternate example of lumped dynamics, consider the rotational dynamics of the same metal bar, hinged on one end as depicted in Fig. 1.1b. Under the assumption that the bar is rigid, the rotational motion of the bar would be equivalent to the rotation of an imaginary point located at the center of gravity of the metal bar with a mass equal to that of the metal bar. Hence, the rotational motion of the metal bar can be described as a lumped dynamical system.

1.2.2 Applications of Dynamical Models

Models of dynamical systems are used in many domains of science and engineering. In each of these domains, dynamical models are used to solve a variety of challenging

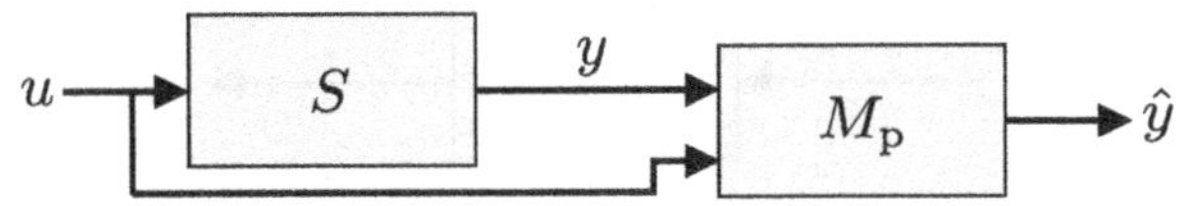

Fig. 1.3 A block-diagram representation of prediction

problems. We illustrate here some of the common utilization scenarios of dynamical models. Note that this is by no means an exhaustive list of applications.

Prediction and Forecasting

A dynamical model M can be used to derive a prediction model M_p that maps input u and output y measured in the past to predicted output data $\hat{y}$ in the future. A prediction model M_p can be used to obtain prediction of the output or a variable of a given system S. This is visualized in the block scheme in Fig. 1.3.

Industrial applications developed around predictions based on dynamical models are ubiquitous. Forecasting [5], prediction of financial trends [30], power generation and load scheduling for power grids [17] and dynamic pricing strategy of services and commodities [9] are some example applications that extensively employ dynamic models to make predictions.

Model-Based Analysis and Design

Models of dynamical systems have enabled us to develop a plethora of analytical tools to interpret and describe the behaviour of a system. Model-based concepts such as Lyapunov's stability, controllability and observability are considered to be fundamental to modern systems and control theory [8]. In civil engineering, models are used to analyse integrity of complex structures [14], optimize material utilization and perform non-destructive testing [36]. In bio-informatics and molecular biology, models developed for biological systems are used to derive inferences [15]. Model-based engineering has been crucial in developing and optimizing the structure and function of modern aeroplanes. In many fields of engineering and design, models have been widely used for virtual prototyping, enabling engineers and designers to test designs before building a physical prototype, leading to enormous savings in time and cost [4].

Control

Control can be described as the manipulation of the behaviour of a system, by interconnecting it with another system (the *controller*), such that the variables of the system (or a subset thereof) follow a desired trajectory [33]. Model-based control methods have enhanced safety, efficiency and reliability of machines significantly. An entire range of control design methods have been proposed over the past few decades. Based on a model, a user can design a controller for a dynamical system that is optimal with respect to a given performance measure, or robust against pre-specified uncertainties in the model [35], or adaptive in order to deal with time-changing phenomena, or even predictive in order to satisfy constraints imposed on the interconnected system [3].

Data-Assimilation

Yet another common use of dynamical models is in the scope of data-assimilation, where a model is used in conjunction with measurements obtained from a physical system, to obtain improved estimates of relevant system variables. Variations of such applications include, for, example, soft-sensing and fault diagnosis.

Rapid detection and diagnosis of faults is vital for engineering systems that must maintain a high production rate to remain economically feasible, or for systems that are safety-critical. Examples of such applications include aeronautical and automotive systems, process industries, power grids and industrial robotics. Besides fault detection, mathematical models are also used for predictive maintenance to enhance the lifetime of systems.

Other applications of data-assimilation include systems that rely on data and models in order to make critical decisions. For example, medical robots must rely on data from multiple cameras and sensors in order to perform a certain task, e.g. see [28]. A similar problem arises in the field of autonomous vehicles, where the path of the vehicle must be optimized based on information obtained from multiple sensors.

1.2.3 Model Representations of Dynamical Systems

As discussed in Sect. 1.2.2, models of dynamical systems are used in a wide array of applications. Depending on the properties of the system, the dynamics may be captured using a variety of mathematical representations. For the purpose of illustration, in this section, we introduce some basic mathematical representations of dynamical systems, with the focus restricted to Single-Input Single-Output (SISO) lumped dynamical systems that are time-invariant (i.e., systems for which the governing mathematical equations do not explicitly depend on time). A more general overview of a number of commonly-used model representations will be provided in Sect. 2.3.

In continuous-time (CT), i.e., when time t resides in a continuous domain $\mathcal{T} \subseteq \mathbb{R}$, an IO representation describes the input-output trajectories that are compatible with the system, and can be expressed as solutions of ODEs. In general, this relationship can be expressed as the following implicit ODE

$$g\left(\frac{\mathrm{d}^{n_y}y(t)}{\mathrm{d}t^{n_y}}, \frac{\mathrm{d}^{n_y-1}y(t)}{\mathrm{d}t^{n_y-1}}, \dots, y(t), \frac{\mathrm{d}^{n_u}u(t)}{\mathrm{d}t^{n_u}}, \frac{\mathrm{d}^{n_u-1}u(t)}{\mathrm{d}t^{n_u-1}}, \dots, u(t)\right) = 0, \qquad (1.2)$$

where $u(t) \in \mathcal{U} \subseteq \mathbb{R}$ is the input, $y(t) \in \mathcal{Y} \subseteq \mathbb{R}$ is the output and $g(\cdot)$ is a function that governs the relationship between input and output, and their higher order derivatives. Signals u and y are ordinary solutions with left compact support. The system representation in (1.2) is typically quite challenging to interpret. In many cases, the system can also be represented by a set of first order ODEs, resulting in a so-called State-Space (SS) representation, as follows

$$\dot{x}(t) = f(x(t), u(t)); \tag{1.3a}$$
$$y(t) = h(x(t), u(t)), \tag{1.3b}$$

where $x(t) \in \mathcal{X} \subseteq \mathbb{R}^n$ is the state vector, $f : \mathcal{X} \times \mathcal{U} \rightarrow \mathcal{X}$ is a function that governs the time-evolution of the states, and $h : \mathcal{X} \times \mathcal{U} \rightarrow \mathcal{Y}$ is a function that maps the input and state to the output. Signals u, x and y are ordinary solutions with left compact support.

In many applications, model-based tools are developed in order to be ultimately implemented in computers or embedded systems that are governed by a discrete clock cycle. Consequently, discrete-time (DT) equivalents of model representations in (1.2) and (1.3) have also been developed. Under suitable conditions on signal sampling and reconstruction, and assuming that the model representation in (1.2) can be made explicit, a DT IO representation of a dynamical system can be expressed as the following difference equation

$$y(k) = g_\mathrm{d}(y(k-1), \ldots, y(k-n_y), u(k), \ldots, u(k-n_u)), \tag{1.4}$$

where $k \in \mathbb{Z}$ is a time index related to the continuous time t as $t = kT_s$ for some sampling time $T_s > 0$, $g_\mathrm{d} : \mathcal{U}^{n_u+1} \times \mathcal{Y}^{n_y} \rightarrow \mathcal{Y}$ is a function that maps past input and output values to the output $y(k)$, and n_u and n_y now represent the maximum number of time-lags in the input and output signals, respectively. Similarly, the DT SS model can be expressed as a set of difference equations

$$x(k+1) = f_\mathrm{d}(x(k), u(k)); \tag{1.5a}$$
$$y(k) = h_\mathrm{d}(x(k), u(k)). \tag{1.5b}$$

The system representations in (1.2)–(1.5) are fairly general, and can be used to represent large classes of dynamical systems.

1.2.4 Obtaining Models of Dynamical Systems

As discussed in Sect. 1.2.2, model-based tools play a vital role in many fields of research and in the industry. Furthermore, as demonstrated in Sect. 1.2.3, a dynamical system may be represented in many different mathematical forms. Depending on the intended application, some model representations may be more favourable than others. This naturally leads to the following challenge: for a given dynamical system, how can we derive or estimate a model that is sufficiently accurate, and furthermore, has structural properties suitable for the intended application?

Over the years, many modelling approaches have been developed to address this challenge. In terms of the usage of prior understanding of the system to-be-modelled, these methods lie in a spectrum of possible modelling approaches, ranging from techniques based on usage of underlying physical laws alone, to those that attempt

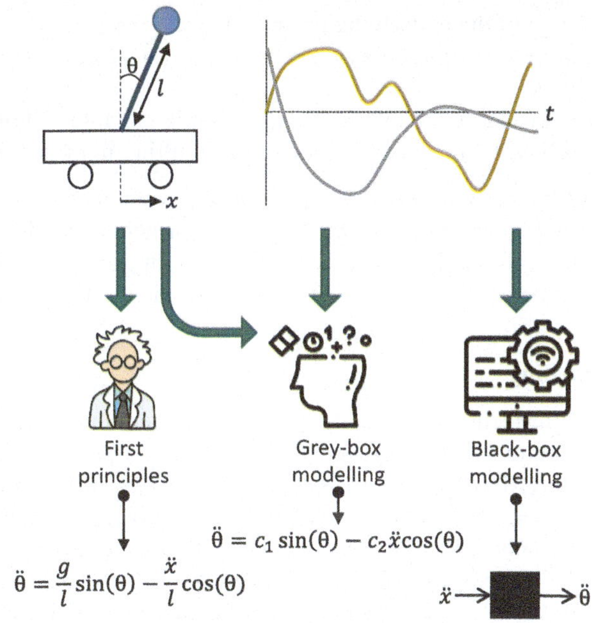

Fig. 1.4 Modelling approaches can be roughly categorized into three types. *First principles modelling* predominantly relies on the use of existing scientific laws, *black-box* modelling predominantly uses measured data obtained from the system, and *gray-box modelling* uses existing scientific laws to construct a partial structure of the model, and used measured data to fine-tune the model to match the given system. Constituent icons made by Freepik from www.flaticon.com

to construct a model solely based on data measured from the system. Roughly, these methods can be classified into three paradigms as depicted in Fig. 1.4: first principles modelling, black-box modelling and gray-box modelling [25]. It should be noted that there exists a large overlap at the borders of these approaches, for instance, Ljung [22] describes a finer characterization of the spectrum of modelling approaches.

First Principle Modelling

The most traditional approach involves making use of prior scientific knowledge in order to build a model of the system. For instance, mechanical motion systems are often modelled using Newton's laws of motion or via the Langrangian equation, and Kirchoff's circuit laws can be used to model simple electronic circuits. This approach to building models is known as *first principles modelling* and results in symbolic or parametric models. The most significant advantage of the first principles modelling approach is that the resulting model is transparent, i.e., each function and parameter in the model can be attributed to a specific component of the dynamical system, making it easier to validate a model, analyse its properties, and employ the model for various applications. On the other hand, modelling complex dynamical systems using first-principles knowledge can be a cumbersome task for the following reasons:

- lack of knowledge of the underlying physical laws,
- inability to extrapolate (interpolate) the known physical laws to a macroscopic (microscopic) level,
- the complexity arising due to interaction of multiple variables within the system,
- the increased reliability on expert's knowledge resulting in increased costs.

These challenges make first-principles modelling a difficult and time-consuming task, even for an expert practitioner. Deriving a first-principles model for a complex dynamical system can sometimes take several months, or even years, thereby diverting the user's attention from the eventual use-case of the model to the derivation of the model itself.

Black-Box Modelling

Black-box modelling techniques were developed as an alternative modelling approach that addresses some of the challenges experienced in first-principles modelling. In black-box modelling, the objective is to estimate a model based only on input-output data measured from the system to-be-modelled. Typically, the user, based on experience and insight gathered from measured data, chooses a model structure that is appropriate for the system to-be-modelled, and subsequently uses data to estimate parameters in the chosen model structure. This alleviates the need to apply scientific knowledge rigorously to build a model. However, choosing an appropriate model structure is not a trivial task, and requires an experienced user to make well-reasoned choices and assumptions that ultimately influence the quality of the estimated model. The challenge of model structure selection will be examined in more details in Chap. 2 for several data-driven modelling approaches. Furthermore, the resulting models are typically not transparent, since it is not possible to relate the estimated parameters to actual physical parameters of the system. As a result, the resulting estimated models are often referred to as *black-box models*.

Gray-Box Modelling

An intermediate approach between first principles modelling and data-driven modelling is the so-called *gray-box modelling* approach. In this approach, a user relies on physical laws to determine a model form, or a part thereof, that is suitable to describe the system to-be-modelled. Unlike first-principles models, gray-box models contain unknown quantities or components, e.g., physical parameters specific to the system or unmodelled dynamics such as friction or backlash. Subsequently, measured data obtained from the system is used to estimate the unknown parameters or components of the model. In practice, gray-box modelling techniques are widely used in many industries since (i) the models are structured in a form that is easy to interpret, (ii) the models are formulated in terms of physical quantities that may be of interest to the user, and (iii) a gray-box model can be tuned to describe the behaviour of a given system taking into account properties that are specific to that system, e.g., taking into account manufacturing errors specific to the system.

1.3 System Identification

System Identification (SI) refers to the task of data-driven modelling of dynamical system. It is an umbrella term that covers a number of different approaches for modelling dynamical systems, ranging from black-box modelling approaches where the model is estimated solely from data, to first principles modelling approaches where certain physical parameters are estimated from measured data. In this section, we introduce the general SI problem and the common steps involved in solving it. In Sect. 1.3.3, we briefly discuss some of the challenges in each of the steps in the identification procedure.

1.3.1 Problem Description

The problem of SI can be informally described as follows. There exists a dynamical system S that can be excited by exogenous input signals $u(k) \in \mathbb{R}^m$, and output signals $y(k) \in \mathbb{R}^q$ can be measured from the system, see Fig. 1.5. Furthermore, there exists an unmeasured exogenous or *noise* signal $\xi(k)$ that may affect the measured response of the system. The unmeasured signal ξ may be caused by unmeasured excitations from the environment, or by measurement noise introduced by the measuring instrument used. The objective is to construct a model M of the system S based on measured data $\mathcal{D}_N = \{u_\mathrm{m}(k), y_\mathrm{m}(k)\}_{k=1}^N$ obtained from the system, such that the model M satisfies a set of specifications given by the user.

At this point, the problem is not yet formalized mathematically, and there exists a number of open questions. What should be the form or structure of model M? What performance measures should be used to quantify the degree of "correctness" of an estimated model. Should there be any conditions imposed on the measured dataset in order to ensure good estimation properties? In order to formulate the SI problem concretely, the user must make a number of assumptions and informed decision dur-

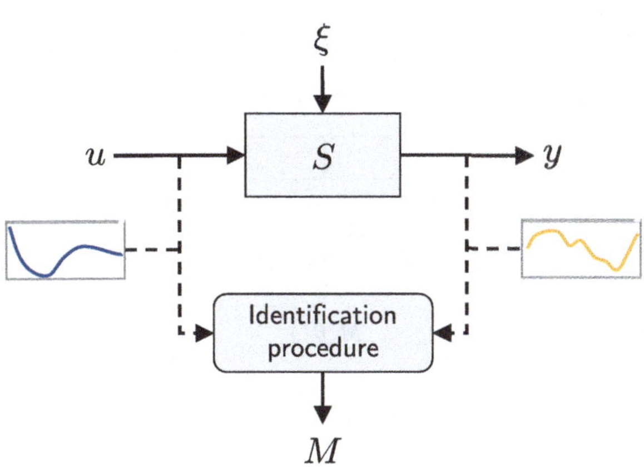

Fig. 1.5 A general schematic of the SI problem

ing the identification procedure. This procedure is referred to as the identification cycle, in which the user iteratively updates the assumptions and decisions made during the identification procedure based on the estimated models. The identification cycle can be roughly categorized into 5 steps [21, Chap. 1]: (i) experiment design, (ii) model structure and complexity selection, (iii) performance measure specification, (iv) model optimization, and (v) model validation. These are described in the following section.

1.3.2 The Typical Identification Cycle

A typical identification cycle is depicted in Fig. 1.6. The steps involved in the procedure are described below.

1. *Experiment design*: Any identification procedure typically begins with experiment design. This includes setting the configuration of the experiment, usually open-loop or closed-loop[1] (see Fig. 1.7), and designing the input excitation signal. The choice of the input signal is influenced by the nature of dynamics to-be-modelled and also by choices yet-to-be-made by the user in the subsequent steps, especially the choice of model structure and complexity.

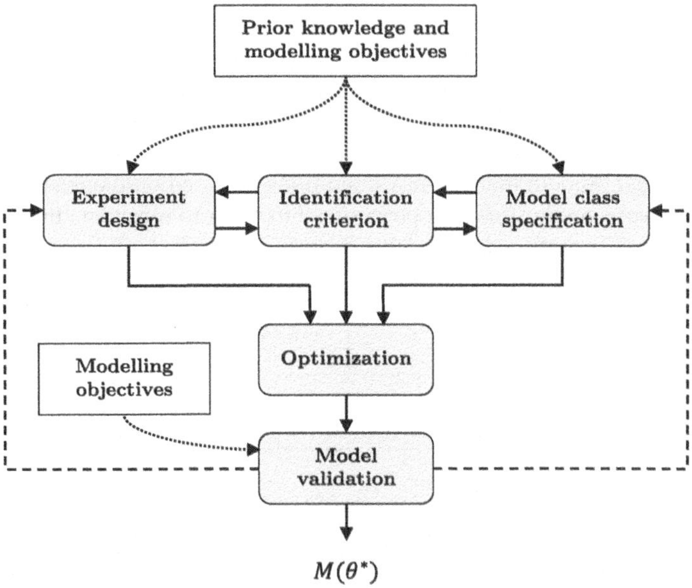

Fig. 1.6 The typical identification cycle (solid and dotted lines indicate flow of information, dashed lines are due to the "cycle") (see [21, Chap. 1])

[1] A closed-loop experiment configuration becomes necessary when the system S is unstable.

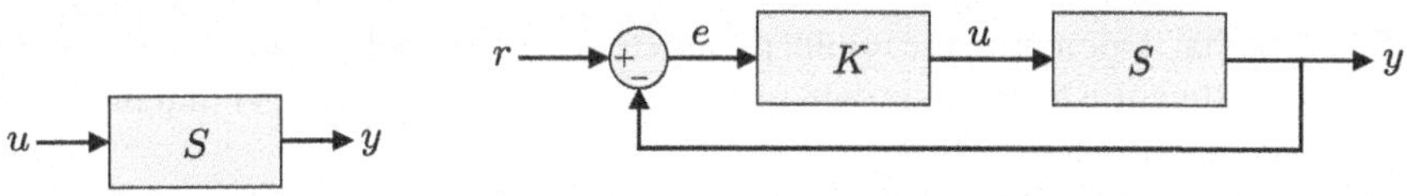

(a) Block-structure representation of an open-loop experiment configuration.

(b) Block-structure representation of a closed-loop experiment configuration, where r is a desired reference trajectory, e is the error between r and y, and K is a stabilizing controller.

Fig. 1.7 Open- and closed-loop configurations for a system identification experiment

2. *Model structure and complexity selection*: One of the more involved choices to be made is the choice of the *model set*, i.e., the set of models to be considered during the identification task. In particular, the user must make decisions related to the following aspects:

   - the structure of the dynamical model—based on experience or experiments, the user must choose a structure for the model M. At this stage, the user may also incorporate considerations from the eventual use-case of the model. In addition, the user must also choose a noise model structure. This includes, among others, the location at which noise enters the experimental setup configuration, i.e., position of noise ξ entering in the block scheme shown in Fig. 1.5, and the statistical and dynamical properties of the noise process. The combined choices of model structure and noise properties leads to the choice of the so-called model class associated with the identification cycle.
   - the complexity of the dynamical model—even when the model class is selected, there exists several degrees-of-freedom within the model class that determines the complexity of a model in the model class. A user must typically also select the degrees-of-freedom or the number of free parameters in the model. This choice is crucial since, if the model structure does not have sufficient degrees-of-freedom, some of the dynamics may remain unmodelled, and conversely, if the model contains too many free parameters, the estimated model may over-fit the data, capturing the effects of noise in the particular dataset which do not generalize to the overall dynamic behaviour of the system. Typically, the choice of model complexity leads to a model structure M that is parameterized by *model parameters* θ that belong to some parameter space Θ. Hence, the choice of model class and model complexity determine the set of all models that will be considered during the identification procedure, a.k.a, the *model set*, given by $\mathcal{M} := \{M(\theta)|\theta \in \Theta\}$.

3. *Identification criterion specification*—After hypothesizing the model structure, the user must define an identification criterion, i.e., the criterion that should be optimized during the estimation of the free parameters of the model. The choice of identification criterion is often determined by the use-case of the model, e.g., short-term prediction, simulation, etc. At this stage, the identification problem has been mathematically defined. Denote the identification criterion as $J : \mathbb{R}^n \to \mathbb{R}$, a

function that depends on the model parameters $\theta \in \mathbb{R}^n$ and the measured data $\mathcal{D}_N$. The identification problem is then to estimate a model $M(\theta) \in \mathcal{M}$ that minimizes the cost function $J(\theta; \mathcal{D}_N)$.

4. *Optimization*—The formulation of the identification problem results in the following mathematical optimization problem

$$\theta^* = \arg\min_{\theta} J(\theta; \mathcal{D}_N). \tag{1.6}$$

   Solving the optimization problem in (1.6) yields the identified model $M(\theta^*)$. The optimization problem can be solved by using an appropriate optimization technique. In some cases, the problem in (1.6) can be reformulated as a convex optimization problem. In the more general setting, however, more sophisticated non-linear optimization techniques must be employed.
5. *Model validation*—The quality of the estimated model $M(\theta^*)$ is influenced by the choices and assumptions made during the identification cycle, e.g., choice of model structure and complexity. Hence, before accepting the estimated model, the decisions made during the identification cycle must be verified. This is commonly done via hypothesis testing on the basis on inferences derived from the model for a given dataset. A dataset used for model validation should be independent of the data that was used in Step 4 of the identification cycle. If the hypothesis is invalidated, then the choices made in Step 1–3 must be re-visited. On the other hand, if the hypothesis cannot be invalidated, then the choices made during the identification cycle are deemed to be correct and the estimated model $M(\theta^*)$ can be accepted as the identified model. In addition to hypothesis testing, in the model validation step, one may also check if the estimated model is suitable for it's eventual use-case, for example, for design of control schemes or simulators.

Many SI approaches have been proposed in the literature, most of which fit within the basic identification cycle described in this section. The diversity in SI methods is primarily due to (i) the diversity of the dynamics to-be-modelled and (ii) the diversity in the intended model structures. For example, the class of Linear Time-Invariant (LTI) models may be represented using Frequency Response Functions (FRFs), IO models as in (1.4), or SS models as in (1.5), and can be modelled using the corresponding data-driven methods presented in [21, 26, 31], respectively. Furthermore, each of these methods have been extended to richer classes of dynamical systems such as Linear Time-Varying (LTV) systems, Linear Parameter-Varying (LPV) systems and Non-Linear (NL) systems, resulting in a vast landscape of methods. A more detailed review of various identification methodologies is reserved for Chap. 2.

1.3.3 Challenges in System Identification: A User's Perspective

A typical identification task involves a number of steps, as laid out in the identification cycle discussed in Sect. 1.3.2. Depending on the identification methodology employed, these steps may require the user to make a number of decisions and assumptions (this will be reviewed in detail in Chap. 2). While the specific nature of user-interaction depends on the identification methodology used, these choices are crucial as they ultimately influence the quality of the identified model. This makes SI a challenging task for an inexperienced user.

Following the identification cycle described in Sect. 1.3.2, some of the challenges for the user are described below.

1. Input design—When designing an input signal for an identification experiment, the user must be cognizant of a number of aspects. Firstly, in an ideal scenario, it is desirable to design an input sequence that is *persistently exciting with respect to* (w.r.t) the chosen model set, i.e., based on the input excitation used, it should be possible to distinguish between any two models in the model set w.r.t the chosen performance measure. Furthermore, for LTI systems, estimating a good model based on a persistently exciting input ensures that the model is accurate even under extrapolation (assuming that the true system belongs to the chosen model set). This is due to the principle of linearity. However, if the given system behaves non-linearly, the user must also take care to design an excitation signal that closely resembles the typical input excitation used for the given system, to ensure model quality in that region of the solution space. If there is no structural error in the chosen model structure, the estimated model will also be valid for a larger class of inputs. However, no such guarantee can be derived when there is a systematic error in the chosen model structure, due to the non-linearity of the system. To determine the nature of dynamics of the system, a user must rely on past experience or knowledge of the system. Alternatively, the user can perform some preliminary experiments in order to develop a better understanding of the dynamics [26]. The choice of input excitation used during the identification experiment also directly affects the variance of the parameter estimates of the model. This knowledge has been used in order to develop input signals that minimize the variance of parameter estimates. Conversely, a poorly designed input sequence can also result in poor variance properties of the parameter estimates.
2. Model structure and complexity selection—The choice of model structure and complexity is primarily motivated by the nature of the dynamics of the system, and the model representation used to describe the dynamics. The class of dynamics of the true system along with the plethora of model representations that can be used to describe the system forms a complex landscape that can be difficult to navigate. A general overview of this landscape would be too extensive to present here, and hence, is reserved for Chap. 2. Here, we provide some examples to merely illustrate the nature of user-intervention required in this step of the identification cycle. The first step, from the user's perspective, is to determine the class of dynamics of the system being modelled. In order to obtain insight into the nature of dynamics of the system, a user may

   (a) perform informative experiments that reveal the nature of the dynamics of the system after appropriate analysis (see [26]), or
   (b) rely of experience or a general understanding of the physical laws that govern the system.

In both cases, determining the dynamic class of the system is a non-trivial task. An overview of classes of dynamical systems, relevant in the context of this dissertation, is presented in Sect. 2.2.
After determining the dynamic class of the system, the user must also choose an appropriate model representation. This choice is often motivated by (i) the ultimate use-case of the model, or (ii) the identification methodology preferred by the user, see Sect. 2.4 for an overview. For example, for the class of LTI systems, the user may choose:

- *Parametric model representations* such as linear IO models or linear SS models. These may be further structured in terms of the process dynamics and noise properties, e.g., linear IO model structures include, among others, Finite Impulse Response (FIR), Auto-Regressive models with eXogenous inputs (ARX) and Box-Jenkins (BJ) structures.
- *Non-parametric model representations*[2] such as FRFs or Gaussian Process models.

The choice of a particular model representation comes with its own merits and demerits. For example, the advantage of using non-parametric models is that the problem of model structure and complexity selection becomes simplified for the user.[3] On the other hand, non-parametric models are not suitable for many model-based methods such as optimal control design and stability analysis and would require an additional modelling or interpolation step. Ultimately, it becomes the user's responsibility to align the choice of model structure and complexity not only with the dynamics of the true system, but also the ultimate use-case of the model.
Finally, based on the inferred dynamic class of the system and the chosen model representation, the user must choose the appropriate model structure and complexity, which is, again, a non-trivial task.

Remark 1.1 For the class of LTI systems, there exist several robust guidelines for a user to follow in order to choose the appropriate model structure and complexity, e.g., see [19, 26] or automated software packages such as [10, 12]. However, for more complex systems such as NL systems, the choice of the model structure becomes far more involved.

3. Identification criterion—In most real-life applications, the identified model is required to satisfy not one but multiple specifications. For example, when estimating a model for the dynamics of the suspension of a vehicle for the purpose

[2] These models are called non-parametric since they do not make use of an explicit finite-dimensional parameter vector to describe the dynamics of the system. This does not imply that non-parametric models do not contain parameters to be estimated. In fact, for non-parametric models like FRFs, the possible number of parameters to be estimated grows with the amount of data available for its estimation.

[3] Nonetheless, the user is still required to specify certain hyper-parameters that effectively results in model structure and/or complexity selection in a more abstract sense.

of Model Predictive Control (MPC) design, the estimated model should have high accuracy and low run-time complexity, simultaneously. This implies that the performance measure $J(\theta; \mathcal{D}_N)$ should be a vector-valued function rather than a real-valued function. In general, optimization of a vector-valued objective function does not yield a single optimal solution, but an entire set of optimal solutions, also called the *Pareto-front*. Hence, optimizing for multiple identification criteria inherently involves making a trade-off between the measures being optimized.

Most identification methods proposed in the literature formulate the identification problem as a single-objective optimization problem. This necessitates the conversion of a multi-objective optimization problem to a single-objective problem. There are essentially three ways in which this can be done. In the first approach, the single performance measure is taken to be a linear combination of the multiple performance measures, weighted by so-called *trade-off parameters*. A user must a-priori select trade-off parameters. If the resulting model does not satisfy all the performance specifications, the identification cycle must be repeated. In the second approach, the trade-off parameters are seen as hyper-parameters that can be optimized to determine the best trade-off value. In the third approach, one of the performance measures is selected as the primary modelling objective, while the other objectives are converted to "reasonable constraints" imposed on the single-objective optimization problem. The user must determine what constitutes as reasonable constraints for the resulting optimization problem.

4. Optimization method—As illustrated in Fig. 1.6, the optimization step in the identification cycle is determined by the choices made in design of experiment, selection of an identification criterion and specification of the model set. A user must take care to ensure that the chosen optimization technique is appropriate for the chosen identification criterion and the structure imposed on the model. An incorrect choice of optimization technique may result in bias (i.e., structural errors) in the model estimates, or high variability in the model estimate.
5. Model validation—This is a crucial step in the identification cycle, as the user must validate (or invalidate) the hypothesis made in the previous steps of the identification cycle. As discussed in Sect. 1.3.2 there are a number of tools available for the user at this step, including correlation analysis and cross-validation. However, the user must be aware of technical details that underpin these techniques, for example, a linear correlations analysis may not be suitable to analyse the properties of residual signals in a non-linear setting.

   The user may also employ information metrics such as Bayesian Information Criterion (AIC) or Bayesian Information Criterion (BIC), to assess the effect of introducing extra degrees-of-freedom (or complexity) to the model. In the linear setting, this can be useful to select an appropriate complexity of the model, since the number of ways in which complexity can be added to the model is restricted. On the other hand, in the non-linear setting, as the complexity of the model increases, the number of ways in which more degrees-of-freedom can be added to the model increases rapidly. This makes it difficult to use information metrics to choose appropriate model complexity in the non-linear setting.

1.4 Goal of the Research

In Sect. 1.3.3, we gave a glimpse of the challenges a user must encounter when applying one of the standard SI approaches available in the literature. For an inexperienced user, these challenges can be daunting. Hence, there is a need to develop an identification framework that simplifies or, whenever possible, eradicates the need for user-interaction during the identification cycle. An identification framework in which the workflow is streamlined and partially automated for the user enables the user to focus on the utilization purpose of the model rather than developing the model itself. This motivates the research direction adopted in this dissertation.

Research Objective

Develop a framework for system identification that, for a given set of measured data and performance measures set by the user, automates the task of model estimation.

The conceptualized identification framework is illustrated in Fig. 1.8. The research objective, as formulated here, is fairly general and ambitious. In the sequel, we make a number of assumptions to narrow down the scope of research presented in this dissertation. Subsequently, we translate the research objective into actionable research questions. These research questions are motivated by the typical challenges experienced by a user during the identification cycle, as discussed in Sect. 1.3.3. These research questions will be positioned w.r.t existing identification methodologies in Chap. 2, where we provide a broad overview of the system identification literature.

Remark 1.2 The identification cycle begins with the design of experiment and collection of data. As mentioned in Sect. 1.3.2, the design of an input signal is influenced by the properties of the system and the utilization purpose of the model. Moreover, diverse application domains dictate a diverse set of constraints on input excitation

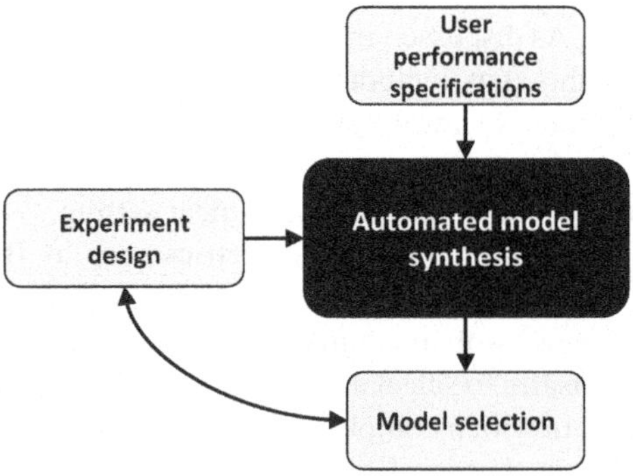

Fig. 1.8 An illustration of the conceptualized identification framework, represented in this figure as 'automated model synthesis'. The desired identification framework automatically estimates models based on performance specifications and measured experimental data provided by the user

signals, e.g., constraints on the amplitude, power, spectral content, length, sign and resolution of the excitation signal. In some application domains, the excitation signal may not be a design variable available to the user, for instance, when modelling stress deformations of mechanical structures due to environmental agents, or when modelling the interaction between biological agents in an organism. Due to this variability in the experiment setting, we begin conceptualizing the desired framework at the point where measured data is already available. This implies that input design is excluded from the scope of research for the purpose of this dissertation. This is done to avoid making application-specific assumptions.

Research Scope

In reality, most physical systems exhibit distributed dynamics, i.e., dynamics w.r.t the spatial domain. However, in terms of measurements obtained at point locations in space, the dynamics of the system are effectively integrated over the spatial domain. Hence, in this dissertation, we consider models of lumped dynamical systems. Furthermore, the primary focus of the research presented in this dissertation will be limited to DT IO representations of SISO dynamical systems. Nevertheless, the identification framework developed in this dissertation is based on general concepts that can be extended to other model representations. Some of the possible extensions will be discussed in Chap. 8.

The objective of "automated system identification" has received a lot of attention in SI literature. In the context of non-parametric system identification, several methodologies have been developed that seek to minimize the number of decisions made by the user, e.g., see [23, 27]. These developments are made possible due to the use of (i) non-parametric model representations like Gaussian Processes, that can be used to describe a large variety of dynamics, and (ii) complex optimization techniques that can be used to automatically estimate the "optimal" complexity of the model. A more detailed overview of such methodologies is reserved for Chap. 2. However, in the setting of parametric models, the research objective of automating system identification remains a relevant challenge. Although some methods have been developed to automate the selection of model terms within a fixed model structure, these approached do not typically extend to other model structures (see Sect. 2.4.2 for an overview). Hence, the scope of this dissertation is restricted to parametric model representations.

Model Structure and Complexity Selection

In Sect. 1.3.3, we discussed the challenges in determining an appropriate model structure and complexity for the identification task. In order to automate this task for the user, the conceptualized identification framework should be able to function across a variety of model structures and complexities. This leads us to the following research question.

Research Question 1

How can we automate the task of model structure and complexity selection in the parametric setting?

Recall that specification of the model set leads to a reformulation of the identification problem as a parameter estimation problem as described in (1.6). Hence, in order to develop a framework where model structure and complexity can be automatically determined, the notion of a model set must be generalized across multiple model structures and varying levels of complexity. While a generalized model set would enhance the general applicability of the desired identification framework, in many use-cases, the user might possess prior knowledge of the system, or application-specific constraints on the model set. These two somewhat contrasting ideas lead to the following two sub-problems.

Research Question 1.1

How can we generalize the notion of model set to span across multiple model structures and model complexities?

Research Question 1.2

How can we introduce meaningful restrictions or prior knowledge to the generalized model set?

Research questions 1.1 and 1.2 are related to the problem of generalizing the model set and adding prior information. However, the problem of selecting the appropriate model structure and complexity remains. This is essentially an optimization problem, and hence, is deferred to Step 4 of the conceptualized identification cycle.

Performance Measure(s)

As discussed in Sect. 1.3.3, a given identification task may involve multiple modelling objectives specified by the user. For an automated identification approach, we cannot rely on a user to reformulate the multiple modelling objectives into a single identification criterion that can be treated with existing identification methodologies (see discussion in Sect. 1.3.3). This motivates the following research question.

Research Question 2

How can we incorporate multiple user-specified performance measures in the identification framework?

There are two key aspects related to this question:

1. In the case of a single modelling criterion, the notion of 'optimality' is well understood. However, in a multi-objective setting, the notion of 'optimality' must be adapted to be able to deal with situations in which the modelling objectives

contradict each other, resulting in not one but a multitude of 'optimal performance values'. While the notion of optimality in a multi-criteria setting has been studied in the field of optimization theory, these concepts have been rarely utilized in the system identification literature.

2. In an automated setting where the model set includes a variety of model structures, there is a greater chance of the conceptualized identification methodology to explore model structures that are *not* appropriate to describe the dynamics of the true system. However, most of the classical identification methodologies use identification criteria that are predicated on the assumption that the true system can be described the chosen model structure (this will be discussed in greater details in Sect. 2.4.2). Hence, in the desired automated setting, we must re-examine the assumptions that are commonly made in traditional identification methodologies.

Remark 1.3 Note that, in classical system identification, the notion of identification criterion, i.e., the cost function to-be-optimized, may differ from the notion of performance measures, which relate to the modelling specifications that an estimated model is required to satisfy. Typically, the identification criterion is a scalar function that measures the "goodness" of a model, and is primarily motivated by the statistical interpretations of the resulting model estimate, while performance measures (or specifications) are used to verify the utility of the estimated model for the intended use-case. However, in this dissertation, we aim to make use of multiple identification criteria, including ones that are typically used as performance measures in classical system identification methods. Hence, in this thesis, we use the terms 'performance measures' and 'identification criteria' interchangeably.

Optimization

The next step in the identification cycle is the optimization step. The generalized problem setting conceptualized by Research Questions 1 and 2 lead naturally to following question.

Research Question 3

How can we optimize model estimates across multiple model structures and multiple identification criteria?

Since the identification problem is formulated across multiple model structures, the optimization method adopted must be capable of robustly solving the model structure selection problem. However, model structure selection is only a part of the identification problem, as any model structure will also contain unknown parameters that must be estimated based on measured data. Finally, the optimization for model structure and parameters must be performed in a multi-objective setting.

1.5 Structure of the Thesis

In the previous section, we formulated a number of research questions motivated by the challenges experienced by a user in a typical identification cycle. We further discuss the relevance of these research questions in the context of existing system identification methodologies in Chap. 2, where we present a general overview of a number of SI methodologies that have been developed in the last few decades. The literature review presented in Chap. 2 is supported by a preliminary discussion on the classes of dynamical systems and some commonly used model representations.

The identification framework developed in this dissertation is built upon three key ingredients: *Evolutionary Computing (EC)*, *Tree Adjoining Grammar (TAG)* and the *multi-objective optimization* framework.

- In Chap. 3, we introduce the preliminaries of EC, which forms the computational basis of the identification framework developed in this thesis, specifically addressing the model structure and complexity selection problem.
- In Chap. 4 we introduce the concept of TAG, a grammar formalism originating in computational linguistics, and develop a generalized notion of a model set based on TAG. The proposed TAG-based method for specifying a model set addresses Research Questions 1.1 and 1.2 proposed in Sect. 1.4.
- In Chap. 5 we introduce the multi-objective framework that will be used to define an ordering on the candidate models based on multiple performance measures. Additionally, we also propose a set of four identification criteria that will be used in the analysis and experimental verification of the proposed identification methodology. Related computational details are also discussed. The ideas presented in Chap. 5 address Research Question 2 proposed in Sect. 1.4.

After introducing the three key ingredients, in Chap. 6 we assemble these ideas together to develop the conceptualized framework for automated identification. The proposed methodology is discussed in detail, and the asymptotic behaviour and computational complexity of the method is analysed. The proposed EC-based identification framework is able to explore models in the generalized model set (based on TAG) and optimize models for multiple identification criteria, thereby addressing Research Question 3, and also the overall research objective.

In Chap. 7, the performance of the proposed identification framework is experimentally evaluated on a collection of identification problems. An academic example is used to empirically study the performance of the proposed framework. Subsequently, a collection of physical case-studies and benchmark problems are used to evaluate the proposed methodology.

Finally, in Chap. 8, we summarize the main contributions of this thesis and reflect on the degree to which the proposed identification framework achieves the research objective set at the beginning of this thesis in Sect. 1.4. We also discuss a number of possible extensions that can be made in order to broaden the scope of the identification concept developed in this thesis.

References

1. Abbott BP, Abbott R, Abbott TD et al (2016) Observation of gravitational waves from a binary black hole merger. Phys Rev Lett 116:061102. https://doi.org/10.1103/PhysRevLett.116.061102
2. Akiyama K, Alberdi A, Alef W et al (2019) First m87 event horizon telescope results. I. The shadow of the supermassive black hole. Astrophys J Lett 875(1):L1
3. Allgöwer F, Zheng A (2012) Nonlinear model predictive control, vol 26. Birkhäuser
4. Boschert S, Rosen R (2016) Digital twin-the simulation aspect. In: Mechatronic futures. Springer, pp 59–74
5. Brockwell PJ, Davis RA (2016) Introduction to time series and forecasting. Springer
6. Close CM, Newell JC, Frederick DK (2002) Modeling and analysis of dynamic systems. Wiley
7. Feynman R (1965) The character of physical law. Cox and Wayman Ltd
8. Franklin GF, Powell JD, Emami-Naeini A, Powell JD (1994) Feedback control of dynamic systems, vol 3. Addison-Wesley Reading, MA
9. Gallego G, Van Ryzin G (1994) Optimal dynamic pricing of inventories with stochastic demand over finite horizons. Manag Sci 40(8):999–1020
10. Gentil S, Barraud A, Szafnicki K (1990) SEXI: an expert identification package. Automatica 26(4):803–809
11. Goel V (2014) Creative brains: designing in the real world. Front Hum Neurosci 8:241
12. Haest M, Bastin G, Gevers M, Wertz V (1990) ESPION: an expert system for system identification. Automatica 26(1):85–95
13. Harari YN (2014) Sapiens: a brief history of humankind. Random House
14. Hibbeler RC, Kiang T (2015) Structural analysis. Pearson Prentice Hall
15. Hood L, Heath JR, Phelps ME, Lin B (2004) Systems biology and new technologies enable predictive and preventative medicine. Science 306(5696):640–643
16. Huang K-Z, Yang H, Lyu MR (2008) Machine learning: modeling data locally and globally. Springer
17. Koutsopoulos I, Tassiulas L (2011) Control and optimization meet the smart power grid: scheduling of power demands for optimal energy management. In: Proceedings of the 2nd international conference on energy-efficient computing and networking. ACM, pp 41–50
18. Leach AR (2001) Molecular modelling: principles and applications. Pearson Education
19. Ljung L (1979) Convergence of recursive estimators. IFAC Proc 12(8):131–144
20. Ljung L (1998) System identification. In: Signal analysis and prediction. Springer, pp 163–173
21. Ljung L (ed) (1999) System identification. Theory for the user, 2nd edn. Prentice Hall PTR. ISBN 0-13-656695-2
22. Ljung L (2010) Perspectives on system identification. Annu Rev Control 34(1):1–12
23. Ljung L, Chen T, Mu B (2019) A shift in paradigm for system identification. Int J Control 1–8
24. Lynch P (2008) The origins of computer weather prediction and climate modeling. J Comput Phys 227(7):3431–3444
25. Nelles O (2002) Nonlinear system identification. IOP Publishing
26. Pintelon R, Schoukens J (2012) System identification: a frequency domain approach. Wiley
27. Rasmussen CE, Williams C (2017) Gaussian processes for machine learning. 2006. Cited on, p 95
28. Ren H, Rank D, Merdes M, Stallkamp J, Kazanzides P (2011) Multisensor data fusion in an integrated tracking system for endoscopic surgery. IEEE Trans Inf Technol Biomed 16(1):106–111
29. Samuelson PA, Scott A (1967) Economics: an introductory analysis, vol 715. McGraw-Hill, New York
30. Treyz GI (2013) Regional economic modeling: a systematic approach to economic forecasting and policy analysis. Springer
31. Van Overschee P, De Moor B (1994) N4SID: subspace algorithms for the identification of combined deterministic-stochastic systems. Automatica 30(1):75–93

32. Willems JC (2007) The behavioral approach to open and interconnected systems. IEEE Control Syst Mag 27(6):46–99
33. Willems JC, Polderman JW (2013) Introduction to mathematical systems theory: a behavioral approach, vol 26. Springer
34. Zaidel DW (2014) Creativity, brain, and art: biological and neurological considerations. Front Hum Neurosci 8:389
35. Zhou K, Doyle JC, Glover K (1996) Robust and optimal control. Prentice-Hall Inc, Upper Saddle River, NJ. ISBN: 0-13-456567-3
36. Zou Y, Tong L, Steven GP (2000) Vibration-based model-dependent damage (delamination) identification and health monitoring for composite structures—a review. J Sound Vib 230(2):357–378

Chapter 2
System Identification: The State-of-the-Art

An original idea? That can't be too hard. The library must be full of them.

Steven Fry

Chapter 1 culminated in the formulation of the research objective pursued in this dissertation: to develop a identification methodology that automates the task of system identification form the user's perspective. The research objective was also dissected into more specific, actionable research questions. The research objective, and the related research question, proposed in Chap. 1 were motivated by the challenges experienced by a user in the typical identification cycle.

In this chapter, we seek to position the proposed research objective, and the constituent research questions, with respect to existing system identification methodologies. In order to achieve this, we provide a general overview of the system identification literature, focusing on some of the commonly-used and well-studied identification methodologies that have been developed over the past decades. The objective is not to provide an exhaustive review, but to shed light on the basic concepts of each of these identification frameworks. This will allow us to understand the nature of decisions a user must make within the context of each of identification frameworks described, and the consequent impact on the quality of estimated models.

2.1 Introduction

The SI problem has been a subject of active research for more than five decades. Many approaches have been developed over the years to address the SI problem. In this chapter, we provide a global overview of the research landscape of SI. The overview presented here is by no means complete. The objective, instead, is to paint a general picture illustrating the various directions in which research in SI has proceeded in the past years. Such a general overview will enable us to position the research direction adopted in this thesis in the context of the SI literature.

D. Khandelwal, *Automating Data-Driven Modelling of Dynamical Systems*,
Springer Theses, https://doi.org/10.1007/978-3-030-90343-5_2

In the context of the research direction adopted in this thesis, two fundamental questions underpin the methodology of any identification technique:

1. What *class(es) of dynamical systems* is the methodology developed for?
2. What *model representation(s)* is the identification technique developed for?

In order to aid the discussion of various identification methods, we first introduce, in Sect. 2.2, a number of ways in which various classes of dynamical systems can be characterized. Subsequently, in Sect. 2.3, we introduce a number of ways in which model structures or representations may be classified. Based on these characterizations, we discuss a number of identification techniques presented in the literature in Sect. 2.4. In Sect. 2.5 we summarize some of the key observations from the presented overview, and connect these with the research questions formulated in Sect. 1.4.

2.2 Classes of Dynamical Systems

The essential challenge of SI is to capture the behaviour of a dynamical system in the form of a mathematical model. Dynamical systems may exhibit a wide range of properties. In this section, we introduce a selection of these properties, restricting our scope to lumped dynamical systems. In order to illustrate these properties, let us assume that a dynamical system can be described as

$$y(t) = f(u(t), t), \tag{2.1}$$

where $u(t) \in \mathcal{U} \subseteq \mathbb{R}$ in the input, $y(t) \in \mathcal{Y} \subseteq \mathbb{R}$ is the output, f is a dynamic operator and time t is in some domain $\mathcal{T}$.

1. Linearity—The system is said to be linear if the following holds

$$f(u_1(t) + u_2(t), t) = f(u_1(t), t) + f(u_2(t), t), \tag{2.2}$$

and

$$f(\alpha u_1(t), t) = \alpha f(u_1(t), t), \tag{2.3}$$

for any $u_1(t), u_2(t) \in \mathcal{U}$ and $\alpha \in \mathbb{R}$. All systems that do not satisfy this property are said to be non-linear.
2. Time-invariance—a system is said to be time-invariant if a finite time-shift $\tau \in \mathcal{T}$ in the input corresponds to an identical delay in the output. i.e., if $y_1(t) = f(u_1(t), t)$, then

$$y_1(t + \tau) = f(u_1(t + \tau), t) \tag{2.4}$$

is true. If the system is time-invariant, then the explicit dependency of (2.1) on time t can be dropped. A system that does not satisfy this property is said to be time-varying.

3. Nature of time-domain—Most physical systems exhibit dynamics that evolve in continuous time, i.e., $\mathcal{T} \subseteq \mathbb{R}$. In contrast, dynamics of digital systems typically evolve in discrete time, which implies $\mathcal{T} \subseteq \mathbb{Z}$. Additionally, hybrid systems exhibit dynamics that evolve in both continuous-time and discrete-time. This may include, for example, dynamics that switch from one mode to another based on time, input excitation or states of the system, or systems in which a physical component interacts with a digital component.
4. Randomness of the cause-effect relationship—When a system is governed by deterministic cause-and-effect relationships between system variables, the system is said to be deterministic. On the other hand, systems may also possess inherent randomness in the cause-and-effect relationships between system variables. Such systems are said to be stochastic.

The behaviour of a given dynamical system may be described by combinations of the properties discussed above. For example, perhaps the most widely studied class of systems are systems that are Linear Time-Invariant (LTI). Systems that are linear, but exhibit time-varying phenomena belong to the class of LTV systems. The class of Non-Linear Time-Invariant (NLTI) systems are far more involved than LTI systems, and may exhibit complex dynamics such as bifurcations, limit cycles and chaos [25]. Switched-Linear (SL) systems, hybrid systems that switch between a collection of LTI systems, may also exhibit unexpected dynamical properties. For example, it is well-known that SL systems can be unstable even when all constituent LTI models are stable [22] (Fig. 2.1).

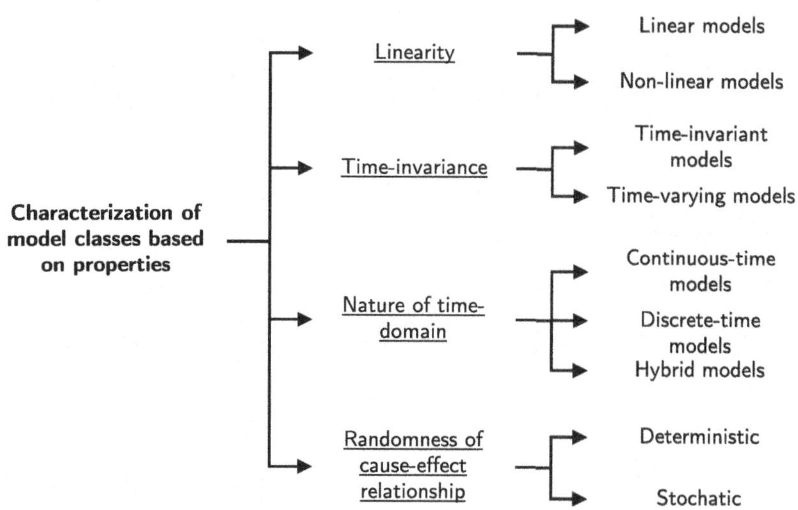

Fig. 2.1 The characterization of model classes based on properties presented in Sect. 2.2

2.3 Characterizations of Model Representations

In order to express various combinations of properties of dynamical systems discussed in the previous section, many model representations have been developed. Model representations can be characterized based on a number of features. In this section, we introduce some of these characterizations of model representations, and examples of model structures therein. The overview is depicted schematically in Fig. 2.3 (Fig. 2.2).

2.3.1 *Deterministic of Probabilistic*

In Sect. 2.2, we introduced stochastic and deterministic relationship between input and output as a characterization of the dynamics of the systems. However, in system identification, this is often seen as a choice of "noise modelling framework", and hence, is relevant in the discussion of model characteristics as well.

Probabilistic models assume that the relationship between the various signals are probabilistic in nature, and probability distributions are required to completely characterize the model. On the other extreme, *deterministic models* assume that the outputs y of a system can be determined completely based on the inputs u given to the

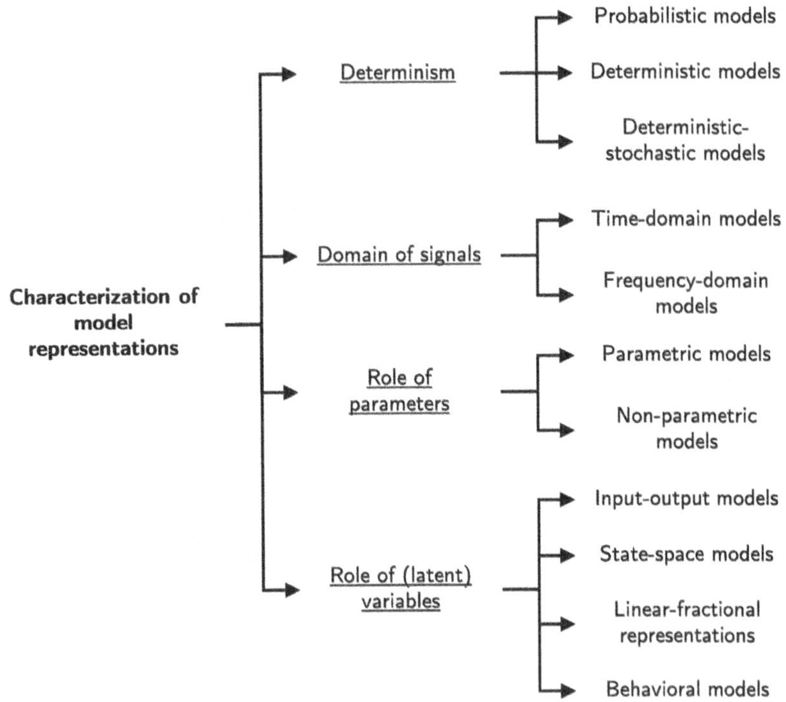

Fig. 2.2 The characterization of model representations presented in Sect. 2.3

system, and do not take into consideration the presence of noise or disturbances. An intermediate approach is to consider *deterministic-stochastic models* that describe the output y as a deterministic consequence of the input u, but affected by stochastic disturbances or noise ξ.

2.3.2 Signal Domain

Models that map input signals to output signals in the time-domain are said to be *time-domain representations* of the system. However, models can also be constructed to map inputs to outputs in transformed domains. A commonly used transformation is the Fourier transform, resulting in *frequency-domain* models.

2.3.3 Parametric or Non-parametric

Non-parametric model sets typically use a large number of parameters, and often, the number of parameters depend on the amount of data available. Due to this flexibility, non-parametric representations of model sets can be used to describe a wide range of dynamics within a particular system class. In contrast, in *parametric model sets*, the dynamics of the system can be mapped to a relatively compact set of model parameters that do not scale with the amount of data measured. As a consequence, for a set of parametric models, the complexity must be tuned w.r.t the dynamics of the system being modelled.

2.3.4 Role of Latent Model Variables

In order to describe the dynamics of any system, the model must, in some way, capture 'the past' of the system. Many model representations have been developed that capture the past of the system with or without using hidden variables. The common model structures include

- *IO* models—These models describe the past directly based on inputs and outputs, and their higher order derivatives (in continuous time) or time-lags (in discrete time).
- *SS* models— The history of the system can also be captured in terms of states of the system. An SS model describes the dynamics of a system using first order derivatives (in continuous time) or a single time-lag (in discrete time) acting on the input and state of the system.
- *Linear Fractional Representation (LFR)* models—These model representations make use of latent variables to separate distinct types of dynamics being described by the model. For example, latent variables can be used to separate the linear part of the model from the non-linear part, see Zhou [89] for more details.

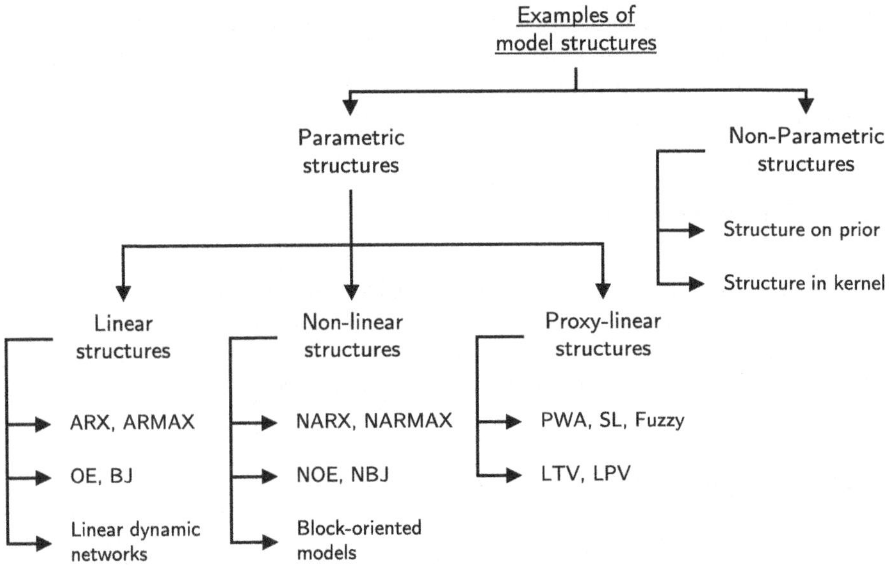

Fig. 2.3 Some examples of structures that can be exposed on models, discussed in Sect. 2.3.5

2.3.5 Examples of Model Structures

In order to accurately describe the dynamics of a system, model representations may be suitably restricted by imposing an *a-priori* structure on the representation. For example, a specific structure may be imposed on the model representation in order to reflect dynamical properties of the system-to-be-modelled, or to reflect the nature of disturbances entering the system and/or measurements. Imposing a meaningful structure on the chosen model representation is an important step in the identification cycle (see discussion in Sect. 1.3.2).

While the imposition of structure is usually associated with parametric models, certain choices made by the user for non-parametric models have a similar effect of imposing desired dynamical properties. In this section we cover some of the commonly used model structures for both parametric and non-parametric models in the SI context.

Parametric models

Broadly speaking, in parametric models, disturbances or noise may be structured in two ways. In the first case, depicted in Fig. 2.4a, noise does not affect the operation of system S directly; noise ξ_u and ξ_y are introduced by the instruments used to measure input u and output y, respectively. This configuration of the identification setup is known as *errors-in-variables* setting and the disturbances are termed as *measurement noise*. More commonly, it is assumed that input u is known exactly, and noise is introduced only in the measurement of output y. This is called the *output-error*

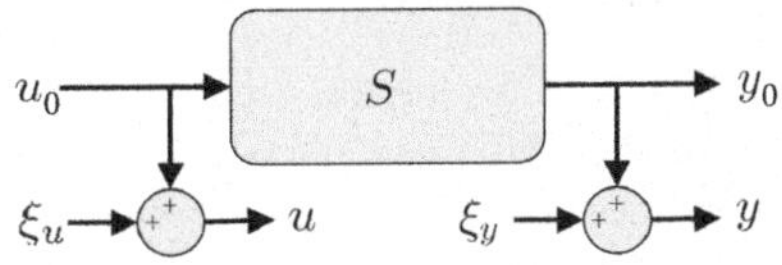

(a) The errors-in-variables noise setting.

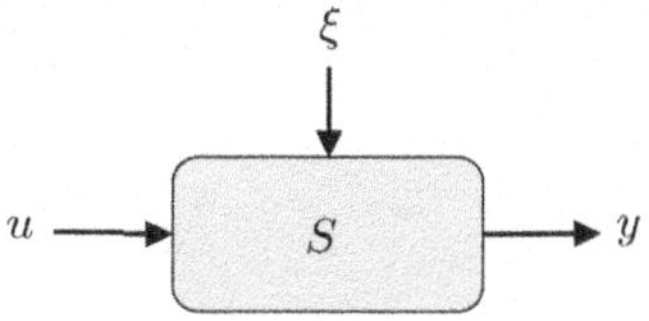

(b) The equation error noise setting.

Fig. 2.4 Commonly adopted noise settings

case.[1] The second configuration is when disturbances directly affect the dynamics of system S, as depicted in Fig. 2.4b. This is usually due to an unknown or unmeasured exogenous excitation. This configuration is termed as the *equation error* setting, and the disturbances are also termed as *process noise*.

Based on this discussion, we introduce a number of commonly used parametric model structures that differ in terms of the dynamics and the noise structure. The material is listed in terms of the intended class of dynamical systems.

1. **LTI model structures**—For the class of LTI systems, one of the more general model structures for an IO representation with equation error is the Auto-Regressive Moving-Average model with eXogenous inputs (ARMAX) structure:

$$y(k) + a_1 y(k-1) + \cdots + a_{n_y} y(k-n_y) = b_0 u(k) + b_1 u(k-1) + \cdots + b_{n_u} u(k-n_u) + \xi(k) + c_1 \xi_{k-1} + \cdots + c_{n_\xi} \xi(k-n_\xi), \quad (2.5)$$

where a_i, b_j and c_k are real coefficients, and n_u, n_y and n_ξ are the maximum lags of the input, output and noise terms respectively, and $\xi[k]$ is a noise process, typically assumed to be an independent and identically-distributed (i.i.d) (i.i.d) random variable distributed as $\mathcal{N}(0, \sigma_\xi)$. Introducing the discrete-time shift operator $q^{-1}u(k) := u(k-1)$, the coefficients in (2.5) can be collected in polynomials $A(q^{-1}) = 1 + a_1 q^{-1} + \ldots a_{n_y} q^{-n_y}$, $B(q^{-1}) = 1 + b_1 q^{-1} + \cdots b_{n_u} q^{-n_u}$ and $C(q^{-1}) = 1 + c_1 q^{-1} + \cdots c_{n_\xi} q^{-n_\xi}$, yielding the short-hand ARMAX representation

$$y(k) = \frac{B(q^{-1})}{A(q^{-1})} u(k) + \frac{C(q^{-1})}{A(q^{-1})} \xi(k). \quad (2.6)$$

The ARMAX model structure can be simplified in order to capture simpler system dynamics with equation error. Setting $C(q^{-1}) = 1$ in (2.6) yields the ARX structure. Furthermore, setting $A(q^{-1}) = 1$ yields the FIR structure.
A general LTI model structure with additive noise on the output is the BJ model

[1] Note that, in the case of linear systems, the errors-in-variables setting reduces to an output-error noise setting, due to the linearity of the system. However, in that case, the noise is correlated with the input, which must be taken into account to accurately estimate a model.

$$
\begin{aligned}
y_O(k) &= \frac{B(q^{-1})}{F(q^{-1})}u(k), \\
\xi_y(k) &= \frac{C(q^{-1})}{D(q^{-1})}\xi(k), \\
y(k) &= y_O(k) + \xi_y(k),
\end{aligned} \tag{2.7}
$$

where $\xi_y(k)$ is a filtered noise process. The BJ model structure is fairly general (in the LTI setting) and includes the models classes of FIR, ARX and ARMAX. The BJ model can also be simplified by setting $C(q^{-1}) = 1$ and $D(q^{-1}) = 1$, yielding the Output Error (OE) (OE) model.
A linear SS representation with linear structure is given by

$$
\begin{aligned}
x(k-1) &= A_S x(k) + B_S u(k) + w(k), \\
y(k) &= C_S x(k) + D_S u(k) + v(k),
\end{aligned} \tag{2.8}
$$

where $u(k) \in \mathbb{R}^m, y(k) \in \mathbb{R}^l$ and $x(k) \in \mathbb{R}^n$ are the input, output and state vectors, respectively, $w(k) \in \mathbb{R}^n$ and $v(k) \in \mathbb{R}^l$ are the process and measurement noise, and A_S, B_S, C_S, D_S are the SS matrices of suitable dimensions.

2. **NL model structures**—The class of NL systems is far more general than that of LTI systems. Determining the appropriate structure for NL models is, as a consequence, a far more involved process.
In this thesis, we make use of parametric IO model structures like Non-linear Auto-Regressive Moving-Average models with eXogenous inputs (NARMAX) or Non-linear Output Error (NOE), that are extensions of their linear counterparts. The NARMAX model is given by

$$
\begin{aligned}
y(k) = f\big(&y(k-1), \ldots, y(k-n_y), u(k), \ldots, u(k-n_u), \xi(k-1), \ldots, \\
&\xi(k-n_\xi)\big) + \xi(k),
\end{aligned} \tag{2.9}
$$

where f is a non-linear function. The Non-linear Auto-Regressive models with eXogenous inputs (NARX) model structure is a special case of the model in (2.9), where the stochastic variables $\xi(k-1), \ldots, \xi(k-n_\xi)$ do not influence the function f. Similarly, the NOE model structure is given by

$$
\begin{aligned}
y_O(k) &= g\left(y_O(k-1), \ldots, y_O(k-n_y), u(k), \ldots, u(k-n_u)\right), \\
y(k) &= y_O(k) + \xi(k),
\end{aligned} \tag{2.10}
$$

where g is a non-linear function. In order to align the model representation to the dynamics of the system, more structure can be imposed on (2.9) and (2.10) by suitably restricting non-linear functions f and g. In the non-linear setting, the Non-linear Box-Jenkins (NBJ) model class, as an extension of NOE, is not well-established.

Non-linear systems can also be described by the NL SS representation

$$\begin{aligned} x(k-1) &= f_S(x(k), u(k), w(k)), \\ y(k) &= g_S(x(k), u(k), v(k)), \end{aligned} \tag{2.11}$$

where f_S and g_S are suitable non-linear functions. Again, more structure can be introduced in the SS representation by imposing suitable restrictions on functions f_S and g_S.
An alternate approach to impose structure in NL models is to use block-oriented models. Block-oriented models are structured interconnections of sub-systems, each of which are described by either LTI models or non-linear functions with or without memory. The sub-systems may be inter-connected in a cascaded, parallel or feedback structure, or combinations thereof. The structure of the interconnection may be inspired from physical interpretation of the system being modelled. For example, a Wiener model can be used to describe an LTI system being measured by an instrument with a non-linear response. See Giri and Bai [21] or Schoukens and Tiels [68] for a more detailed treatment of block-oriented model structures.

3. **Pseudo-linear model structures**—In a number of application domains such as control design and systems analysis, a number of tools have been developed for the class of linear systems that do not extend to the class of non-linear systems. Nevertheless, most real physical systems demonstrate non-linear behaviour. As a result, a number of model structures have been developed to extend modelling capabilities beyond the class of linear systems, while still retaining properties of linear systems, in some sense. This permits the extension of linear techniques to non-linear systems that can be described by these "proxy-linear" model structures. Some commonly used proxy-linear model structures are introduced below.

   - LTV and LPV models—LTV models describe the dynamics of linear systems with time-varying behaviour, see Sect. 2.2. However, LTV models can also be used to describe non-linear dynamical systems, e.g., Kuijstermans [31, Chaps. 6–8] used LTV models to analyse and design non-linear circuits. Several model forms have been used for the identification of LTV models, including frequency-domain representations [34], SS representations [83] and IO representations [37].
     In LPV models, the dynamics (between input and output) are assumed to be linear but dependent on an exogenous signal called the "scheduling signal". LPV models can be considered as extensions to LTV models, and can be used to model non-linear systems, for example, see Balas [5]. LPV techniques have been used widely in control design, see Mohammadpour and Scherer [50] for an overview. Many identification techniques have been proposed for LPV models, see Tóth [74] and Cox [15] for an overview. Furthermore, existing NL models can also be realized as LPV models, see Schoukens and Tóth [69].
   - Piece-Wise Affine (PWA), SL and fuzzy models—Certain classes of non-linear dynamical systems can be described by a collection of linear (or affine)

models. PWA models describe the dynamics of a system as linear (or affine) models in a partitioned input and state domain (for SS representations) or input and output domain (for IO representation). PWA models are commonly used for design of control schemes and analysis of system properties such as stability.

SL models are extensions of PWA models, where the (linear) dynamics of the system is dependent on "switching signals". The switching signal may switch from one mode to the other based on input, state or output values, time, or the past values of the switching signal itself.

Similar to PWA and SL models, fuzzy dynamic models describe non-linear dynamics by partitioning the input, state or output spaces, and using linear models in each partition to describe the dynamics of the system. In the case of fuzzy systems, the partitioning is done based on a set of *if-then* rules.

A survey of identification methods for PWA and SL systems can be found in Garulli et al. [20]. Identification techniques for dynamic fuzzy models, and several applications, can be found in Nelles [53].

Non-parametric structures

For parametric model representations, structure can be explicitly imposed on the functional relationship described by the model. In the case on non-parametric representations, structure is imposed on the model in a more implicit way. Some examples for the LTI and NLTI cases are listed below.

1. **LTI model structures**—In the non-parametric setting, LTI systems can be described by *Markov parameters* (in the time-domain) or FRFs (in the frequency-domain). In a system identification context, the complexity of the model (i.e., number of Markov parameters or the resolution of the FRF estimate) is determined by the experiment design settings, such as sampling properties and length of the dataset.
2. **NL model structures**—Non-linear dynamics of a system can also be described using non-parametric representations, for example, joint probability distributions of the measured outputs of a system, or so-called kernel-based representations of the underlying function. Again, while such descriptions of the system are very flexible, for use in system identification, some structure is imposed on the nature of relationship between the measured signals. These will be discussed in more details in Sect. 2.4.4.

2.4 Identification Methodologies

The SI problem has been studied under various lenses, including statistical inference, systems and control and machine learning. A number of identification methodologies have been developed to infer dynamical models from data. These methodologies are typically developed for a specific class of systems, as discussed in Sect. 2.2, and

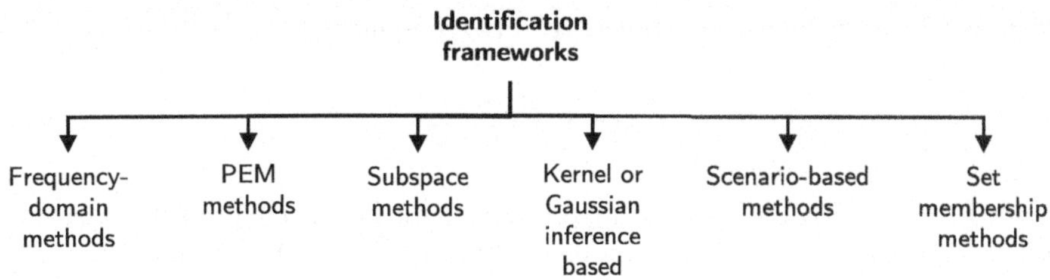

Fig. 2.5 The identification frameworks presented in Sect. 2.4

for a specific model structure representation, as discussed in Sect. 2.3. For a given class of dynamical systems, a number of model representations and structures can be used to describe the system. Invariably, the user determines the model structure or representation that is suitable for his or her intended use-case. Alternatively, the user may choose a familiar identification methodology, that ultimately leads to a choice of model structure representation.

In this section, we introduce some of the main identification methodologies that have been proposed in the literature over the past decades (see Fig. 2.5), and connect them to the system properties and model structures introduced in Sects. 2.2 and 2.3, respectively. The primary purpose of this review is to highlight the various decisions and well-reasoned assumptions a user must make within existing identification methodologies, and the corresponding impact on the resulting model estimates.

2.4.1 Frequency-Domain Methods

Frequency-domain methods were some of the first approaches developed for SI, and are widely applied in many engineering problems such as modal analysis and vibration analysis in structural engineering and acoustics. Frequency-domain methods are used by experienced engineers to gain insight into the system behaviour since they reveal invariant properties of the system such as resonance properties. Frequency-domain insights are also used by control engineers to design simple yet effective Proportional-Integral-Derivative (PID) controllers via loop-shaping for a wide range of applications.

System class and model types

Frequency-domain identification was initially developed for LTI class of systems, and are used to estimate non-parametric input-output models. Consider the true system to be given by.

$$Y(z) = G_0(z)U(z) + \Xi(z), \tag{2.12}$$

where $U(z)$, $Y(z)$ and $\Xi(z)$ are the z-transforms of the input, output and additive measurement noise, and $G_0(z)$ is the z-transform of the impulse response of the system. Based on measurements $\mathcal{D}_N$, one can describe the system in terms of its

Empirical Transfer-Function Estimate (ETFE), computed as

$$\hat{G}(e^{j\omega_k}) = \frac{Y(e^{j\omega_k})}{U(e^{j\omega_k})}, \tag{2.13}$$

where $Y(\cdot)$ ($U(\cdot)$) is the Discrete-time Fourier Transform (DFT) of the measured output (input), and ω_k is a discrete point on the frequency grid determined by experimental conditions (sampling rate and length of the dataset).

Frequency-domain methods have been extended to more general noise structure assumptions such as the errors-in-variables setting [62]. Beyond the class of LTI systems, frequency-domain analysis can also be used to detect 'non-linear distortions' or time-varying dynamics present in the data [34, 62]. In terms of describing non-linear or time-varying dynamics, frequency-domain methods have been extended to estimate non-parametric proxy-linear model structures such as LTV [34] and LPV models [17], and non-parametric non-linear model structures like Generalized Frequency Response Functions (GFRFs) [71]. Frequency-domain methods are also used to initialize parametric identification methods, for example, in Paduart et al. [57].

Identification Criterion and Optimization

The choice of the identification criteria used is typically based on the assumed noise properties of the system and a statistical interpretation of the resulting estimates. For a system with zero-mean additive measurement noise on the output, a linear-least squares estimator $\theta[\text{LS}]$ provides unbiased estimates for the least-squares cost function

$$J(\theta, \mathcal{D}_N) = \frac{1}{N}\sum_{\omega}\left(Y(\omega) - \hat{Y}(\omega,\theta)\right)^2, \tag{2.14}$$

where $\hat{Y}(\omega, \theta)$ is the model output. Furthermore, if the additive noise is i.i.d Gaussian random variable, the least-squares estimate corresponds to the Maximum Likelihood (ML) estimate, see Pintelon and Schoukens [62] for more details. Based on the noise properties, the identification criteria can also be adapted by including a weighting factor $W(\omega)$ over the frequency range. For a noise process with a known covariance matrix, the uncertainty in parameter estimates can be reduced by choosing a suitable weighting factor for the weighted least squares cost function.

When both inputs and outputs are measured with additive Gaussian noise, the least-squares estimator no longer corresponds to the ML estimate. In this case, more complex estimators such as TLS of IV estimator must be used to obtain unbiased estimates, see Pintelon and Schoukens [62] for a more detailed overview.

Discussion

Recall that the purpose of the review presented in this chapter is to highlight the role of the user within the identification methodology. Hence, in this section, we discuss some of the important decisions a user must make in the frequency-domain identification framework. Note that the user-made decisions discussed here are not necessarily specific only to this method.

In order to choose a suitable model structure, a user must be aware of the nature of dynamics being modelled and the properties of the noise process. Certain dynamic properties of the system may be revealed through specially designed experiments. Analysis of measurements obtained from such experiment can allow an experienced user to determine the dynamic properties of the system, and hence, select an appropriate model structure. However, spectral analysis of data inferred from a dynamical system can be sensitive to the experimental conditions. A poorly designed experiment (or pre-processing of the resulting data) can introduce artefacts such as spectral leakage and errors due to transients. Furthermore, improper choice of sampling frequency and experiment length affects the resolution and bandwidth of the resulting estimated model. Finally, since most systems are measured in discrete-time, the user must also be cognizant of implicit assumptions made on the inter-sample behaviour of the dynamical system. Hence, the choices made by the user during experiment design, and the subsequent analysis of measurements play a crucial role in determining a suitable model structure for the system. As discussed earlier, an improper choice of model and noise structure may lead to biased model estimates.

2.4.2 Parametric Prediction Error Minimization

Parametric Prediction Error Minimization (PPEM) method, often referred to as Prediction Error Minimization (PEM) method, came into the foray in the 1960s and 1970s. The need for PEM methods originated from industrial applications originating from engineering domains such as process industries and electro-mechanical production systems [4]. While frequency-domain analysis was enabling design of simple control architectures in these application domains, there was an emerging need for *parametric* methods for more complex control schemes. Initially, several parametric identification methods were developed for various application cases. These methods were subsequently collected under a unifying framework, called PEM method [4]. Since the origins of PEM methods are closely related to the need for parametric models, the term 'PEM' is often implicitly linked with parametric model structures. To avoid confusion, we refer to these as PPEM methods.

System class and model types

PPEM methods have been developed for many classes of dynamical systems, including LTI systems, NL systems and hybrid systems. PPEM methods typically make use of parametric IO model structures. In general, the choice of model structure, and its complexity (i.e. degrees of freedom), is treated as a hypothesis made by the user. Hence, in Step 2 of the identification cycle, based on expert knowledge, experience, and informative experiments, the user must choose a suitable structure and complexity of the dynamical and noise properties of the system. This choice leads to the choice of a model set $\mathcal{M}$ parameterized by a parameter vector θ. Subsequently, in Step 5 of the identification cycle, the user must validate the model structure hypotheses made in the earlier step. At this stage, a number of tools can be used to validate the hypothesis or determine the generalizability of the estimated model, e.g., cross validation using an independent dataset, information metrics such as AIC or BIC to estimate the generalization error of the model, and correlation analysis of residual signals. For more details, see Ljung [40].

For the case of LTI systems, users may impose equation-error type model structures, such as ARX or ARMAX (see (2.6)), or output-error type model structures, such as OE or BJ (see (2.7)). More complex LTI systems may be described as a network of simpler LTI systems [76].

For the case of NL systems, the user may employ non-linear extensions of parametric models structures such as NARMAX in (2.9) or NOE in (2.10). To estimate such models, the user must prescribe the structure and complexity of non-linear functions f and g, which is a non-trivial task. A common approach is to restrict functions f and g to specific classes of non-linear functions such as polynomial or wavelet functions [10]. Polynomial NARMAX models have been widely used in the literature due to their favourable approximation properties for continuous functions over closed domains.

Alternatively, for NL systems, structure can also be imposed in the form of block-oriented models [21]. The primary advantage of block-oriented models is its transparency; the structure of the interconnection may be inspired from physical interpretation of the system being modelled.

Choosing a suitable model structure can be a daunting task for an inexperienced user. In the case of LTI systems, there exist some robust guidelines for user, e.g., see Ljung [40]. However, for more complex classes of dynamics, the task of model structure and complexity selection is far more involved, and no general guidelines exist.

Identification Criterion and Optimization

As the name suggests, PPEM methods use prediction error to define an optimization criterion. The model structures presented in the previous section can be used to infer a parametric *prediction model*, i.e., a model that computes a prediction of the output of the system, based on past measured data and given model parameters. Assume that the prediction model is defined as

$$\hat{y}(k \mid \theta) = f_{\mathrm{p}}(k, \mathcal{D}_{k-1}, u(k) \mid \theta), \tag{2.15}$$

where $\hat{y}(k)$ is the predicted output at time-instant k, $\mathcal{D}_{k-1}$ is the measured data upto time-instant $k-1$, θ is a vector containing the unknown parameters of the model, and f_{p} is the prediction model. More specifically, the model in (2.15) is called a one-step-ahead prediction model since, based on data measured upto $k-1$ time samples and input $u(k)$, the model predicts the output at time sample k. This notion can be generalized to multi-step-ahead prediction models. Prediction error ε is defined as the difference between the measured and the predicted output, i.e.

$$\varepsilon(k) = y(k) - \hat{y}(k \mid \theta). \tag{2.16}$$

A general prediction error identification criterion takes the form

$$J(\theta, \mathcal{D}_N) = \sum_{k=1}^{N} l(y(k) - \hat{y}(k \mid \theta)), \tag{2.17}$$

where the $l(\cdot)$ is a positive real-valued function, called the *loss function*. A common choice for the loss function is the mean squared loss function.

Using the mean squared one-step-ahead prediction error to determine the quality of the model has been widely popular in the identification community. This can be attributed to two reasons. Firstly, under the assumption that the underlying noise process is zero-mean and i.i.d Gaussian process, and that the true system can be described by the chosen model structure, the least-squares estimate corresponds to the ML estimate of the parameters. Secondly, for several classes of systems such as ARX, ARMAX and polynomial NARMAX, relatively simple and reliable numerical optimization techniques such as (linear) Least Squares (LS) and Iterative Least Squares (ILS) can be used to compute the optimal model parameters efficiently. For more complex parameterization, a gradient-descent method is typically used to estimate model parameters.

Discussion

Despite the variety of LTI model structures available, there exists effective and robust guidelines enabling users to choose appropriate model structures for their identification task [39]. These guidelines also provide reliable heuristics to choose the appropriate model complexity, i.e., degrees of the $A(q^{-1})$, $B(q^{-1})$, $C(q^{-1})$, $D(q^{-1})$ and $F(q^{-1})$ polynomials in (2.6 and 2.7). Most commonly, frequency-domain analysis is used to guide the user in this selection process. Using non-parametric analysis, a user can determine the number of poles in the system, or equivalently, the order of polynomial $A(q^{-1})$ (or $F(q^{-1})$) in (2.6) (or (2.7)). Moreover, with specially designed experiments, the same non-parametric analysis can be extended to determine the complexity of the noise model. Details on non-parametric modelling and analysis techniques can be found in Pintelon and Schoukens [62]. While there exist many tools to validate (or invalidate) the model structure hypothesized, e.g., via residual analysis, they can only be applied *a-posteriori*, i.e., after having estimated the model.

Unlike in the LTI case, there are not many robust guidelines for the user to select appropriate structure and complexity for NARMAX models. Instead, the common approach is to restrict functions f and g in (2.9 and 2.10) to a specific class of non-linear functions, and use numerical algorithms that essentially solve a combinatorial optimization problem using greedy search techniques [12], l_1 regularization or shrinkage methods [65]. In general, the resulting models may not be immediately interpretable for a user. However, research efforts have been directed to developing non-parametric analysis techniques based on identified NARMAX models, making the models more open to human interpretation. More details on the 'NARMAX philosophy' and related numerical and analysis techniques can be found in Billings [10].

For block-oriented modelling, some guiding principles have been developed to discriminate between some simple block-oriented structures. These principles are based on non-parametric analysis of data generated from the system using specialized experiments, see Schoukens et al. [67] for details. However, for more complex block-oriented models, there is no general guiding framework for the user. Furthermore, determining the appropriate complexity of each of the sub-systems in a given block-oriented model can be a challenging task for the user. Instead, similar to the case of NARMAX models, the complexity is determined by numerical algorithms that essentially solve a combinatorial optimization problem.

2.4.3 Subspace Methods

PPEM methods discussed in Sect. 2.4.2 are primarily developed for IO representations, as in (2.5)–(2.10). In the subjects of filtering and control, a number of techniques were developed for SS representations of systems. In the case of SISO LTI systems, realization theory can be utilized to infer a canonical SS realization from an identified input-output model. However, for MIMO systems, still in the LTI setting, conversion to SS form can be challenging. Hence, there was a need to estimate SS models directly and efficiently from data. This lead to the development of subspace methods for SI. Seminal works that introduced subspace methods to SI include [33, 77, 81].

A second motivation that lead to the development of subspace identification is the need for robust numerical techniques. PPEM methods must often rely on non-linear optimization techniques that lead to locally optimal solutions that may possibly be far away from the global optima. Subspace techniques utilize numerical linear algebra techniques such as QR factorization and Singular Value Decomposition (SVD) that are more robust than, for example, gradient-based algorithms for optimization. In fact, in state-of-the-art implementations of PPEM methods, subspace identification is often used to derive an initial LTI estimate of the model.

Model structures

Subspace methods for SI were originally developed for the class of LTI systems to be described by SS models. The typical DT LTI SS form is given by

$$\begin{aligned} x(k-1) &= A_{\mathrm{S}}x(k) + B_{\mathrm{S}}u(k) + w(k), \\ y(k) &= C_{\mathrm{S}}x(k) + D_{\mathrm{S}}u(k) + v(k), \end{aligned} \tag{2.18}$$

where $u(k) \in \mathbb{R}^m$, $y(k) \in \mathbb{R}^l$ and $x(k) \in \mathbb{R}^n$ are the input, output and state vectors, respectively, $w(k) \in \mathbb{R}^n$ and $v(k) \in \mathbb{R}^l$ are the process and measurement noise, and A_{S}, B_{S}, C_{S}, D_{S} are the SS matrices of suitable dimensions. Typically, the process and measurement noise are assumed to be zero-mean, with positive-definite covariance

$$\operatorname{cov}(w(k), v(k)) = \begin{bmatrix} Q & S \\ S^\top & R \end{bmatrix}, \tag{2.19}$$

where matrices Q, S, R have appropriate dimensions.

Algorithms have been developed for identification of deterministic models (matrices A_S, B_S, C_S, D_S) (e.g. [43, 51]), stochastic models (matrices A_S, C_S, Q, S, R) (e.g. [7]), and combined stochastic-deterministic models (e.g. [33, 58, 77]).

Due to the nature of computations involved (to be discussed in the sequel), extension of subspace techniques to non-linear model structures is typically not straightforward. However, these methods have been successfully extended to specific classes on non-linear and proxy-linear model structures, such as bilinear systems [18], Wiener systems [86], Hammerstein systems [82] and LPV systems [80]. In a deterministic setting, Noël and Kerschen [55] extended subspace methods to estimate non-linear mechanical systems.

Numerical methods

Unlike frequency-domain or PPEM methods, subspace methods do not optimize for models based on a given identification criterion. Instead, subspace methods rely on linear algebra techniques and realization theory. The three commonly used subspace identification methods are *N4SID* (numerical algorithms for subspace state-space system identification) (N4SID), *MOESP* (multivariable output-error state-space) (MOESP) and *CVA* (canonical variate analysis) (CVA). Van Overschee and De Moor [78] proved that, under certain conditions, the three algorithms can be unified in a single framework and are similar to each other upto two weighting matrices. The key assumptions for this result include (but not limited to) (i) infinitely long dataset (for asymptotic statistical analysis), and (ii) statistical independence of input $u(k)$ with process noise w_k and measurement noise $v(k)$. However, the insight that led to the development of these three algorithms differ, and so do the underlying numerical techniques. Hence, when any of the assumptions in Van Overschee and De Moor [78] is not satisfied, the results may vary significantly.

The typical sequence of operations in subspace methods is illustrated in Fig. 2.6. Here, we provide a brief description of the steps involved in the N4SID algorithm, computational details can be found in, e.g., Van Overschee and De Moor [79]. The subspace approach is set up by first splitting the dataset in two parts—a *past* dataset, denoted by (U_p, Y_p), and *future* dataset, denoted by (U_f, Y_f). Then, using linear projections of the future output Y_f onto the past data (U_p, Y_p) and future inputs U_f, one can compute an estimate of the extended observability matrix $\hat{\Gamma}_i := (C, CA, \ldots, CA^{i-1})$, and state vector estimates, denoted by $\hat{X}_i$, where i (greater than n, the dimension of the system) is determined by the partitioning of the data. Subsequently, the SVD of $\hat{\Gamma}_i$ is used to determine a suitable model order selection[2]—the number of 'dominant' singular values is chosen as the order n of the model. For finite datasets, all singular values are expected to be non-zero, making this a non-trivial choice for the user. Based on this choice, the state vector estimates $\hat{X}_n$ can be computed. Finally, the system matrices in (2.18) and covariance matrices in (2.19) can be computed based on $\hat{\Gamma}_i$ and $\hat{X}_n$.

[2] Pre- and post-multiplication of $\hat{\Gamma}_i$ with suitable weighting matrices W_1 and W_2 yield MOESP or CVA algorithms under suitable assumptions. See details in Van Overschee and De Moor [78].

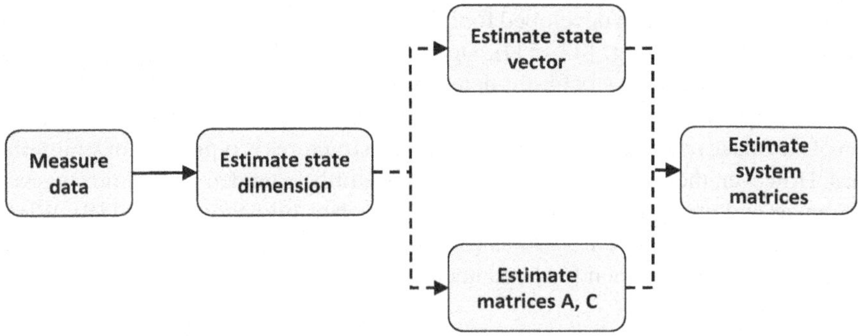

Fig. 2.6 The typical subspace identification workflow [78]. Dashed lines indicate alternate routes, depending on the algorithm

Discussion

From the user's perspective, subspace methods require little interaction with the user. An implicit assumption here is that the user is interested in obtaining an LTI[3] SS model (deterministic or stochastic) for the given system. Perhaps the most crucial choice made by the user is the dimensionality of the state vector. Depending on the particular algorithm used, the user may be required to choose the dimensionality of the state vector based on, for example, singular values of certain matrices or variance of certain residual signals estimated from data [6]. When the dataset is finite and noisy, or when the data-generating system does not satisfy assumptions made on system propoerties (such a linearity), the estimate of state dimensionality may be inaccurate. Furthermore, the estimation of the system order typically requires the partition of measured data into a *past* and *future* dataset. Bauer [6] demonstrated that order estimation in finite datasets can be sensitive to a number of user-specified parameters, including the partitioning of the datasets. Dahlén et al. [16] demonstrated experimentally that, when certain underlying assumptions on the data-generating system are not satisfied, stochastic subspace methods can fail drastically.

2.4.4 Kernel Methods and Related Approaches

Kernel methods were proposed in the study of solutions of integral equations (e.g., see Mercer [45]), and also in the study of general classes of functions, e.g., in Bergman and Schiffer [9]. A general theory of Kernel functions and regularization in the related function spaces can be found in Aronszajn [2]. For ease of illustration, we consider a static data generating system governed by the following equation

[3] As mentioned earlier, while subspace methods can be extended to specific classes of non-linear systems, the extension is usually more involved and require a specially designed numerical methods.

$$y = f_0(u) + \xi, \tag{2.20}$$

where, $u \in \mathscr{U} \subseteq \mathbb{R}$ and $y \in \mathbb{R}$ and ξ is a random variable in $\mathbb{R}$. The basic idea of kernel-based methods is to estimate a continuous function f in some generalized space of functions $\mathscr{H}$ that minimizes some regularized optimization criterion formulated on the basis of sampled data obtained from (2.20). More details will be provided in the sequel. The main advantages of Kernel methods are:

- When the space of functions $\mathscr{H}$ is carefully chosen, continuous functions can be arbitrarily well approximated.
- The definition of the function space $\mathscr{H}$ automatically provides a suitable regularization term that allows to tune the complexity of the function estimates in a smooth manner while ensuring well-posedness of the solution.
- Since this is a non-parametric approach, the user must not pre-determine a suitable model structure and complexity.

An alternate perspective for function estimation from sampled data is provided by the framework of Bayesian estimation of stochastic processes. In this setting, a prior probability distribution is postulated for function $f(\cdot)$, and is denoted by $p(f)$. Additionally, a probability distribution $p(\xi)$ is assumed for the noise. The objective is to compute the posterior distribution of the function $p(f \mid Y_N)$ conditioned on the sampled data $Y_N = (y_1 \cdots y_n)^\top$. A common approach in Bayesian inference is to assume that the noise in (2.20) is a i.i.d Gaussian random variable and that the prior distribution on f is a Gaussian process.[4,5] In this setting, Kimeldorf and Wahba [28] pointed out a correspondence between statistical inference using kernels and Bayesian estimation of Gaussian Processes. This correspondence will be highlighted in the sequel.

While Kernel methods and Gaussian processes became popular in the machine learning community to estimate (distributions over) functions from sampled data, these methods have been suitably adapted for identification of dynamical systems (see Pillonetto and De Nicolao [59], Pillonetto et al. [60, 61]).

Function Space

Unlike PPEM method, the Kernel method does not require the user to specify a model structure and complexity. However, the user must specify the space of functions $\mathscr{H}$. In particular, Kernel-based methods utilize the so-called Reproducing-Kernel Hilbert Space (RKHS) of functions. An RKHS $\mathscr{H}$ is a possibly infinite-dimensional Hilbert space of functions defined over some domain $\mathscr{U}$ with the associated inner-product $\langle f, g\rangle$ defined for all $f, g \in \mathscr{H}$ and the induced norm $\|f\|_{\mathscr{H}} := \langle f, f\rangle$. Additionally, an RKHS is associated with a positive-definite symmetric kernel function $K : \mathscr{U} \times \mathscr{U} \to \mathbb{R}$ that satisfies the property that for any $u \in \mathscr{U}$ the kernel satisfies

$$\langle K(u, \cdot), f\rangle = f(u), \ \forall f \in \mathscr{H}. \tag{2.21}$$

[4] A Gaussian Process is a generalization of the Gaussian distribution over functions with a continuous domain, see Rasmussen and Williams [64] for details.

[5] These assumptions provide significant computational advantage, as will be discussed in the sequel.

For any fixed $u \in \mathscr{U}$ the kernel function $K(u, \cdot) \in \mathscr{H}$ is a linear functional in terms of the second argument and is called the Kernel section $K_u(\cdot)$ centred at u.

There are a number of ways in which a user can choose or construct an RKHS, or the corresponding kernel function $K(\cdot, \cdot)$, see [66, Chap. 2] for a detailed discussion. Examples of commonly used Kernel functions include the homogeneous polynomial kernel

$$K_\mathrm{p}(x_1, x_2) = (u_1^\top u_2)^d, \tag{2.22}$$

for $u_1, u_2 \in \mathscr{X} \subseteq \mathbb{R}_n$ and $d \in \mathbb{N}$, the Gaussian kernel

$$K_\mathrm{G}(u_1, u_2) = \exp\left(-\frac{u_1 - u_2^2}{2\sigma^2}\right), \tag{2.23}$$

for $\sigma \in \mathbb{R}_{>0}$, and cubic spline kernels

$$K_\mathrm{CS} = \begin{cases} \frac{u_1^2}{2}\left(u_2 - \frac{u_1}{3}\right), & u_1 \geq u_2 \\ \frac{u_2^2}{2}\left(u_1 - \frac{u_2}{3}\right), & u_2 < u_1. \end{cases} \tag{2.24}$$

For the purpose of system identification, Pillonetto et al. [61] illustrated the importance of introducing suitable constraints such as stability in the kernel functions. As a consequence, Pillonetto and De Nicolao [59] introduced the stable spline kernel

$$K_\mathrm{SS}(u_1, u_2; \beta) = K_\mathrm{CS}\left(e^{-\beta u_1}, e^{-\beta u_2}\right), \tag{2.25}$$

where $\beta \in \mathbb{R}_{>0}$ is a hyper-parameter.

In the Bayesian setting for inference over Gaussian processes, the user must specify the prior distribution $p(f)$. A common choice of prior $p(f)$ is a zero-mean Gaussian process with a covariance function $K_\mathrm{COV} = \mathbb{E}(f(u_1), f(u_2))$, which is taken as the Kernel function. The choice of covariance function K_COV can bear a strong influence on the result of the inference [64, Chap. 5].

Kernel methods and Gaussian process have been proposed in the literature for modelling of FIR models [59], Volterra series models [11], Wiener systems [38], state space models [19], and many other model types and structures. A recent review can be found in Chiuso and Pillonetto [14].

Identification Criterion

An advantage of using Kernel methods is the automatic selection of model complexity that achieves an optimal trade-off between bias an variance of the estimate. This is achieved through a regularization term in the optimization problem. Regularization was popularized by Tikhonov and Arsenin [73] to solve inverse problems that are ill-posed. A typical regularized function estimation criterion with a sum-of-squares loss function can be written as

$$V_N(f) = \sum_{i=1}^{N}(y_i - f(u_i))^2 + \lambda\|f\|^2_{\mathcal{H}}, \qquad \lambda \in \mathbb{R}, \tag{2.26}$$

where $\{(u_i, y_i)\}_{i=1}^N$ are sampled data obtained from (2.20) and $f \in \mathcal{H}$. The regularization term $\|f\|^2_{\mathcal{H}}$ penalizes the complexity of the function f, as measured in the induced norm of the RKHS $\mathcal{H}$. The hyper-parameter λ achieves a trade-off between the quality of fit (measured in terms of the the sum of squared errors in (2.26)) and the complexity of the function. Besides the squared-error loss function used in (2.26), a user may choose other loss functions such as 0–1 misclassification, absolute error or logistic error loss function (see [66]), depending on the ultimate use-case of the estimated model.

In contrast to Kernel methods, where the objective is to estimate a function $f \in \mathcal{H}$, the objective in the Bayesian setting is to provide a complete probabilistic description of the estimated function, denoted by $p(f \mid U_N, Y_N)$, where $U_N = (u_1 \cdots u_n)^\top$ and $Y_N = (y_1 \cdots y_n)^\top$ are the sampled data points. The celebrated Bayes' rule establishes the following relationship

$$p(f \mid U_N, Y_N) = \frac{p(Y_N \mid U_N, f)p(f)}{p(Y_N \mid U_N)}, \tag{2.27}$$

where $p(Y_N \mid f)$ is the likelihood function and $p(Y_N)$ is the marginal likelihood, a normalizing constant that does not depend on f. The posterior $p(f \mid U_N, Y_N)$ effectively combines the prior information encoded in $p(f)$ along with the evidence provided in the measurements, encoded in the likelihood function $p(Y_N \mid U_N, f)$. Often, the prior over functions is parameterized in terms of some parameter vector θ. In this case, the Bayes' rule can be written as

$$p(f;\theta \mid U_N, Y_N) = \frac{p(Y_N \mid U_N, f;\theta)p(f;\theta)}{p(Y_N \mid U_N)}, \tag{2.28}$$

where $p(f;\theta)$ denotes the probability distribution of function f parameterized by vector θ.

Optimization

Solving the optimization problem in (2.26) in a possibly infinite-dimensional space $\mathcal{H}$ is made possible by the acclaimed Representer Theorem [27], which states that any minimizer of (2.26) in $\mathcal{H}$ can be expressed in the following form

$$\hat{f}(u) = \sum_{i=1}^{N} \hat{\alpha}_i K(u_i, u), \qquad \hat{\alpha}_i \in \mathbb{R}. \tag{2.29}$$

The implication of the Representer theorem is that while $\mathcal{H}$ might be an infinite-dimensional space, the optimizer of (2.26) can be expressed as combination of kernel sections centered at each data-point x_1. The Representer theorem has been extended to other loss functions and regularization terms, see Schölkopf et al. [66]. For the criterion in (2.26) and for a known λ, an estimate of the coefficients $\hat{\alpha} = (\hat{\alpha}_1, \ldots, \hat{\alpha}_N)^\top$ can be computed algebraically as

$$\hat{\alpha} = \left(\hat{K} + \lambda I_N\right)^{-1} Y_N, \tag{2.30}$$

where $\hat{K}$ is a matrix with elements $\hat{K}_{ij} = K(u_i, u_j)$. The results in (2.29) and (2.30) have been extended to system identification, e.g. in Pillonetto and De Nicolao [59] for estimating FIR models.

Surprisingly, Bayesian solution to the inference problem using Gaussian processes leads to the same solution as in (2.29) and (2.30), under some conditions. It can be shown that, if the hyper-parameters (i.e., noise variance σ_ξ^2) is known, the posterior distribution $p(f \mid Y_N, U_N, \sigma)$ is distributed as $\mathcal{N}(m_\mathrm{p}, \Sigma_\mathrm{p})$, where m_p is the posterior mean function and Σ_p is the posterior covariance function (details can be found in [64, Chap. 2]). Since $p(f \mid Y_N, U_N, \sigma)$ is Gaussian, m_p is also the maximum *a-posteriori* (MAP) estimate of $p(f \mid Y_N, U_N, \sigma)$. Kimeldorf and Wahba [28] demonstrated that when $\lambda = \sigma_\xi^2$ and $K = K_\mathrm{COV}$ in (2.29) and (2.30), the kernel solution coincides with the MAP estimate obtained by Bayesian estimation. More details of this comparison can be found in Rasmussen and Williams [64, Chap. 6].

Discussion

When the kernel function is carefully chosen and for relatively short data-sets, Chen et al. [13] demonstrated that kernel methods can out-performed PPEM methods in well-studied identification problems such as estimation of FIR models. However, Chen et al. [13] also demonstrated that the results obtained using kernel methods significantly depend on how well the kernel function is adapted to the estimation problem at hand. Similar to kernel methods, the performance of Bayesian inference for a given finite dataset depends on the selection and tuning of the prior's covariance function [64, Chap. 5]. Hence, for the Bayesian or kernel-based identification approaches, the primary challenge for the user is to select an appropriate prior or kernel function. This choice, in an abstract sense, is akin to the choice of model structure or complexity in the PPEM framework.

In terms of computations, the solutions presented here can be computed analytically when (i) the hyper-parameters (noise variance σ_ξ^2, parameter related to kernel specification θ and regularization parameter λ) are known to the user, (ii) the noise scenario is assumed to be additive white Gaussian, and (iii) there are no hidden variables such as state variables. However, this is usually not the case. The problem of selecting hyper-parameters in Bayesian or kernel methods have been likened to the problem of selecting model structures in PPEM method [61]. When the hyper-parameters are unknown, the problem must be solved via the optimization of the marginal likelihood. In a simple noise setting (i.e., additive white Gaussian noise on the output) and in the absence of hidden variables, this can be solved analytically. In other cases, more complex optimization approaches must be employed by the user, e.g., the Markov Chain Monte-Carlo (MCMC) algorithm [54]. Hence, while Gaussian processes or kernel methods appear to circumvent the problem of model structure and complexity selection, from the user's perspective, the challenge is diverted to the design of the kernel and hyper-parameter selection. Although the Bayesian setting offers a new paradigm for optimizing hyper-parameters via the marginal likelihood, it is, in general, not necessarily an easier problem for the user to solve.

2.4.5 *Neural Networks*

Rapid improvement in computational techniques (such as backpropagation-based gradient-descent) and computing capabilities has led to the meteoric rise in popularity of Neural Networkss (NNs). Development of NN-based techniques have revolutionized several research fields such as computer vision, speech recognition, by out-competing existing techniques overwhelmingly. The initial success of NNs attracted widespread research interest, and several NN architectures were developed, for examples, Recurrent Neural Network (RNN) and Convolutional Neural Network (CNN). Development of scalable and numerically efficient stochastic gradient-based optimization techniques, e.g., see Kingma and Ba [29], led to scaling up of NN architectures, resulting in the paradigm of deep learning [35].

In the domain of system identification, NN-based methodologies have also been developed for identifying dynamical systems from data. Initially, NN-based identification techniques made use of NNs as an alternate approach to approximate non-linearities, and effectively combined NNs with existing identification frameworks such as PPEM. A review of such approaches can be found in Sjöberg et al. [70], Narendra and Parthasarathy [52], Suykens et al. [72].

With the advent of deep learning, several deep architectures have been developed for the purpose of identification of non-linear dynamical systems. The use of deep learning in non-linear system identification is currently a subject of active research, and cannot be said to constitute a general framework for non-linear system identification yet. In the following, we discuss some of the methodologies that have been recently proposed in the literature. It should be noted that this is by no means an exhaustive review.

One of the deep NN architectures that have received significant attention in SI literature is auto-encoders (see [23] for details). Masti and Bempoad [44] developed a deep auto-encoder-based approach to estimate (i) a mapping from high-dimensional past input (and possibly output) data to a low-dimensional vector that serves as a state estimate of the system, (ii) a mapping from the low-dimensional state estimate to the output at the next time step, and (iii) a mapping from the current state estimate and input to the state at the next time step. In order to estimate the three mappings simultaneously, the authors propose a multi-objective optimization problem which is converted to a single-objective optimization problem by introducing trade-off parameters. Auto-encoders have also been used to estimate *Koopman eigen-function* representations of non-linear systems, for example, in Lusch et al. [41], Mardt et al. [42], Otto and Rowley [56]. Koopman eigen-functions serve as non-linear coordinates that enable the description of the dynamics of the system using linear functions, which makes it possible to study various system-theoretic properties of non-linear systems, such as stability and limit cycles. While these methods use NN to describe a different representation of a non-linear system, the underlying concept is similar to that in Masti and Bemporad [44].

Other NN architectures that have been used for SI include Long Short-Term Memory (LSTM) [85] and temporal CNN [1]. In Andersson et al. [1], the authors discussed similarities, in representational capabilities, between NN architectures, such as LSTMs and temporal CNN, on one hand, and traditional non-linear system representations, such as Volterra series and cascaded parallel Wiener models, on the other, thereby establishing the relevance of such NN architectures in context of SI.

Discussion

In order to use auto-encoder-based learning methods, the user must choose several hyper-parameters of the algorithm including the number of hidden layers of the architecture, the activation functions, and crucially, the number of states (i.e., number of neurons in the central layer of the auto-encoder). These choices are similar, in effect, to choices related to model structure and complexity in other identification methodologies such as subspace and PPEM methods. Similar hyper-parameter selection must also be done for other NN architectures. Due to the abstract nature of these hyper-parameters, and lack of general guidelines, selecting the specifications of the NN architecture is a non-trivial task for a user.

These methods also require relatively large sizes of training data to learn the dynamics of the system (except temporal CNN, see Andersson et al. [1]). Finally, while NN-based techniques can be used to describe complex non-linear dynamics, the interpretability and generalizability of the models is, in general, lacking. While, in recent years, research focus has been directed towards learning interpretable and generalizable NN, much work is yet to be done [88].

2.4.6 Evolutionary Algorithms

EAs are a class of global optimization techniques that make use of biologically-inspired heuristics to solve complex non-linear optimization problems. Due to the flexible nature of these optimization techniques, they have been successfully used in a large number of application domains, for example, see Lehman et al. [36] and references therein. In the context of system identification, Evolutionary Algorithmss (EAs) have been primarily used in two capacities: (i) in estimation of parameters of a fixed model structure for a given, typically non-convex, identification criteria, and (ii) in determining a suitable model structure and complexity in a combinatorial landscape of possible model structures. Evolutionary Algorithms-based SI approaches do not form a general SI methodology unique from the other SI frameworks described in this chapter. Instead, EAs are typically used as powerful optimization approaches to solve challenging problems occurring in existing methodologies, such as the PPEM method.

EAs play a prominent role in the identification approach developed in this thesis. Hence, a more detailed discussion of EAs, especially in the context of SI, is deferred to Chap. 3.

2.4.7 Set-Based Methods

The set-membership approach to system identification distinguishes itself from other traditional system identification methodologies such as PPEM or frequency-domain methods based on the treatment of disturbances in the system or measured data. Traditional SI methods treat disturbances in a statistical framework, and results such as uncertainty bounds on parameter estimates are dependent on assumptions made on the statistical properties of the noise process (e.g., see Ljung [40]). Such results are no longer valid when the underlying assumptions are violated, for example, in the scenario of approximate modelling, the residuals signals may be correlated with the input, which violates one of the standard assumptions in the PPEM framework. On the other hand, in set-membership method for SI, disturbances are assumed to be unknown but bounded in some signal norm, typically L_1, L_2 or L_∞, and no statistical properties of the noise process is assumed. This allows the development of estimators that provide theoretical uncertainty bounds on parameter (or state) estimates obtained from finite data, without any statistical assumption of noise properties, e.g., see Witsenhausen [87] and review in Walter and Piet-Lahanier [84]. Furthermore, set-theoretic methods do not estimate a single *optimal* model, but a *feasible set* of models (or parameters) that have not been invalidated by the measured data. The estimated feasible set essentially consists of models that could have produced the measured data given the bounds considered on the disturbances, see Milanese and Vicino [48].

Set-membership methods typically consider parametric models affected by additive disturbances, such that tight uncertainty bounds (or approximations thereof) can be computed. Such techniques have been developed for linear systems, for example, in Kacewicz [26], Belforte and Tay [8] and Gutman et al. [24], non-linear systems, for example in [46], and time-varying systems, for example, for linear dynamics in Kuntsevich [32]. A review of set-membership techniques for system identification can be found in Milanese and Vicino [48] and Milanese et al. [49].

Set-membership identification techniques, have been used in several applications, most notable in the design of robust controllers, see Van den Boom et al. [75] and Kosut [30], and in fault detection and diagnosis [63].

Similar to traditional methods like PPEM and subspace methods, set-membership identification method also requires the user to postulate a model structure for the system, however, in this case, model structure does not include specification of the statistical properties of the noise. Furthermore, in cases where the uncertainty bound on parameter or state estimates cannot be exactly computed, or relies on computationally burdensome computations, the user must make use of approximation techniques. The conservativeness of the approximation techniques can have an adverse effect on the utility of the model in its ultimate utilization scenario, e.g., robust control. Although several approximation techniques have been proposed in the literature, e.g., in Arruda and Favier [3] and Milanese and Taragna [47], choosing a suitable approximation technique for the identification task is not a trivial task for the user.

2.5 Discussion

In this chapter, we reviewed some of the essential features of a number of identification methodologies that have been developed over the past decades. The methodologies discussed here differ in terms of

(i) the class of dynamical systems that can be considered,
(ii) the model representations used,
(iii) the identification criteria adopted, and
(iv) the optimization technique they use.

In Sect. 2.4 we discussed, for each of these identification frameworks, the choices a user must make during the identification cycle. Some of the key observations are summarized here.

- Many of the identification frameworks have been extended to address a wide variety of dynamical classes (see Sect. 2.2 for a discussion on classes of dynamical systems). This is primarily achieved by adopting various model representations that are suitable to describe the dynamics of the system, and by adopting various numerical optimization techniques that are suitable for the corresponding model representations. Form the user's perspective, the challenge is to be able to navigate the variety of model representations and optimization techniques available within an identification framework, while having little or no insight into the nature of dynamics of the true system.
- One of the main challenges for a user is to choose a suitable model structure (and complexity). For some frameworks this choice must be made explicitly, e.g., choosing a parameterized model structure in PPEM method. For others, this choice is made on a more abstract level, e.g., choosing the network architecture in a neural-network-based method or a kernel. In both cases, the choices made by the user a crucial as they ultimately determine the quality of the model estimate.
- Most identification methodologies estimate models for a single identification criterion. The choice of the identification criterion is typically motivated by statistical arguments based on the chosen model structure, or is based on the capability of the optimization technique being used. On the other hand, in many engineering applications, model estimates must satisfy multiple performance requirements.

Based on the observations listed above, we can distil the features required to achieve automated system identification:

- The identification methodology to-be-developed should be able to function across multiple classes of dynamical systems. This is necessary to ensure that the concep-

tualized identification framework can be applied to a variety of application with minimal changes.
- Instead of requiring the user to make decisions related to model structure and complexity, the desired identification methodology should be able to infer the appropriate structure required in the model based on measured data obtained from the system-to-be-modelled. The ability to automatically determine appropriate model structures from a generalized model set makes it possible to minimize the number of crucial decisions made by the user during the identification cycle.
- In many applications, it is possible that the user has a prior understanding of the dynamics of the system, or a preference for the desired model representation or structure. In such a scenario, it should be possible to suitably restrict the scope of the conceptualized identification algorithm based on this information. This will not only make the estimated models better suited for the ultimate use-case, but also make the desired identification algorithm more efficient.
- It should be possible to incorporate multiple user-specified identification criteria, such that models are optimized for performance measures that are relevant to the ultimate use-case of the model.

The list of features suggest a diverse range of specifications for the desired identification framework. In Chaps. 3, 4 and 5, we introduce a number of concepts that serve as the basic ingredients required to develop a recipe for an identification framework that possesses the list of features presented here. As hinted in Sect. 2.4.6, the first of these ingredients is evolutionary algorithms.

References

1. Andersson C, Ribeiro AH, Tiels K, Wahlström N, Schön T (2019) Deep convolutional networks in system identification. *arXiv preprint* arXiv:1909.01730
2. Aronszajn N (1950) Theory of reproducing kernels. Trans Am Math Soc 68(3):337–404
3. Arruda L, Favier G (1992) A review and a comparison of robust estimation methods. IFAC Proc Vol 25(15):247–253
4. Åström KJ, Eykhoff P (1971) System identification—a survey. Automatica 7(2):123–162
5. Balas GJ (2002) Linear, parameter-varying control and its application to a turbofan engine. Int J Robust Nonlinear Control 12(9):763–796
6. Bauer D (2001) Order estimation for subspace methods. Automatica 37(10):1561–1573
7. Bauer D, Deistler M, Scherrer W (1999) Consistency and asymptotic normality of some subspace algorithms for systems without observed inputs. Automatica 35(7):243–1254
8. Belforte G, Tay T (1992) Recursive estimation for linear models with set membership measurement error. IFAC Proc Vol 25(15):241–246
9. Bergman S, Schiffer M (1948) Theory of kernel functions in conformal mapping. Appl Math Ser 18:199
10. Billings SA (2013) Nonlinear system identification: NARMAX methods in the time, frequency, and spatio-temporal domains. Wiley, New York
11. Birpoutsoukis G, Marconato A, Lataire J, Schoukens J (2017) Regularized nonparametric volterra kernel estimation. Automatica 82:324–327
12. Chen S, Billings SA, Luo W (1989) Orthogonal least squares methods and their application to non-linear system identification. Int J Control 50(5):1873–1896

13. Chen T, Ohlsson H, Ljung L (2012) On the estimation of transfer functions, regularizations and gaussian processes-revisited. Automatica 48(8):1525–1535
14. Chiuso A, Pillonetto G (2019) System identification: a machine learning perspective. Robotics, and Autonomous Systems, Annual Review of Control
15. Cox PB (2018) Towards efficient identification of linear parameter-varying state-space models. Ph.D. thesis, Eindhoven University of Technology
16. Dahlén A, Lindquist A, Mari J (1998) Experimental evidence showing that stochastic subspace identification methods may fail. Syst Control Lett 34(5):303–312
17. de Rozario R, Oomen T (2018) Frequency response function identification of lpv systems: a global approach with application to mechanical systems. IFAC-Papers OnLine 51(15):108–113
18. Favoreel W, De Moor B, Van Overschee P (1999) Subspace identification of bilinear systems subject to white inputs. IEEE Trans Autom Control 44(6):1157–1165
19. Frigola R, Lindsten F, Schön TB, Rasmussen CE (2013) Bayesian inference and learning in gaussian process state-space models with particle mcmc. In: Advances in neural information processing systems, pp 3156–3164
20. Garulli A, Paoletti S, Vicino A (2012) A survey on switched and piecewise affine system identification. IFAC Proc Vol 45(16):344–355
21. Giri F, Bai E-W (2010) Block-oriented nonlinear system identification, vol 1. Springer, Berlin
22. Goebel R, Sanfelice RG, Teel AR (2009) Hybrid dynamical systems. IEEE Control Syst Mag 29(2):28–93
23. Goodfellow I, Bengio Y, Courville A (2016) Deep learning
24. Gutman P-O, Baril C, Neumann L (1994) An algorithm for computing value sets of uncertain transfer functions in factored real form. IEEE Trans Autom Control 39(6):1268–1273
25. Hirsch MW, Smale S, Devaney RL (2012) Differential equations, dynamical systems, and an introduction to chaos. Academic, New York
26. Kacewicz B (1996) Solving linear problems in the presence of bounded data perturbations. In: Bounding approaches to system identification, pp 29–42. Springer, Berlin
27. Kimeldorf G, Wahba G (1971) Some results on Tchebycheffian spline functions. J Math Anal Appl 33(1):82–95
28. Kimeldorf GS, Wahba G (1970) A correspondence between Bayesian estimation on stochastic processes and smoothing by splines. Ann Math Stat 41(2):495–502
29. Kingma DP, Adam JB (2014) A method for stochastic optimization. In: 3rd International Conference for Learning Representation (ICLR)
30. Kosut RL (1988) Adaptive robust control via transfer function uncertainty estimation. In: 1988 American Control Conference. IEEE, , pp 349–354
31. Kuijstermans FCM (2004) Design of nonlinear circuits: The linear time-varying approach
32. Kuntsevich V (1996) Set-valued estimation of state and parameter vectors within adaptive control systems. In: Bounding approaches to system identification. Springer, pp 239–259
33. Larimore WE (1990) Canonical variate analysis in identification, filtering, and adaptive control. In: 29th IEEE conference on decision and control. IEEE, pp 596–604
34. Lataire J, Pintelon R, Louarroudi E (2012) Non-parametric estimate of the system function of a time-varying system. Automatica 48(4):666–672
35. LeCun Y, Bengio Y, Hinton G (2015) Deep learning. Nature 521(7553):436–444
36. Lehman J, Clune J, Misevic D, Adami C, Altenberg L, Beaulieu J, Bentley PJ, Bernard S, Beslon G, Bryson DM et al (2018) The surprising creativity of digital evolution: a collection of anecdotes from the evolutionary computation and artificial life research communities. *arXiv preprint* arXiv:1803.03453
37. Li Y, Wei H-L, Billings SA (2010) Identification of time-varying systems using multi-wavelet basis functions. IEEE Trans Control Syst Technol 19(3):656–663
38. Lindsten F, Schön TB, Jordan MI (2013) Bayesian semiparametric wiener system identification. Automatica 49(7):2053–2063
39. Ljung L (1998) System identification. In: Signal analysis and prediction. Springer, Berlin, pp 163–173

40. Ljung L (ed) System identification, 2nd ed. Theory for the User. Prentice Hall PTR. ISBN 0-13-656695-2
41. Lusch B, Kutz JN, Brunton SL (2018) Deep learning for universal linear embeddings of nonlinear dynamics. Nat Commun 9(1):4950
42. Mardt A, Pasquali L, Wu H, Noé F (2018) Vampnets for deep learning of molecular kinetics. Nat Commun 9(1):5
43. Markovsky I, Willems JC, Rapisarda P, De Moor BL (2005) Algorithms for deterministic balanced subspace identification. Automatica 41(5):755–766
44. Masti D, Bemporad A (2018) Learning nonlinear state-space models using deep autoencoders. In: IEEE Conference on Decision and Control (CDC). IEEE, pp 3862–3867
45. Mercer J (1909) Functions of positive and negative type, and their connection the theory of integral equations. Philos Trans R Soc Lond Seri A 209(441–458): 415–446 (containing papers of a mathematical or physical character)
46. Milanese M, Novara C (2004) Set membership identification of nonlinear systems. Automatica 40(6):957–975
47. Milanese M, Taragna M (2002) Optimality, approximation, and complexity in set membership H/sub/spl infin//identification. IEEE Trans Autom Control 47(10):1682–1690
48. Milanese M, Vicino A (1991) Optimal estimation theory for dynamic systems with set membership uncertainty: an overview. Automatica 27(6):997–1009
49. Milanese M, Norton J, Piet-Lahanier H, Walter É (2013) Bounding approaches to system identification. Springer Science & Business Media
50. Mohammadpour J, Scherer CW (2012) Control of linear parameter varying systems with applications. Springer Science & Business Media
51. Moonen M, Ramos J (1993) A subspace algorithm for balanced state space system identification. IEEE Trans Autom Control 38(11):1727–1729
52. Narendra KS, Parthasarathy K (1990) Identification and control of dynamical systems using neural networks. IEEE Trans Neural Networks 1(1):4–27
53. Nelles O (2002) Nonlinear system identification. IOP Publishing
54. Ninness B, Henriksen S (2010) Bayesian system identification via Markov chain Monte Carlo techniques. Automatica 46(1):40–51
55. Noël J-P, Kerschen G (2013) Frequency-domain subspace identification for nonlinear mechanical systems. Mech Syst Signal Process 40(2):701–717
56. Otto SE, Rowley CW (2019) Linearly recurrent autoencoder networks for learning dynamics. SIAM J Appl Dyn Syst 18(1):558–593
57. Paduart J, Lauwers L, Swevers J, Smolders K, Schoukens J, Pintelon R (2010) Identification of nonlinear systems using polynomial nonlinear state space models. Automatica 46(4):647–656
58. Picci G, Katayama T (1996) Stochastic realization with exogenous inputs and 'subspace-methods' identification. Signal Process 52(2):145–160
59. Pillonetto G, De Nicolao G (2010) A new kernel-based approach for linear system identification. Automatica 46(1):81–93
60. Pillonetto G, Chiuso A, De Nicolao G (2011) Prediction error identification of linear systems: a nonparametric gaussian regression approach. Automatica 47(2):291–305
61. Pillonetto G, Dinuzzo F, Chen T, De Nicolao G, Ljung L (2014) Kernel methods in system identification, machine learning and function estimation: a survey. Automatica 50(3):657–682
62. Pintelon R, Schoukens J (2012) System identification: a frequency domain approach. Wiley, New York
63. Puig V (2010) Fault diagnosis and fault tolerant control using set-membership approaches: application to real case studies. Int J Appl Math Comput Sci 20(4):619–635
64. Rasmussen CE, Williams C (2017) Gaussian processes for machine learning, p 95
65. Rojas CR, Tóth R, Hjalmarsson H (2014) Sparse estimation of polynomial and rational dynamical models. IEEE Trans Autom Control 59(11):2962–2977
66. Schölkopf B, Smola AJ, Bach F et al (2002) Learning with kernels: support vector machines, regularization, optimization, and beyond. MIT Press, Cambidge

67. Schoukens J, Pintelon R, Rolain Y, Schoukens M, Tiels K, Vanbeylen L, Van Mulders A, Vandersteen G (2015) Structure discrimination in block-oriented models using linear approximations: a theoretic framework. Automatica 53:225–234
68. Schoukens M, Tiels K (2017) Identification of block-oriented nonlinear systems starting from linear approximations: a survey. Automatica 85:272–292
69. Schoukens M, Tóth R (2018) Linear parameter varying representation of a class of mimo nonlinear systems. IFAC-Papers OnLine 51(26):94–99
70. Sjöberg J, Hjalmarsson H, Ljung L (1994) Neural networks in system identification. Linköping University
71. Stoddard J, Birpoutsoukis G (2017) Gaussian process regression for generalized frequency response function estimation. *arXiv preprint* arXiv:1710.09828
72. Suykens JA, Vandewalle JP, de Moor BL (2012) Artificial neural networks for modelling and control of non-linear systems. Springer Science & Business Media
73. Tikhonov AN, Arsenin VI (1977) Solutions of ill-posed problems, vol 14. Vh Winston
74. Tóth R (2010) Modeling and identification of linear parameter-varying systems, vol 403. Springer
75. Van den Boom T, Klompstra M, Damen A (1991) System identification for H_∞-robust control design. IFAC Proc Vol 24(3):971–976
76. Van den Hof PM, Dankers A, Heuberger PS, Bombois X (2013) Identification of dynamic models in complex networks with prediction error methods-basic methods for consistent module estimates. Automatica 49(10):2994–3006
77. Van Overschee P, De Moor B (1994) N4sid: subspace algorithms for the identification of combined deterministic-stochastic systems. Automatica 30(1):75–93
78. Van Overschee P, De Moor B (1995) A unifying theorem for three subspace system identification algorithms. Automatica 31(12):1853–1864
79. Van Overschee P, De Moor B (2012) Subspace identification for linear systems: theory-implementation-applications. Springer Science & Business Media
80. Van Wingerden J-W, Verhaegen M (2009) Subspace identification of bilinear and LPY systems for open-and closed-loop data. Automatica 45(2):372–381
81. Verhaegen M, Dewilde P (1992) Subspace model identification part 2. Analysis of the elementary output-error state-space model identification algorithm. Int J Control 56(5):1211–1241
82. Verhaegen M, Westwick D (1996) Identifying MIMO Hammerstein systems in the context of subspace model identification methods. IFAC Proc Vol 29(1):4092–4097
83. Verhaegen M, Yu X (1995) A class of subspace model identification algorithms to identify periodically and arbitrarily time-varying systems. Automatica 31(2):201–216
84. Walter E, Piet-Lahanier H (1990) Estimation of parameter bounds from bounded-error data: a survey. Math Comput Simul 32(5–6):449–468
85. Wang Y (2017) A new concept using lstm neural networks for dynamic system identification. In: American Control Conference (ACC). IEEE, pp 5324–5329
86. Westwick D, Verhaegen M (1996) Identifying MIMO Wiener systems using subspace model identification methods. Signal Process 52(2):235–258
87. Witsenhausen H (1968) Sets of possible states of linear systems given perturbed observations. IEEE Trans Autom Control 13(5):556–558
88. Zhang C, Bengio S, Hardt M, Recht B, Vinyals O (2017) Understanding deep learning requires rethinking generalization. In: 5th International Conference for Learning Representation (ICLR)
89. Zhou K, Doyle JC, Glover K (1996) Robust and optimal control. Prentice-Hall Inc., Upper Saddle River. 0-13-456567-3

Chapter 3
Preliminaries—Evolutionary Algorithms

I have called this principle, by which each slight variation, if useful, is preserved, by the term of Natural Selection.

Charles Darwin

In Chap. 2, we presented, based on a general overview of the system identification literature, a set of desired features for the identification framework conceptualized in this thesis. To aid the development of the conceptualized framework, we must first introduce several concepts that will be ultimately fused together to realize the research objective stated in Chap. 1.

In this chapter, we introduce *evolutionary algorithms*—one of the concepts that will play a vital role in addressing the problem of automated selection of model structure and complexity (see Research Question 3 proposed in Chap. 1). Evolutionary algorithms are a class of global optimization techniques that make use of heuristics inspired by Darwin's Theory of Evolution. In this chapter, we pay special attention to *genetic programming*, which is one of the optimization methodologies that belongs to the class of evolutionary algorithms.

In the field of evolutionary algorithms, the problem of estimating models, of unknown structure and complexity, from measured data is often referred to as the *symbolic regression* problem. Although the symbolic regression problem is similar, in concept, to the system identification problem, there are crucial differences in the types of systems and models considered in the two fields. Nonetheless, a number of developments proposed in the evolutionary algorithm literature to address symbolic regression are also relevant in the system identification context. Hence, these developments are also discussed in this chapter.

D. Khandelwal, *Automating Data-Driven Modelling of Dynamical Systems*, Springer Theses, https://doi.org/10.1007/978-3-030-90343-5_3

3.1 Introduction

We discussed in Sect. 2.5 that the conceptualized methodology for automatic system identification should be able to determine suitable model structures that capture the functional relationship realized in the inputs and output measurements obtained from the system. In order to achieve this, we need an optimization methodology that can explore a variety of model structures from some abstract model set, and subsequently converge to an "optimal" model structure, where the optimality is determined by the chosen identification criteria. The Evolutionary Computing (EC) community has studied this problem, known as the so-called Symbolic Regression (SR) *problem*, and many Evolutionary Algorithms (EAs) have been developed for this purpose. In this chapter, we introduce the preliminaries of EC, and more specifically, Genetic Programming (GP), which is one of the key ingredients of the identification methodology that will be developed in the course of this dissertation.

This chapter is structured as follows. We introduce the concept of EC in Sect. 3.1.1. In Sect. 3.1.2, we introduce the problem of SR, which originates in the EC literature, and shed light on the link between SR and system identification. A number of EA-based identification methodologies have been proposed in the literature, and will be reviewed in Sect. 3.1.3. In Sect. 3.2.2, we introduce a specific variant of EA known as genetic programming, and describe the basic GP algorithm. Finally, in Sect. 3.3, we discuss some of the challenges commonly encountered in solving the SR problem using EAs, and the corresponding solutions that have been proposed in the literature.

3.1.1 Evolutionary Computing and Algorithms

EC is a global optimization technique inspired from biological theories and observations, more specifically, Darwin's Theory of Evolution. EC attempts to replicate the process of *evolving a population* of species via *natural selection* in order to solve optimization problems. The basic tenets of natural evolution that are replicated in EC are the following:

1. A population, consisting of a number of *individuals*, exists in an *environment*.
2. The *fitness* of an individual is determined by its *environment*.
3. Individuals with higher fitness are likelier to *survive and reproduce*.
4. New individuals, or *offsprings*, can be produced via reproduction from existing individuals, or *parents*. An offspring inherits traits from its parents.

Following these basic principles, a population evolves over time, through the processes of reproduction and mutation, and since the fittest individuals are likelier to survive, the population as a whole, on average, improves in fitness.

In EC, these principles are replicated in the following way. The *optimization problem* forms the environment, and a *candidate solution* is treated as an individual.

A collection of such candidate solutions forms the population, and reproduction is replicated as mathematical operations that generate new candidate solutions based on existing ones. Selection and propagation of fit candidate solutions, where the fitness is determined by the problem specification, ensures survival of the fittest. On average, the fitness of the candidate solutions increase w.r.t the optimization problem.

EC is computationally implemented as EAs, a class of meta-heuristic search algorithms. Some examples of EAs are Genetic Algorithms [25], Evolutionary Programming [14], Evolutionary strategies [54], Genetic Programming [31] and Gene Expression Programming [11]. These algorithms chiefly differ in terms of biological inspirations they are derived from, the data structures used, and the evolutionary reproduction operators applied. EAs are relatively domain independent, as can be seen from the broad range of successful applications [17, 22, 37, 62].

At the core of EC is the idea to solve problems by utilizing a *generate-and-evaluate* strategy, i.e., candidate solutions are first generated using heuristic operations, and subsequently evaluated for their ability to solve the given problem. This makes EC an inefficient optimization tool for problems that are relatively simple to solve, and for which efficient algorithms already exist. However, for challenging combinatorial or multi-modal optimization problems for which no efficient solution techniques are known (e.g., the class of NP problems), EC can be used as a generic global optimization technique to obtain 'useful' solutions. Additionally, the generate-and-validate strategy often yields solutions that are unconventional with respect to standard solution techniques (for example see [26]), or even simply surprising, as demonstrated by the collection of anecdotes in [36].

3.1.2 Symbolic Regression

Symbolic Regression refers to the problem of inferring a symbolic mathematical function, of unknown form, from data generated by the function. While in typical regression techniques, the (parametric) form of the function is determined by the user, in SR, the objective is to infer the structure of the function from fundamental mathematical entities (like operators and operands) using computational methods.

The formulation of the SR problem originates in the EC literature. Koza [31] was the first to employ EA, more specifically GP, to solve the SR problem. The basic idea of the GP approach to SR is depicted in Fig. 3.1. The problem is set up by collecting data from the function to be estimated, and providing a set of modular 'lego' blocks that can be combined to form a mathematical function. For example, the set of blocks may consist of operations such as addition, multiplication and exponential, and operands (e.g., x, y in Fig. 3.1). Based on the given set of modular blocks, the GP is initialized with a randomly generated set of candidate solutions. The solutions can be compared to measured data using a user-specified criterion, e.g., the Euclidian norm. Individuals with higher comparative fitness are likelier to reproduce and survive during the iterations of GP. The evolution of the population of candidate solutions continues until some stopping criterion is satisfied, typically

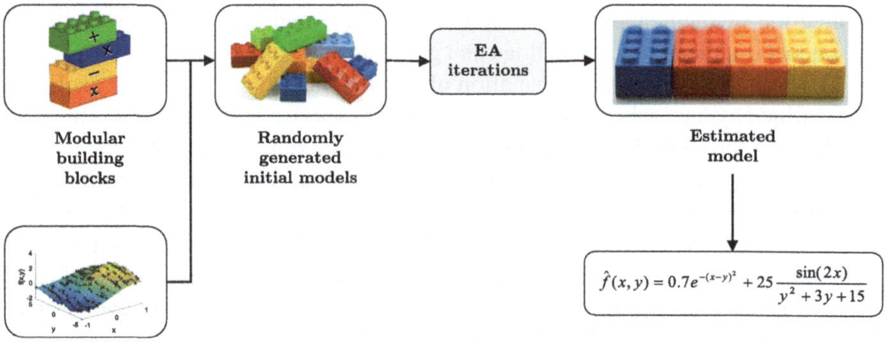

Fig. 3.1 An illustration of EC approach applied to a static SR. Some constituent images were obtained from Wikipedia Commons

the maximum number of iterations. Finally, the candidate solution, built from the modular modelling blocks, that achieves the best performance is returned as the function estimate. An illustration of the GP iterations is depicted in Fig. 3.2. More details on GP is provided in Sect. 3.2.2.

On the surface, the SR problem is closely related to SI. Both problems involve estimating models from measured data. However, most SR techniques do not consider

- dynamic dependencies between inputs and outputs, and
- disturbances acting on the system, or noisy measurements,

which are crucial considerations for any SI method.

Since [31], many EA-based techniques have been developed for SR—Keijzer [29], Ferreira [13], Peng et al. [48], Smits and Kotanchek [61], Uy et al. [64], Vladislavleva et al. [65], Hoai et al. [23], Zelinka et al. [70], McConaghy [39] to name a few. However, most techniques for SR have *not* been developed for the considerations listed above. On the other hand, in the SI literature, some EA-based techniques have been proposed for estimating non-linear models from noisy data. These will be reviewed in Sect. 3.1.3.

3.1.3 Evolutionary Computing in System Identification: An Overview

In this section, we provide an overview of some of the EC-based SI methodologies proposed in the literature. In the context of SI, EC have primarily been used in two capacities:

1. For the estimation of model parameters for a known (and parameterized) model structure, for example, see [69], where the authors used EAs to estimate model

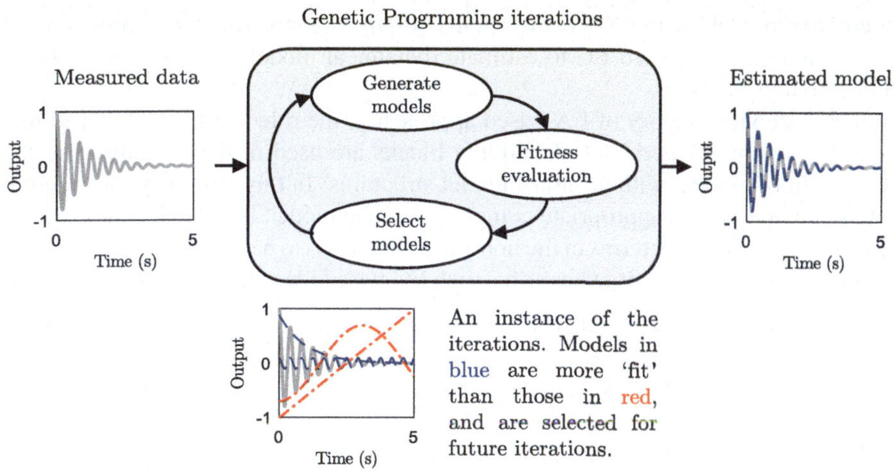

Fig. 3.2 Illustration of GP iterations for an SR problem

parameters for complex non-linear benchmark problems, or [3], for a review of the use of EAs in estimating model and design parameters in aerospace applications.
2. To determine appropriate model structures, possibly in conjunction with estimating model parameters, for example, see [15, 20].

The first mode of usage of EA in the SI context is not of significantly relevance for this dissertation (since EA is used merely as a numerical optimization tool), and hence, is not treated in this review. Interested readers are referred to [1] for a review. The second scenario, in which EAs are used to determine appropriate model structures, bears a closer resemblance to the SR setting, and is more relevant in the context of *automated* system identification. Hence, we focus our review on the second scenario.

Evolutionary Computing-based SI approaches, in which EAs are used for model-structure determination, can be categorized as follows:

(i) approaches that choose a fixed class of dynamical system (and its representation) and use EC to determine the appropriate model complexity (or model terms), and
(ii) approaches that use EC to explore model structure and model complexity.

In the first category of EA-based approaches, the basic building blocks of an EA are chosen such that only models with a specific model structure can be generated, leaving model complexity as the remaining degree of freedom. Typically, these approaches cannot be extended to other model structures without significant modifications. This approach can be found in [16, 55, 56], where the authors use EAs to perform term selection within a chosen model structure. This approach is also used in [32], where the authors use Genetic Algorithms (GAs) to estimate pole-zero locations for ARMAX models, and in [21], where authors used a variant of Particle swarm optimization (PSO) to select model terms in a polynomial Non-linear Auto-

Regressive models with eXogenous inputs (P-NARX) structure. In a similar vein, Naitali and Giri [45] used EC to estimate dynamical models with a fixed Wiener-Hammerstein structure.

In the second category of EA-based approaches, the role of EC is similar to that in an SR setting. Generic set of building blocks are used in the EA, allowing the generation of models with arbitrary model structures. In this case, EAs are used to determine not just the appropriate complexity of the model, but also the appropriate model structure (e.g., in terms of the non-linear functions to be included in the model). In general, there are two drawbacks for such EC-based identification methodologies:

- Unrestrained generation of arbitrary model structures using EA may result in models that are not well-posed, e.g., models with discontinuities, non-causality, or finite escape-time. Typically, these problems are avoided by using arbitrary ad-hoc solutions, e.g., setting all discontinuities to 0.
- Prior knowledge of the dynamical system or user-preference for desired model structures cannot be incorporated systematically in the identification procedure.

In [38], the authors use GP to identify NARMAX models that may contain arbitrary non-linearities. While the authors are interested in models that are linear-in-the-parameters, GP may return models that do not belong to that class. Consequently, the authors use an ad-hoc solution to ensure that the candidate model structures generated by GP are linearly parameterized. In [20], the authors used GP to construct linear or non-linear models from basic elements like SIMULINK blocks and static non-linearities. The combination of various SIMULINK blocks cannot be systematically structured to avoid ill-posed models, or to incorporate prior knowledge or user-preference. In [33], the authors used a stack-based GA, to obtain compact dynamical models of wind turbines. The proposed stack-based GA always produces syntactically correct models, however, this is achieved by ignoring part of the genotype of the candidate model, which may lead to undesirable behaviour of the search algorithm. Furthermore, there exists no systematic way to introduce prior knowledge.

Besides the methodologies described above, a number of SR techniques have also been extended and applied to identify dynamical systems, such as [52, 59]. While these approaches are capable of determining complex dynamical model structures starting from basic building blocks using measured data, they typically (i) do not take into account stochastic properties of systems or measured data, and (ii) implicitly assume that measurements of all the relevant variables of the system are available. In realistic identification applications, both of these aspects play an important role and must be taken into consideration.

3.2 Genetic Programming

Genetic programming is a variant of EA proposed by Koza [31]. GP encodes candidate solutions in the form of tree structures. GP forms an integral part of the identification methodology presented in this thesis, and hence, is introduced in this section. In Sect. 3.2.1, we introduce some preliminary notions, and in Sect. 3.2.2, we introduce the standard GP algorithm and discuss each step of the algorithm. Over the years, many extensions have been proposed to the standard GP algorithm, depending on the problem being solved. Some of these extensions, relevant for the SR problem, are discussed in Sect. 3.3.

3.2.1 GP Preliminaries

A candidate solution in GP can be generated from elements of two sets—a *set of functions* $\mathfrak{N}$ and a *set of terminals* $\mathfrak{T}$. The combined set $\mathfrak{S} = \mathfrak{N} \cup \mathfrak{T}$ can be viewed as a dictionary of building blocks, provided by the user, that determine the set of all possible solutions that can be generated by GP. Regardless of the problem domain, a candidate solution in GP is encoded as a finite labelled tree, defined as follows.

Definition 3.1 (*Finite tree* [28]) A *finite tree* is a directed graph, denoted by $\psi = \langle V, E, r \rangle$, where, V is the set of vertices, E is the set of edges, and $r \in V$ is the root node, such that

- ψ contains no cycles,
- $r \in V$ has in-degree (number of incoming edges) 0,
- All $\nu \in V \setminus \{r\}$ have in-degree 1,
- Every $\nu \in V$ is accessible from r,
- A vertex with out-degree (i.e., number of outgoing edges) 0 is a leaf.

Definition 3.2 (*Labelling*) Let A_1, A_2 be a set of disjoint alphabets. The *labelling* of vertices and edges of a graph $\psi = \langle V, E \rangle$ over (A_1, A_2) is given by the mappings $l : V \to A_1$ and $g : E \to A_2$, respectively.

Let $\Lambda(\psi)$ denote the set of all leaves in a finite tree ψ. The depth of a finite-tree is defined as follows.

Definition 3.3 (*Tree-depth*) The *tree-depth* $d_t(\psi)$ of a finite tree $\psi = \langle V, E, r \rangle$ is defined as

$$d_t(x) := \max_{\nu \in \Lambda(\psi)} n_e(\nu, \psi) + 1, \tag{3.1}$$

where $n_e(\nu, \psi)$ is the number of edges between a node ν and the root r of tree ψ.

In GP the finite-tree representation of a solution is often referred to as the *genotype* of the individual. The solution itself may have a different case-dependent form, for example, a computer code or a mathematical expression. The final form of the solution is said to be part of the *phenotype* of the individual.

Let m_d be the maximum tree-depth to be explored in GP. For the given set of functions $\mathfrak{N}$ and terminals $\mathfrak{T}$, the set of candidate solution genotypes $L_T(\mathfrak{S})$ is defined as follows.

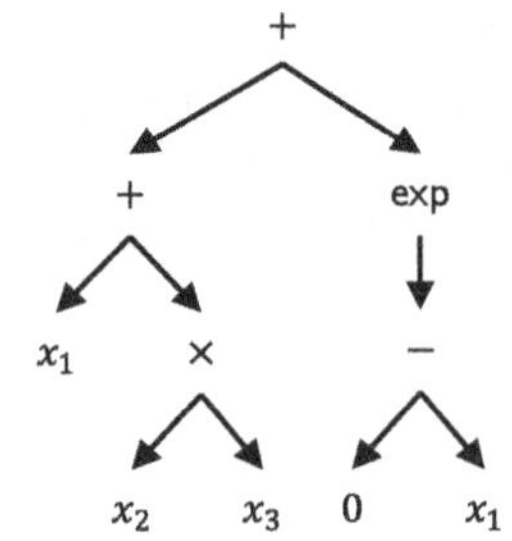

Fig. 3.3 Expression tree for $x_1 + x_2x_3 + \exp(-x_1)$

Definition 3.4 The set of candidate solution genotypes $L_T(\mathfrak{S})$ consists of all finite labelled trees $\psi = \langle V, E, r\rangle$ that satisfy

- $l(v) \in \mathfrak{N}$ for all $v \in V \setminus \Lambda(\psi)$,
- $l(v) \in \mathfrak{T}$ for all $v \in \Lambda(\psi)$,
- $d_t(\psi) < m_d$.

For notational convenience, the argument in $L_T(\mathfrak{S})$ will be dropped, unless it leads to ambiguities. Let Φ denote the mapping from the tree-representation of the individual to the final form of the candidate solution. The mapping Φ is usually evident from the choice of encoding used for the solutions. The set of all candidate solutions $\mathcal{M}$ can then be defined as $\mathcal{M} := \{\Phi(\psi) \mid \psi \in L_T\}$.

In the context of SR, the sets $\mathfrak{N}$ and $\mathfrak{T}$ consist of mathematical operations and operands, respectively, that are chosen for the SR problem. In this setting, the genotype is represented as an expression tree, for example, see Fig. 3.3.

3.2.2 *The Genetic Programming Algorithm*

The standard GP, as proposed by Koza [31], is an iterative search algorithm, with at most L iterations or *generations*. The search scheme is initialized with a *population* of n_s randomly generated candidate solutions $X^{(0)} = \{x_i^0\}_{i=1}^{n_s}$. Each individual in a generation of solutions is denoted as x_i^j, where $i \in [1, n_s]$ is the index of the individual in a generation, and $j \in [0, L]$ is the index of the generation. A maximum tree-depth of m_d is imposed on the individuals.

In each iteration, *variation operators* are used to propose new candidate solutions, that are subsequently evaluated using the given performance measure. The *selection operator* is used to select the solutions that are propagated to the next iteration.

The basic GP algorithm described in [31] is presented in Algorithm 1. Each step in the algorithm is described in the following.

Algorithm 1 Genetic Programming

Require: population size $n_s > 0$, number of iterations $L > 0$, set of functions $\mathfrak{N}$ and terminals $\mathfrak{T}$, crossover rate p_c, mutation rate p_m, maximum tree-depth m_d

1: *Initialize* population $X(0)$, $l = 0$
2: *Evaluate* fitness of models in $X(l)$
3: **repeat**
4: Propose new population $\tilde{X}^{(l+1)}$ using *variation operations*
5: *Evaluate* fitness of population $\tilde{X}^{(l+1)}$
6: *Select* population $X^{(l+1)}$
7: $l \leftarrow l + 1$
8: **until** $l \leq L + 1$ **return** $X(L)$

Initialization

The population $X^{(0)}$ is initialized in Step 1 of Algorithm 1 as a set of randomly generated trees $\{x_i^0 \in L_\mathrm{T}(\mathfrak{S})\}_{i=1}^{n_s}$. Koza [31] proposed three approaches to generate the initial population $X^{(0)}$—*grow*, *full* and *ramped half-and-half*. The 'grow' approach is described[1] in Algorithm 2. The 'grow' initialization results in trees with arbitrary shape and depth in the range of $[2, m_d]$. In contrast, the 'full' initialization results in trees that are full, i.e., all leaves are at maximum depth m_d. The 'full' initialization follows the same steps as the 'grow' initialization in Algorithm 2 with a modification in Step 9: the labels for the newly added nodes are selected from $\mathfrak{N}$ (instead of $\mathfrak{N} \cup \mathfrak{T}$) randomly with uniform distribution.

The ramped half-and-half initialization is a combination of the 'grow' and 'full' initializations. The population $X^{(0)}$ is divided into $m_d - 1$ groups, each with a corresponding maximum tree-depth of $2, 3, \ldots m_d$. In each group, half of the individuals are initialized using the 'grow' initialization, and the other half using the 'full' initialization.

Fitness evaluation

A candidate solution x can be evaluated based on its phenotype $\Phi(x)$. Let $f : \mathcal{M} \rightarrow \mathbb{R}$ be the user-specified performance criterion that assigns an individual its fitness value. Unless specified, we will assume that the performance criterion is to be minimized. The given performance measure is used to compute the fitness of each individual in the population (see Steps 2, 5 of Algorithm 1).

In order to evaluate the fitness of an individual, it is crucial that each genotype $x \in L_\mathrm{T}$ is mapped to a phenotype $\Phi(x)$ that is a 'valid' candidate solution, where the validity is determined by the problem domain. For example, if the phenotype of an individual is a computer program, then the program must not contain syntactic or semantic errors. This is referred to as the *closure* property. When the tree-based encoding of the individuals does not possess the closure property, one must use ad-hoc strategies to 'fix' the individual such that it can be evaluated. Alternatively, more

[1] It is assumed that each function in $\mathfrak{N}$ has arity (i.e., number of arguments) 2. Algorithm 2 can easily be extended to the case where each function in $\mathfrak{N}$ has an arbitrary but known arity.

Algorithm 2 The 'grow' initialization

Require: set of functions $\mathfrak{N}$ and terminals $\mathfrak{T}$, maximum tree-depth m_{d}

Initialize $x \leftarrow \langle V, E, r\rangle$, where $V = \{r\}$, $E = \phi$, and $l(r) \in \mathfrak{N}$ is randomly selected with uniform distribution.

$\Lambda_{\mathrm{F}} \leftarrow \{v \in \Lambda(x) \mid l(v) \in \mathfrak{N}\}$ ▷ Set of leaves with labels that belong to $\mathfrak{N}$

$d \leftarrow 1$ ▷ depth of the tree

$i \leftarrow 1$

while $d \leq m_{\mathrm{d}}$ **do** ▷ Grow tree until maximum depth m_{d}

  while $\Lambda_{\mathrm{F}} \neq \phi$ **do** ▷ Visit each leaf in Λ_{F}, and add new nodes

    Select any $v \in \Lambda_{\mathrm{F}}$

    if $d < m_{\mathrm{d}}$ **then**

      $V \leftarrow V \cup \{v_{i+1}, v_{i+2}\}$, where $l(v_{i+1}), l(v_{i+2}) \in (\mathfrak{N} \cup \mathfrak{T})$ is randomly selected with uniform distribution

    else

      $V \leftarrow V \cup \{v_{i+1}, v_{i+2}\}$, where $l(v_{i+1}), l(v_{i+2}) \in \mathfrak{T}$ is randomly selected with uniform distribution

    $E \leftarrow E \cup \{\langle v, v_{i+1}\rangle, \langle v, v_{i+2}\rangle\}$,

    $\Lambda_{\mathrm{F}} \leftarrow \Lambda_{\mathrm{F}} \setminus \{v\}$

    $i \leftarrow i + 2$

  $x \leftarrow \langle V, E, r\rangle$

  $\Lambda_{\mathrm{F}} \leftarrow \{v \in \Lambda(x) \mid l(v) \in \mathfrak{N}\}$

  $d \leftarrow d + 1$

return $\langle V, E, r\rangle$

advanced tree-based encodings can be used, for example, strongly-type GP [43] or grammar-based GP [66].

Typically, fitness evaluation of individuals is the most time-consuming step in Algorithm 2. However, computing the fitness of individuals in a population can usually be parallelized, thereby making the step scalable to larger population sizes.

Variation operators

In each iteration of GP, variation operators such as *crossover* and *mutation* are used to propose a new generation of solutions from the existing generation (see Step 4 of Algorithm 1). These operators are described below:

- The crossover operator, denoted as $\kappa : L_{\mathrm{T}} \times L_{\mathrm{T}} \rightarrow L_{\mathrm{T}} \times L_{\mathrm{T}}$, exchanges sub-trees of two individuals $x^j_{i_1}$ and $x^j_{i_2}$ to create two new individuals $\tilde{x}^{j+1}_{i_1}$ and $\tilde{x}^{j+1}_{i_2}$, for some $i_1, i_2 \in [1, n_{\mathrm{s}}]$ and some $j \in [0, L - 1]$. Many crossover operators have been proposed and studied in the literature, e.g., in [49]. Perhaps the most common one, proposed in [31] is the *standard sub-tree crossover*. The standard sub-tree crossover, depicted in Fig. 3.4, involves the following steps:
  1. Two individuals $x^j_{i_1}, x^j_{i_2}$ are selected randomly from the population $X^{(j)}$. The probability distribution for the random selection may be dependent on the relative fitness of the individuals w.r.t the population (as proposed in [31]). However, many other strategies have been proposed in the literature, e.g., see review in [35]. Crossover is applied to the selected pair with a probability of p_c.

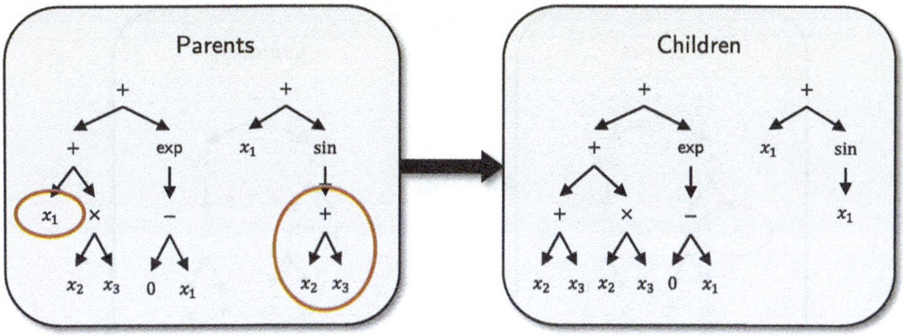

Fig. 3.4 Standard crossover operator illustrated

2. Two nodes ν_1, ν_2 are selected randomly from individuals $x_{i_1}^j, x_{i_2}^j$, respectively, with uniform distribution.[2]
3. The two sub-trees in $x_{i_1}^j, x_{i_2}^j$ rooted at ν_1, ν_2, respectively, are swapped. This results in two new candidate solutions $\tilde{x}_{i_1}^{j+1}, \tilde{x}_{i_2}^{j+1}$.

- The mutation operator, denote as $\mu : L_T \rightarrow L_T$, modifies a sub-tree of an individual x_i^j to create x_i^{j+1}. The standard sub-tree crossover, depicted in Fig. 3.5, involves the following steps:

1. An individual x_i^j is selected randomly from the population with a probability determined by the relative fitness of the individual. The individual is mutated with probability p_m.
2. A node ν in x_i^j is selected randomly with uniform distribution. The sub-tree rooted at ν is swapped with a newly generated tree (using the same techniques as in the initialization step).

While only the standard crossover and mutation operators have been introduced here, there exist a large number of proposed strategies for crossover and mutation, for example, in [9, 34, 49, 50, Chap. 5].

Selection

In [31], no clear distinction was made between the variation operators of crossover and mutation, and the selection operator. In its original form, the selection operator, denoted as $\varsigma : L_T \rightarrow L_T$ simply copies an individual x_i^j to the next generation as $\tilde{x}_i^{j+1}$ with a probability of p_s. The individual x_i^j is selected randomly from the population with a probability that is determined by the relative fitness of the individual w.r.t the population. In this setting, the new population for the next iteration $X^{(j+1)}$ is the same as the proposed set of individuals $\{\tilde{x}_i^{j+1}\}_{i=1}^{n_s}$ obtained from the crossover, mutation and selection operators.

[2] It is common to select nodes for crossover with a uniform distribution over all nodes except leaves of the tree.

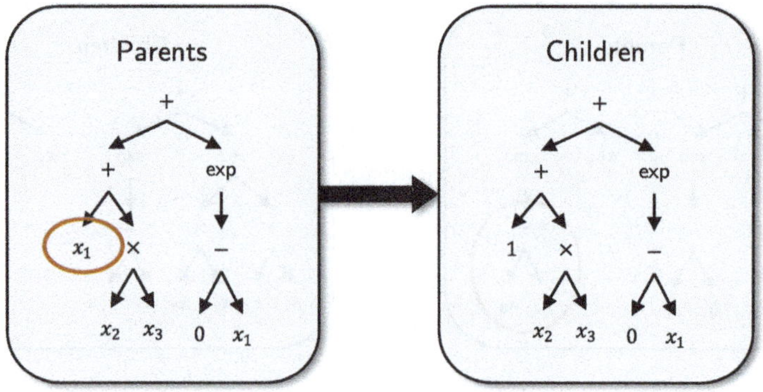

Fig. 3.5 Standard mutation operator illustrated

In most recent approaches for GP, it is far more common to separate the roles of the variation operators and the selection operator. In this setting, the variation operators are used to propose a new set of solutions, and the selection operator is used to select individuals from the current population $X^{(j)}$ and the proposed set of solutions $\{\tilde{x}_i^{j+1}\}_{i=1}^{n_s}$ to form the new population $X^{(j+1)}$. Many strategies have been proposed, such as crowding-based selection [41], rank-based selection and tournament selection [19].

3.3 Genetic Programming for Symbolic Regression—Challenges and Extensions

Genetic programming, and other EAs, have been applied to a number of SR problems over the years. Depending on the challenges posed by the SR test problems, a number of shortcomings have been identified in standard GP. Correspondingly, a number of extensions have also been proposed to enable better performing EAs for SR. Due to the similarity between SR and SI, the challenges posed by the SR problem, and the corresponding extensions of GP, are also relevant in the context of SI. Hence, in this section, we discuss some of the commonly identified challenges and the corresponding possible GP extensions. For detailed reviews, refer to O'Neill et al. [46], Dabhi and Chaudhary [7], Poli et al. [50, Part 2] and Eiben and Smith [10, Part 2].

Estimation of numerical constants

Symbolic regression is inherently a two-part problem: (i) the search for an appropriate model structure, and (ii) the estimation of model coefficients. The search for an appropriate model structure is a combinatorial optimization problem since, in EC, the model is built from fundamental building blocks (see Fig. 3.1). The estimation of model coefficients is usually a continuous optimization problem, the complexity of which is dependent on the problem domain. The estimation of model coefficients is a challenging task for the standard GP algorithm [58]. This stems from the fact that standard GP typically resorts to combining various numerical constants, that were present in the initial set of solutions, to propose a new constant [31]. This is a very slow process with no guarantees of finding the correct model coefficients. The inability of GP to find apt numerical constants has a more severe effect on the combinatorial part of the SR problem. Consider, for example, the true data-generating function to be $f_0(x) = \sin(x) + 1000$. For this example, a candidate solution $f_1(x) = \sin(x) + 1$ would have a very low fitness score, even though the solution is structurally similar to the true function. Thus the candidate solution $f_1(x)$ is highly likely to be dropped out of the current generation of solutions. While the GP algorithm could potentially re-introduce the sine function at a later stage of the search, the inability to find appropriate numerical constants makes GP highly inefficient for searching in continuous domains.

This inefficiency has often been observed in scientific literature [7], and several extensions to GP have been proposed. A commonly used approach is to hybridize the GP algorithm with a *local search* technique that estimates model coefficients more efficiently. Examples of such local search techniques include gradient-descent [30, 63], randomized algorithms and the Nelder-Mead simplex method [42], and other evolutionary algorithms [27]. Such hybrid algorithms are commonly referred to as Memetic Algorithmss (MAs) [44]. Ferreira [13] proposed an approach, based on an alternate representation, to estimate model coefficients during the evolutionary search.

Over-fitting and bloating

A common flaw in standard GP is that, as the number of iterations increase, the individuals in the population tend to grow in complexity (e.g., in terms of size of the tree), without any significant improvement in the fitness of the individuals. This is a symptom of the evolution of candidate solutions that over-fit a finite amount of measured data. In EC terminology, this is referred to as *bloating*.

A variety of techniques have been proposed in the literature to avoid bloating. One approach to controlling bloat is to modify the variation or selection operators, such that the operations function on individuals of similar complexities, for example crowding-based selection [41] and size-fair crossover [6]. A second approach to deal with over-fitting is to cross-validate the candidate models using an independent dataset in the selection step [18]. Perhaps the most direct approach to control bloating is to use complexity as an additional performance measure in a Multi-Objective Evolutionary Algorithm (MOEA), for example, Fonseca and Fleming [15], Deb et al. [8], and Zitzler et al. [72].

Efficiency of the algorithm

EAs are often required to search for solutions in combinatorial search landscapes, which can be challenging task [5, 24]. Analysis into the performance of EAs via the *schemata theory* [25] only provides a partial description of why EAs work, and has been criticized for being too simplistic [2]. On the other hand, convergence analysis for EAs are often established in an asymptotic computational time setting [57]. While analysing the limit behaviour of the algorithm provides relevant insights, it does not reveal the non-asymptotic performance of the search.

In order to improve the efficiency of EAs, developments have been made in several directions, some of which are listed below.

- **Solution space and representations**: In the EC community, it is well known that the performance of EAs can be improved by incorporating domain-specific prior knowledge into the algorithm [53, 67]. One way to incorporate domain-specific knowledge is to restrict the solution space or modify the representation of solutions based on prior knowledge. Grammar-based approaches have been widely used to restrict search spaces in a declarative sense, see [40], or adaptively, see [68]. Many authors have used solution representations specially catered to the problem domain, a good illustration of this idea can be found in [4]. In this paper, the authors also illustrate how the choice of representation may influence the fitness landscape by introducing or removing local optima. EAs have also been extended to multiple-gene representations for combinatorial problems [12], continuous optimization problems [13], and graph-based representations for network-based applications [51].
- **Modified variation operators**: Schemata theory implicitly assumes that individuals with relatively high fitness produce offsprings with relatively high fitness [2]. This is in general not true for standard crossover and mutation operators, and is dependent on the nature of the fitness landscape. Modified variational operators have been proposed to improve the fitness of offsprings, for examples see [9, 12, 47, 49]
- **Modified selection operator**: Similar to variation operators, selection operators have also been modified to introduce favourable traits in EAs, see [19] for a comparative analysis of some commonly-used selection mechanisms. The importance of selection operation with regards to the global convergence properties of the search algorithm was pointed out in [57]. Crowding-based selection, see [41], provide a selection mechanism that trades-off performance versus complexity of individuals, and has been used in MOEAs [8] to maintain diversity in the population and prevent early convergence.
- **Fitness evaluation**: Evaluating the fitness of individuals is most often the most time-consuming step of any EA. Efforts have been made to reduce time-complexity of the fitness evaluation step by predicting the fitness of individuals [60] or by active control of the length of the data-set [71].

References

1. Abd Samad M (2014) Evolutionary computation in system identification: review and recommendations. Int Rev Autom Control (IREACO) 7(2):208–216
2. Altenberg L (1995) The schema theorem and Price's theorem. In: Foundations of genetic algorithms, vol 3. Elsevier, pp 23–49

3. Arias-Montano A, Coello CAC, Mezura-Montes E (2012) Multiobjective evolutionary algorithms in aeronautical and aerospace engineering. IEEE Trans Evol Comput 16(5):662–694
4. Ashlock D, McGuinness C, Ashlock W (2012) Representation in evolutionary computation. In: IEEE world congress on computational intelligence. Springer, pp 77–97
5. Colorni A, Dorigo M, Maffioli F, Maniezzo V, Righini G, Trubian M (1996) Heuristics from nature for hard combinatorial optimization problems. Int Trans Oper Res 3(1):1–21
6. Crawford-Marks R, Spector L (2002) Size control via size fair genetic operators in the PushGP genetic programming system. In: Proceedings of the 4th annual conference on genetic and evolutionary computation. Morgan Kaufmann Publishers Inc., pp 733–739
7. Dabhi VK, Chaudhary S (2015) Empirical modeling using genetic programming: a survey of issues and approaches. Nat Comput 14(2):303–330
8. Deb K, Pratap A, Agarwal S, Meyarivan T (2002) A fast and elitist multiobjective genetic algorithm: NSGA-II. IEEE Trans Evol Comput 6(2):182–197
9. D'haeseleer P (1994) Context preserving crossover in genetic programming. In: Proceedings of the first IEEE conference on evolutionary computation. In: IEEE world congress on computational intelligence. IEEE, pp 256–261
10. Eiben AE, Smith JE (2003) Introduction to evolutionary computing, vol 53. Springer
11. Ferreira C (2001) Gene expression programming: a new adaptive algorithm for solving problems. Complex Syst 13(2):87–129
12. Ferreira C (2002) Combinatorial optimization by gene expression programming: inversion revisited. In: Proceedings of the Argentine symposium on artificial intelligence, pp 160–174
13. Ferreira C (2003) Function finding and the creation of numerical constants in gene expression programming. In: Advances in soft computing. Springer, pp 257–265
14. Fogel LJ, Owens AJ, Walsh MJ (1966) Artificial intelligence through simulated evolution. Wiley
15. Fonseca CM, Fleming PJ (1993) Genetic algorithms for multiobjective optimization: formulation, discussion and generalization. In: Proceedings of the 5th international conference on genetic algorithms. Morgan Kaufmann Publishers Inc., pp 416–423
16. Fonseca CM, Fleming PJ (1996) Non-linear system identification with multiobjective genetic algorithms. In: Proceedings of 13th IFAC world congress, pp 1169–1174
17. Frowd CD (2002) EvoFIT: a holistic, evolutionary facial imaging system
18. Gagné C, Schoenauer M, Parizeau M, Tomassini M (2006) Genetic programming, validation sets, and parsimony pressure. In: European conference on genetic programming. Springer, pp 109–120
19. Goldberg DE, Deb K (1991) A comparative analysis of selection schemes used in genetic algorithms. In: Foundations of genetic algorithms, vol 1. Elsevier, pp 69–93
20. Gray GJ, Murray-Smith DJ, Li Y, Sharman KC, Weinbrenner T (1998) Nonlinear model structure identification using genetic programming. Control Eng Pract 6(11):1341–1352
21. Hafiz F, Swain A, Mendes EM (2019) Two-dimensional (2D) particle swarms for structure selection of nonlinear systems. Neurocomputing 367:114–129
22. Hill T, Lundgren A, Fredriksson R, Schiöth HB (2005) Genetic algorithm for large-scale maximum parsimony phylogenetic analysis of proteins. Biochim Biophys Acta (BBA) Gen Subj 1725(1):19–29
23. Hoai NX, McKay RI, Essam D, Chau R (2002) Solving the symbolic regression problem with tree-adjunct grammar guided genetic programming: the comparative results. In: Proceedings of the 2002 congress on evolutionary computation. CEC'02 (Cat. No. 02TH8600), vol 2. IEEE, pp 1326–1331
24. Hoffman A, Wolfe J, Garfinkel R, Johnson D, Papadimitriou C, Gilmore P, Lawler E, Shmoys D, Karp R, Steele J et al. (1986) The traveling salesman problem: a guided tour of combinatorial optimization. Wiley
25. Holland JH (1975) Adaptation in natural and artificial systems: an introductory analysis with applications to biology, control, and artificial intelligence. University of Michigan Press
26. Hornby GS, Lohn JD, Linden DS (2011) Computer-automated evolution of an x-band antenna for Nasa's space technology 5 mission. Evol Comput 19(1):1–23

27. Howard LM, D'Angelo DJ (1995) The GA-P: a genetic algorithm and genetic programming hybrid. IEEE Expert 10(3):11–15
28. Kallmeyer L (2009) A declarative characterization of different types of multicomponent tree adjoining grammars. Res Lang Comput 7(1):55–99
29. Keijzer M (2003) Improving symbolic regression with interval arithmetic and linear scaling. In: European conference on genetic programming. Springer, pp 70–82
30. Kommenda M, Kronberger G, Winkler S, Affenzeller M, Wagner S (2013) Effects of constant optimization by nonlinear least squares minimization in symbolic regression. In: Proceedings of the 15th annual conference companion on genetic and evolutionary computation. ACM, pp 1121–1128
31. Koza JR (1992) Genetic programming: on the programming of computers by means of natural selection, vol 1. MIT Press
32. Kristinsson K, Dumont GA (1992) System identification and control using genetic algorithms. IEEE Trans Syst Man Cybern 22(5):1033–1046
33. La Cava W, Danai K, Spector L, Fleming P, Wright A, Lackner M (2016) Automatic identification of wind turbine models using evolutionary multiobjective optimization. Renew Energy 87:892–902
34. Langdon WB (1999) Size fair and homologous tree genetic programming crossovers. In: Proceedings of the 1st annual conference on genetic and evolutionary computation, vol 2. Morgan Kaufmann Publishers Inc., pp 1092–1097
35. Larranaga P, Kuijpers CMH, Murga RH, Inza I, Dizdarevic S (1999) Genetic algorithms for the travelling salesman problem: a review of representations and operators. Artif Intell Rev 13(2):129–170
36. Lehman J, Clune J, Misevic D, Adami C, Altenberg L, Beaulieu J, Bentley PJ, Bernard S, Beslon G, Bryson DM et al (2018) The surprising creativity of digital evolution: a collection of anecdotes from the evolutionary computation and artificial life research communities. arXiv preprint arXiv:1803.03453
37. Li Y, Ng KC, Murray-Smith DJ, Gray GJ, Sharman KC (1996) Genetic algorithm automated approach to the design of sliding mode control systems. Int J Control 63(4):721–739
38. Madár J, Abonyi J, Szeifert F (2005) Genetic programming for the identification of nonlinear input-output models. Ind Eng Chem Res 44(9):3178–3186
39. McConaghy T (2011) FFX: fast, scalable, deterministic symbolic regression technology. In: Genetic programming theory and practice IX. Springer, pp 235–260
40. Mckay RI, Hoai NX, Whigham PA, Shan Y, O'neill M (2010) Grammar-based genetic programming: a survey. Genet Progr Evolvable Mach 11(3–4):365–396
41. Mengshoel OJ, Goldberg DE (2008) The crowding approach to niching in genetic algorithms. Evol Comput 16(3):315–354
42. Molina D, Lozano M, García-Martínez C, Herrera F (2010) Memetic algorithms for continuous optimisation based on local search chains. Evol Comput 18(1):27–63
43. Montana DJ (1995) Strongly typed genetic programming. Evol Comput 3(2):199–230
44. Moscato P (1999) Memetic algorithms: a short introduction. New ideas in optimization. McGraw-Hill Ltd., UK, pp 219–234
45. Naitali A, Giri F (2016) Wiener-Hammerstein system identification—an evolutionary approach. Int J Syst Sci 47(1):45–61
46. O'Neill M, Vanneschi L, Gustafson S, Banzhaf W (2010) Open issues in genetic programming. Genet Progr Evolvable Mach 11(3–4):339–363
47. Pawlak TP, Wieloch B, Krawiec K (2014) Semantic backpropagation for designing search operators in genetic programming. IEEE Trans Evol Comput 19(3):326–340
48. Peng Y, Yuan C, Qin X, Huang J, Shi Y (2014) An improved gene expression programming approach for symbolic regression problems. Neurocomputing 137:293–301
49. Poli R, Langdon WB (1998) On the search properties of different crossover operators in genetic programming. Genet Progr 293–301
50. Poli R, Langdon WB, McPhee NF, Koza JR (2008) A field guide to genetic programming. http://www.lulu.com/product/file-download/a-field-guide-to-genetic-programming/2502914

51. Poli R et al (1997) Evolution of graph-like programs with parallel distributed genetic programming. In: ICGA. Citeseer, pp 346–353
52. Quade M, Abel M, Shafi K, Niven RK, Noack BR (2016) Prediction of dynamical systems by symbolic regression. Phys Rev E 94(1):012214
53. Radcliffe NJ, Surry PD (19995) Fundamental limitations on search algorithms: evolutionary computing in perspective. In: Computer science today. Springer, pp 275–291
54. Rechenberg I (1971) Optimierung technischer Systeme nach Prinzipien der biologischen Evolution
55. Rodríguez-Vázquez K, Fleming PJ (2000) Use of genetic programming in the identification of rational model structures. In: Proceedings of European conference on genetic programming. Springer, pp 181–192
56. Rodriguez-Vazquez K, Fonseca CM, Fleming PJ (2004) Identifying the structure of nonlinear dynamic systems using multiobjective genetic programming. IEEE Trans Syst Man Cybern Part A: Syst Hum 34(4):531–545
57. Rudolph G (1994) Convergence analysis of canonical genetic algorithms. IEEE Trans Neural Netw 5(1):96–101
58. Ryan C, Keijzer M (2003) An analysis of diversity of constants of genetic programming. In: European conference on genetic programming. Springer, pp 404–413
59. Schmidt M, Lipson H (2009) Distilling free-form natural laws from experimental data. Science 324(5923):81–85
60. Schmidt MD, Lipson H (2008) Coevolution of fitness predictors. IEEE Trans Evol Comput 12(6):736–749
61. Smits GF, Kotanchek M (2005) Pareto-front exploitation in symbolic regression. In: Genetic programming theory and practice II. Springer, pp 283–299
62. To CC, Vohradsky J (2007) A parallel genetic algorithm for single class pattern classification and its application for gene expression profiling in streptomyces coelicolor. BMC Genomics 8:49
63. Topchy A, Punch WF (2001) Faster genetic programming based on local gradient search of numeric leaf values. In: Proceedings of the genetic and evolutionary computation conference (GECCO-2001), vol 155162
64. Uy NQ, Hoai NX, O'Neill M, McKay RI, Galván-López E (2011) Semantically-based crossover in genetic programming: application to real-valued symbolic regression. Genet Progr Evolvable Mach 12(2):91–119
65. Vladislavleva EY et al (2008) Model-based problem solving through symbolic regression via Pareto genetic programming. Citeseer
66. Whigham PA et al. (1995) Grammatically-based genetic programming. In: Proceedings of the workshop on genetic programming: from theory to real-world applications, vol 16, pp 33–41
67. Wolpert DH, Macready WG et al (1997) No free lunch theorems for optimization. IEEE Trans Evol Comput 1(1):67–82
68. Wong ML, Mun T (2005) Evolving recursive programs by using adaptive grammar based genetic programming. Genet Progr Evolvable Mach 6(4):421–455
69. Worden K, Manson G (2012) On the identification of hysteretic systems. Part I. Fitness landscapes and evolutionary identification. Mech Syst Signal Process 29:201–212
70. Zelinka I, Oplatkova Z, Nolle L (2005) Analytic programming—symbolic regression by means of arbitrary evolutionary algorithms. Int J Simul Syst Sci Technol 6(9):44–56
71. Zhang B-T, Cho D-Y (1998) Genetic programming with active data selection. In: Asia-Pacific conference on simulated evolution and learning. Springer, pp 146–153
72. Zitzler E, Laumanns M, Thiele L (2001) SPEA2: improving the strength Pareto evolutionary algorithm. TIK-report 103

Chapter 4
Tree Adjoining Grammar

What I cannot create, I do not understand.

Richard Feynman

In Chap. 3, we introduced the preliminaries of GP, a generic optimization tool that can generate and search for optimal solutions in a general solution space. GP is the first key ingredient that will be used to realize the conceptualized identification methodology. Recall that in the context of automated SI, the desired solution space would consist of all model structures (with varying degrees of model complexity) that are relevant for the identification task at hand. Furthermore, when the user has prior knowledge of, or preference for the desired model structure, it should be possible to suitably adapt the search space of the desired SI method. These ideas connect to Research Sub-questions 1.1 and 1.2 proposed in Chap. 1.

In this chapter, we introduce the concept of TAG, a grammar formalism for generating trees with desired structures. Although TAG was developed to describe syntactic structures in linguistic sentences, we demonstrate in this chapter that the same concept can be applied to describe structure in dynamical models. Based on this idea, we generalize the notion of a model set, which ultimately enables us to shape the solution space explored by GP. We formalize these concepts by developing a TAG to describe the structure of a class of IO models, specifically the P-NARMAX model class, and demonstrate that the proposed grammar also describes a number of other model structures such as ARMAX and truncated Volterra series. We also demonstrate that the proposed concept can be extended to a larger class of IO models, and to other model representations such as state-space models. Parts of this chapter were published in [11].

D. Khandelwal, *Automating Data-Driven Modelling of Dynamical Systems*,
Springer Theses, https://doi.org/10.1007/978-3-030-90343-5_4

4.1 Introduction

One of the primary challenges to automating the task of system identification is to address the problem of model structure determination. In most identification methodologies proposed in the literature, the user must either make an explicit choice in terms of the structure and complexity of the model (e.g., in PEM methods, see Sect. 2.4.2), or make a choice on a more abstract level that eventually determines the model set (e.g., in Bayesian identification, see Sect. 2.4.4). In order to alleviate the dependence on a user's choice, the conceptualized identification framework must be able to function across multiple model structures and complexities. This argument led to the formulation of Research Question 1: "How can we automate the task of model structure and complexity selection in the parametric setting?", and Sub-questions 1.1 and 1.2. These questions will be addressed in this chapter.

In Chap. 3 we introduced EC, and discussed how algorithms such as GP may be used to solve the problem of model structure selection. In order to apply GP in a system identification context, one must ensure that

(i) The set of desired dynamical model structures can be expressed in terms of expression trees that are generated using a set of functions and terminal symbols, since GP uses these functions and terminals as basic building blocks to construct candidate solutions for the given problem.
(ii) Any expression tree constructed using the set of functions and terminal symbols must be a well-posed model that belongs to the set of desired model structures. This is necessary since the GP algorithm should be able to evaluate models corresponding to the expression trees generated in each iteration. If the set of terminals and functions do not always generate a well-posed model structure, one must make use of a number of ad-hoc correction mechanisms in order to obtain a model structure that can be numerically evaluated. By design, these ad-hoc correction mechanisms would be specific to the chosen set of functions and terminals, thereby making it difficult to apply the conceptualized identification methodology across different model structures or representations.

Typically, requirement (i) can be satisfied since mathematical expressions can be represented by expression trees. Ensuring the satisfaction of requirement (ii) can be more involved, see discussion in Sect. 3.3. Furthermore, GP in itself does not provide tools to impose constraints on the structure of models generated during the evolution of candidate solutions.

In order to satisfy requirement (ii), we need to specify a set of rules that can accurately describe the structure of well-posed dynamical models, more specifically, parametric IO models. Surprisingly, a fundamentally similar problem arises in the field of computational linguistics, where one would like to describe the structure of words that constitute syntactically-correct sentences in a grammar. In this comparison, a well-posed model structure is similar to a syntactically-correct sentence, and the desired model set is akin to the grammatical language. Continuing along the lines of this analogy, we are interested in determining the grammar rules that generate well-posed model structures.

Whigham et al. [18] was the first to use generative grammars to shape the set of candidate models explored in GP. In this chapter we explore the possibility of using a formalism of generative grammar called TAG to generate model structures with the desired properties. TAG was originally developed to study the structure of sentences in natural languages. In this chapter we use TAG to generate structured parametric dynamical models that belong to a desired model set. We also discuss how the model set, i.e., the underlying language, can be modified to incorporate prior knowledge and constraints on the model structure.

In Sect. 4.2 we introduce generative grammars, illustrate the basic notions of TAG on a linguistic example and formally define the relevant notions. In Sect. 4.3 we introduce the notion of a model set specified by a TAG and develop a TAG that specifies a class of parametric IO models. We also demonstrate how prior information can be incorporated into the model set by imposing restrictions in the TAG. In Sect. 4.3.3 we demonstrate the use of TAG to describe other types of IO dynamic models, and in Sect. 4.4 we develop a TAG for non-linear SS model representations.

4.2 Formal Grammars and TAG

4.2.1 Introduction to Generative Grammars

A *formal grammar* can be described as a set of rules for generating strings. The resulting set of strings is called the *language* generated by the grammar. Formal grammars find applications in development of programming languages and parsers, theoretical linguistics and in the study of mathematical logic [14]. The most common uses for formal grammars and languages are in the context of (i) *generative grammars*, which can be considered as systems of rules that generate only the combination of words that exist in a given formal language, or (ii) *recognizers*, which determine whether a given string belongs to a language.

The notion of generative grammars was introduced by Chomsky [2]. Chomsky formulated a grammar, more specifically a *phrase structured grammar*, as a tuple comprising

(i) an alphabet of non-terminals $\mathfrak{N}$,
(ii) an alphabet of terminals $\mathfrak{T}$,
(iii) a distinguished start symbol S in the non-terminal alphabet, and
(iv) a set of production rules P that generate (or rewrite) strings that belong to the grammar.

Let $\mathfrak{N}^*$ denote the set of all strings that can be made using zero or more symbols in alphabet $\mathfrak{N}$. Similarly, let $\mathfrak{N}^+$ denote the set of all strings that can be made using one or more symbols in alphabet $\mathfrak{N}$. A single production rule of a grammar can take the from

$$\sigma_1 \rightarrow \sigma_2, \tag{4.1}$$

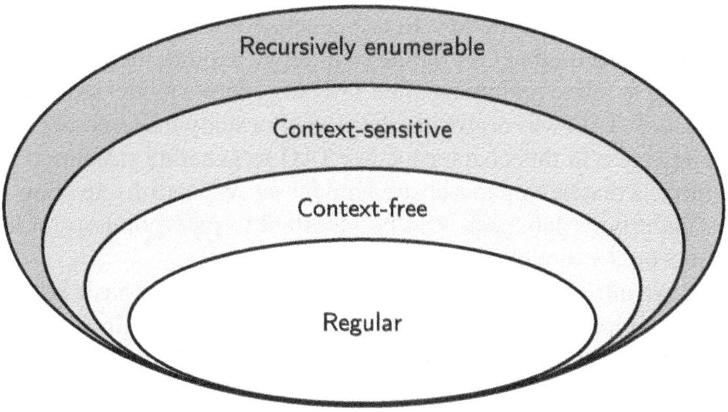

Fig. 4.1 Illustration of the Chomsky hierarchy

where sub-strings σ_1, σ_2 satisfy $\sigma_1 \in (\mathfrak{N} \cup \mathfrak{T})^+$ and $\sigma_2 \in (\mathfrak{N} \cup \mathfrak{T})^*$. Equation (4.1) can be interpreted as follows: if a string σ, that belongs to the grammar, contains a substring that matches σ_1, the substring can be replaced with σ_2 in order to obtain a new sentence σ' that also belongs to the grammar.

Based on the structure of σ_1 and σ_2 in (4.1), phrase structured grammars can be categorized into four types—regular, context-free, context-sensitive and unrestricted. Regular grammars are the most structured phrase-structured grammars. The production rules in a regular grammar can take the form of (4.1), where $\sigma_1 \in \mathfrak{N}$ and $\sigma_2 \in \mathfrak{T}$ or $\sigma_2 = \sigma_3\sigma_4$ with $\sigma_3 \in \mathfrak{T}$ and $\sigma_4 \in \mathfrak{N}$. Context-free grammars are less structured than regular grammars since σ_1, σ_2 in (4.1) are required to satisfy $\sigma_1 \in \mathfrak{N}$ and $\sigma_2 \in (\mathfrak{T} \cup \mathfrak{T})^*$. Similarly, context-sensitive grammars form a more general class of grammars with σ_1, σ_2 in (4.1) satisfying $\sigma_1, \sigma_2 \in (\mathfrak{N} \cup \mathfrak{T})^+$. As a consequence, the four categories of phrase structured grammars form a hierarchy since the set of regular languages are properly included in the set of context-free languages, which in turn is properly included in the set of context-sensitive languages, and so on. This is referred to as the Chomsky hierarchy, and is depicted in Fig. 4.1. For a more detailed discussion, see Parkes [15].

Until 1970s, context-free grammars were commonly used to describe semantics of natural languages. However, Shieber [16] demonstrated that context-free grammars were inappropriate to describe certain features of natural languages. On the other hand, context-sensitive languages were generally considered as too big a class of grammars, making the resulting algorithms, for example for parsing, computationally complex and non-scalable [9]. As a result, several formulations were proposed in order to extend the class of context-free languages such that relevant context-sensitive features of natural languages could be incorporated (see [9]). One of the first *mildly context-sensitive grammar* formalisms was TAG developed by Joshi et al. [7]. In this dissertation, we put forth the idea to use the generative capacity of TAG to generate stochastic dynamical models. In Sects. 4.2.2 and 4.2.3 we introduce TAG both informally and formally.

4.2.2 TAG—An Informal Introduction

Unlike formal grammars that generate strings, TAG describes a set of rules for generating trees. We introduce the concept of TAG through an example derived from Joshi and Schabes [6].

Consider the sentence "a man saw Mary". Simple grammatical constructs can be used to decompose the given sentence into its basic components. For example, the sentence consists of articles ("a"), nouns ("man", "Mary") and verbs ("saw"). Other underlining structures, such as subjects and predicates, can also be observed in the sentence. The sentence, together with the underlying grammatical structure can be represented in a tree form as shown in Fig. 4.2. The tree depicted in Fig. 4.2 is called a *derived tree*. The *yield* of a derived tree are the labels associated with the leaves of the tree. Hence, the yield of the derived tree in Fig. 4.2 is "A man saw Mary".

The given derived tree can be obtained by combining basic building blocks that are constituents of the TAG. Figure 4.3 depicts the set of *initial trees* $\mathcal{I}$ and *auxiliary trees* $\mathcal{A}$, collectively known as *elementary trees*, that can be combined in specific ways to produce the derived tree in Fig. 4.2. The set of initial trees $\mathcal{I}$ can be informally described as a set of non-recursive replacement rules that can be used to generate a set of trees. The set of auxiliary trees $\mathcal{A}$ can be described as a set of recursive replacement rules. Consequently, each auxiliary tree has a terminal node with the same label as that of its root node.

The TAG framework defines two operations that can be used to combine trees from sets $\mathcal{I}$ and $\mathcal{A}$—*substitution* and *adjunction*. A substitution operation can be used to substitute an initial tree into, for instance, another initial tree, if and only if the latter has a leaf with a label that matches the label of the root node of the prior. The downward arrows symbol $\downarrow$ in Fig. 4.3 represents nodes in initial trees that are available for substitution. Adjunction can be loosely described as the operation of inserting an auxiliary tree into an existing tree (for example, an initial tree or a tree generated from sets $\mathcal{I}$ and $\mathcal{A}$ using substitution and adjunction operations). Adjunction of an auxiliary tree can take place on a non-terminal node of a tree if and only if the node has a label that matches the label of the root node of the auxiliary tree to be adjoined. The star symbol $\star$ in Fig. 4.3 marks the node in auxiliary tree α_1 that participates in an adjunction operation.

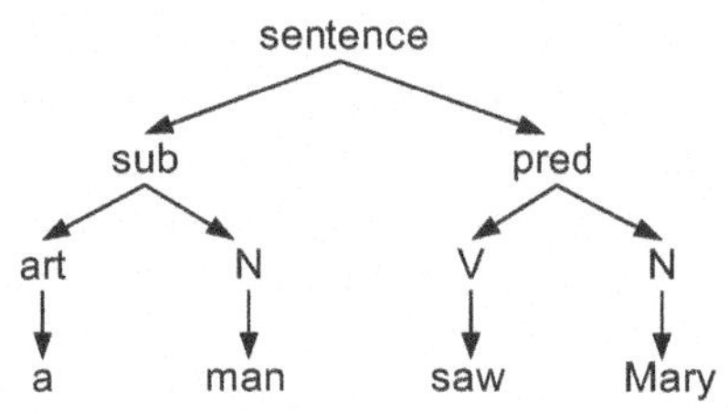

Fig. 4.2 A derived tree with the yield "A man saw Mary" [11]. The tree depicts the grammatical constructs that are evident in the structure of the sentence—a subject (sub) and a predicate (pred), an article (art), a verb (V) and nouns (N)

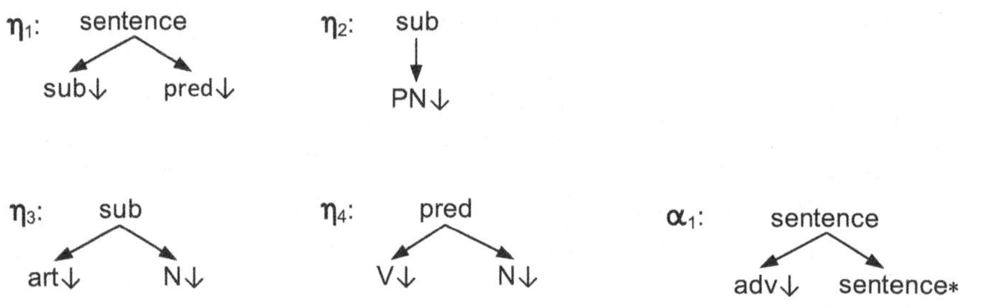

(a) Set of *initial trees* ($\mathcal{I}$). **(b)** Set of *auxiliary trees* ($\mathcal{A}$).

Fig. 4.3 Initial ($\mathcal{I}$) and auxiliary trees ($\mathcal{A}$) for the example "A man saw Mary" [11]. The sets $\mathcal{I}$ and $\mathcal{A}$ serve as building blocks of the tree set of a TAG

(a) Derivation tree for "a man saw Mary". **(b)** Derivation tree for "yesterday a man saw Mary".

Fig. 4.4 Derivation tree representation [11]—dashed lines represent substitutions, solid lines represent adjunction, and labels on the edges represent the Gorn addresses (a method to assign a label to a node in a tree structure, see Gorn [4]) of the nodes participating in substitution or adjunction

We can now construct the sentence "a man saw Mary" through the following sequence of operations. The initial tree η_3 is substituted in η_1 at the location of the "sub" node. Let's denote the resulting tree as ψ_1. The tree ψ_1 is an example of a *syntactic tree*, a tree obtained by applying an arbitrary number of substitution and adjunction operations to a given initial tree. Next, the initial tree η_4 is substituted to the syntactic tree ψ_1 at the location of the "pred" node. Let the result be denoted as ψ_2. In fact, ψ_2 is exactly the tree structure given in the example in Fig. 4.2, upto the last level of the derived tree, where specific articles, nouns and verbs are substituted in the tree to obtain the yield "a man saw Mary".

Substitution can be performed on a initial tree or syntactic tree for as long as there exist nodes available for substitution, marked by $\downarrow$. A *derived tree* is a syntactic tree in which none of the terminal nodes (leaves) are available for substitution. The initial and auxiliary trees provide an alternative representation, the *derivation tree*, as shown in Fig. 4.4a. Based on the TAG in Fig. 4.3, more complex sentences can also be generated. For example, the auxiliary tree α_1 can be adjoined to the root node of ψ_2 since both root nodes have the label "sentence". This operation effectively adds an adverb before the sentence, yielding the sentence "yesterday a man saw Mary". The resulting derivation tree is depicted in Fig. 4.4b.

The set of all derived trees that can be obtained, by starting from a given start symbol, say 'sentence', and applying an arbitrary number of adjunctions and/or substitutions using elementary trees is called the *tree language* of the corresponding TAG. The string yield of all trees in the tree set is called the *string language* of the corresponding TAG.

4.2.3 TAG—Formal Definitions

The formal definitions of TAG and related concepts can be found in Joshi and Schabes [6] and Kallmeyer [8]. These definitions are reproduced here for completeness. To introduce the basic notions of a TAG, assume $\mathfrak{N}$ and $\mathfrak{T}$ to be disjoint alphabets that serve as the labels of non-terminal and terminal vertices, respectively.

Definition 4.1 (*Syntactic Tree*) A *syntactic tree* is an ordered, labelled tree $\langle V, E, r\rangle$ such that the label $l(\nu)$ satisfies $l(\nu) \in \mathfrak{N}$ for each vertex ν with out-degree at least 1 and $l(\nu) \in (\mathfrak{N} \cup \mathfrak{T} \cup \epsilon)$ for each leaf ν.

Definition 4.2 (*Auxiliary Tree*) An *auxiliary tree* is a syntactic tree $\langle V, E, r\rangle$ such that there is a unique leaf f, marked as *foot node*, with $l(f) = l(r)$. An auxiliary tree is denoted as $\langle V, E, r, f\rangle$.

Definition 4.3 (*Initial Tree*) An *initial tree* is a non-auxiliary syntactic tree.

Remark 4.1 Recall that a syntactic tree is obtained by combining initial and auxiliary trees using the substitution and adjunction operations, respectively. If a syntactic tree $\psi = \langle V, E, r\rangle$ is formed by substituting (or adjoining) initial (or auxiliary) trees to an auxiliary tree $\psi' = \langle V', E', r, f\rangle$, then the syntactic tree also contains the foot node f, and is denoted as $\psi = \langle V, E, r, f\rangle$.

With the basic ideas defined, we can now define TAG, and the related operations.

Definition 4.4 (*Tree Adjoining Grammar*) A TAG is a tuple $G = \langle \mathfrak{N}, \mathfrak{T}, S, \mathcal{I}, \mathcal{A}\rangle$, where

- $\mathfrak{N}$, $\mathfrak{T}$ are disjoint sets of non-terminals and terminals,
- $S \in \mathfrak{N}$ is a start symbol,
- $\mathcal{I}$ is a finite set of initial trees and $\mathcal{A}$ is a finite set of auxiliary trees.

The set of trees $\mathcal{I} \cup \mathcal{A}$ is called *elementary trees*. Elementary trees can be combined using substitution and adjunction operations to generate more complex trees.

Definition 4.5 (*Substitution*) Let $\psi = \langle V, E, r\rangle$ be a syntactic tree and $\psi' = \langle V', E', r'\rangle$ be an initial tree and $\nu \in V$. The result of substituting ψ' into ψ at node ν, denoted as $\psi[\nu, \psi']$, is defined as follows

- If ν is not a leaf or ν is a foot node or $l(\nu) \neq l(r')$, then $\psi[\nu, \psi']$ is not defined,

- otherwise, $\psi[\nu, \psi'] = \langle V'', E'', r\rangle$ with

$$V'' = V \cup V' \setminus \{\nu\}, \tag{4.2}$$

and

$$E'' = (E \setminus \{\langle \nu_1, \nu\rangle \mid \nu_1 \in V\}) \cup E' \cup \{\langle \nu_1, r'\rangle \mid \langle \nu_1, \nu\rangle \in E\}. \tag{4.3}$$

The substitution operation is illustrated in Fig. 4.5a.

Definition 4.6 (*Adjunction*) Let $\psi = \langle V, E, r\rangle$ be a syntactic tree and $\psi' = \langle V', E', r', f\rangle$ be an auxiliary tree and $\nu \in V$ with out-degree at least 1. The result of adjoining ψ' into ψ at node ν, denoted as $\psi[\![\nu, \psi']\!]$, is defined as follows

- if $l(\nu) \neq l(r')$ or ν is a leaf, then $\psi[\![\nu, \psi']\!]$ is undefined,
- else $\psi[\![\nu, \psi']\!] = \langle V'', E'', r''\rangle$ with

$$V'' = V \cup V' \setminus \{\nu\}, \tag{4.4}$$

and

$$\begin{aligned} E'' = (E \setminus \{\langle \nu_1, \nu_2\rangle \mid \nu_1 = \nu \text{ or } \nu_2 = \nu\}) \cup \\ E' \cup \big\{\langle \nu_1, r'\rangle \mid \langle \nu_1, \nu\rangle \in E\big\} \cup \\ \{\langle f, \nu_2\rangle \mid \langle \nu, \nu_2\rangle \in E\}. \end{aligned} \tag{4.5}$$

The adjunction operation is illustrated in Fig. 4.5b.

More complex trees in a TAG can be produced by combining multiple elementary trees using substitution and adjunction operations. When combining trees using substitution or adjunction, we make use of *instances* of elementary trees such that any

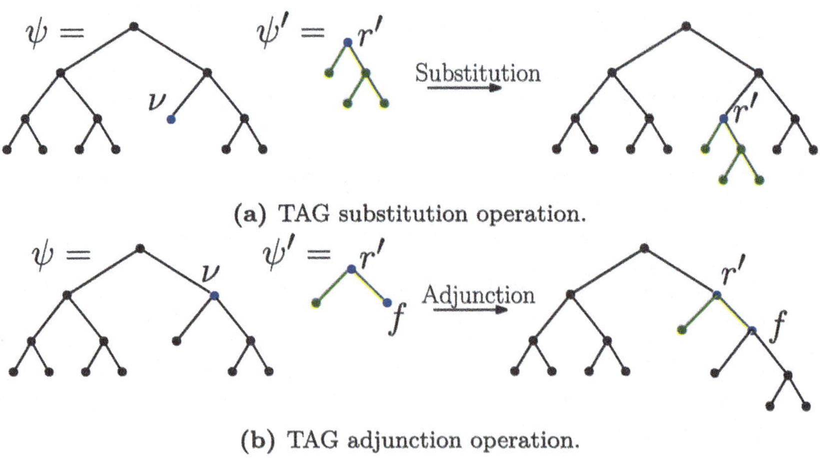

Fig. 4.5 Illustration of the TAG operations [10]

two trees used to generate a complex tree in a given TAG are pairwise disjoint (in terms of the set of vertices and edges). Any tree generated from a TAG can be represented in two forms—derived tree and derivation tree. A tree obtained by performing an arbitrary number of valid substitution and adjunction operations to an initial tree $\psi = \langle V, E, r \rangle \in \mathcal{I} \cup \mathcal{A}$ with $l(r) = S$ is called a *derived tree* (for example, as in Fig. 4.2). The substitution and adjunction operations performed can be represented in a tree representation called *derivation tree* (for example, as in Fig. 4.4), where

- the vertices of the derivation tree represent the various elementary trees being substituted or adjoined,
- edges of the derivation tree represent substitution or adjunction of the elementary tree associated to the child vertex to the elementary tree associated with the parent vertex, and
- the edges are labelled with the Gorn address of the vertex in the elementary tree associated with the parent node where adjunction or substitution takes place. The Gorn addressing scheme is as follows. The root node of a tree has address 0. The n children of the root node have addresses $1, \dots, n$, respectively. For any other node in the tree with address a, the m children of the node have addresses $a.1, \dots, a.m$, respectively.

A derived tree is said to be *saturated* if all leaves of the derived tree belong to the set $\mathfrak{T}$ and cannot be further substituted. The corresponding derivation tree is also said to be *saturated*. The *yield* of a saturated derived tree ψ is the string of terminal symbols associated with the leaves of ψ read from left to right.

Definition 4.7 (*Tree Language and String Language*) Let $G = \langle \mathfrak{N}, \mathfrak{T}, S, \mathcal{I}, \mathcal{A} \rangle$ be a TAG. The *tree language* $L_{\mathrm{T}}(G)$ of grammar G is defined as the set of all saturated derived trees $\psi = \langle V, E, r \rangle \in G$ with $l(r) = S$.

The *string language* $L(G)$ of G is the set of yields of the trees in $L_{\mathrm{T}}(G)$.

For a given TAG G and derivation tree ρ, the derived tree can be computed as $\psi = \Delta_G(\rho)$, where Δ_G denotes the mapping from derivation trees to derived trees by performing the required substitution and adjunction operations. Furthermore, we define

$$L_{\mathrm{D}}(G) := \{\rho \mid \Delta_G(\rho) \in L_{\mathrm{T}}(G)\} \tag{4.6}$$

as the set of all derivation trees constructed using grammar G such that the corresponding derivation tree $\Delta_G(\rho)$ satisfies $\Delta_G(\rho) \in L_{\mathrm{T}}(G)$.

4.3 TAG Representation of IO Models

In Sect. 4.2 we introduced the basic notion of TAG. In this section, we demonstrate how TAG can be used to generate parametric stochastic dynamical models. In Sect. 4.3.1 we introduce a new notion of model set based on the generative capacity of TAG. To showcase the capability of a TAG-based approach to specifying the structure of dynamical models, in Sect. 4.3.2 we develop a TAG for a general class of non-linear IO models—the polynomial Non-linear Auto-Regressive Moving-Average models with eXogenous inputs (P-NARMAX) model class. The proposed TAG for P-NARMAX models contains, as special cases, grammars that describe more structured model classes, such as FIR, ARX and truncated Volterra series. In Sect. 4.3.2 we also provide illustrative examples to demonstrate the use of TAG to generate model structures. While we make use of the P-NARMAX model class to establish formal results and to develop illustrative examples, the scope of TAG-based specification of model sets is by no means limited to the P-NARMAX class. In Sect. 4.3.3 we present TAG for model structures that go beyond the class of P-NARMAX models.

4.3.1 Grammar-Based Model Set Specification of IO Models

Consider the following DT IO representation of a non-linear dynamical model

$$y(k) = f\left(u(k), \ldots, u(k-n_u), y(k-1), \ldots, y(k-n_y), \xi(k-1), \ldots, \xi(k-n_\xi)\right) \\ + \xi(k) \tag{4.7}$$

where $u(k), y(k) \in \mathbb{R}$ are the input and output signals at time-instant k, $\xi(k) \sim \mathcal{N}(0, \sigma_\xi^2)$ is a noise signal independent of input u, constants n_u, n_y and n_ξ are the corresponding maximum time-lags and the non-linear function $f(\cdot)$ belongs to an arbitrary set of functions $\mathcal{M}$. In PPEM, the set of functions $\mathcal{M}$, also known as the *model set*, along with a specified choice for n_u, n_y and n_ξ, is determined by a user based on expert knowledge, prior information and informative experiments. It will be demonstrated in Sect. 4.3.2 that TAG can be used to generate trees that yield non-linear functions $f(\cdot)$ with desirable structural properties and varying choices of arguments (time lags of the involved u, y and ξ signals).

In order to formalize this concept, we introduce a function $\Pi_f(u, y, \xi, k)$ that maps from function f to the right-hand-side expression in (4.7) (in string form). We can now define a new notion of model set, based on TAG, defined as follows.

Definition 4.8 ([11]) For a given TAG G, the corresponding *model set* $\mathcal{M}(G)$ is defined as the set of models in the form of (4.7) such that $\Pi_f(u, y, \xi, k) \in L_T(G)$.

Remark 4.2 Recall that while standard GP, introduced in Sect. 3.2.2, makes use of expression trees that can be parsed as mathematical expressions, in TAG, the derived tree yields the model expression as a string. Hence, we employ mapping Π_f to map from function f to the model expression.

Note that the notion of model set introduced in Definition 4.8 is more generalized than that used in PPEM. In PPEM, a model set is typically determined by choosing a fixed model structure along with a suitable parameterization (i.e. model complexity). On the other hand, in this work, the choice of initial and auxiliary trees of a TAG automatically determines the model set. The advantage of such a declaration of a model set is that, when no prior information is available, the model set can be chosen to span a number of commonly used model classes without a prior specification of the model complexity. On the other hand, when prior information on the structure or complexity of the model is available, the grammar can be suitably refined to restrict the model set. In Sect. 4.3.2 we introduce a general non-linear IO model class—the P-NARMAX class, and in Sect. 4.3.2, we propose a TAG for the P-NARMAX model class, and demonstrate that the resulting model set spans a number of commonly used parametric model structures.

4.3.2 TAG Representation of P-NARMAX Models

The NARMAX model class is a flexible class on NL IO dynamical models [12]. The P-NARMAX model class is the set of all NARMAX models where the non-linear relationships are of the polynomial kind. P-NARMAX models can be used to describe a large variety of system dynamics since any continuous function on a closed space can be approximated arbitrary well using polynomial functions (based on Weierstrass theorem, see [17]). Furthermore, the family of P-NARMAX models includes, as special cases, other commonly used model classes such as FIR and ARMAX. It will be shown that these models can be generated by suitably restricting the TAG presented here.

P-NARMAX Models

A DT SISO P-NARMAX model can be represented as (see [1])

$$\begin{aligned} y(k) &= \theta_0 + \sum_{i_1=1}^{n} \theta_{i_1} x_{i_1}(k) + \sum_{i_1=1}^{n} \sum_{i_2=i_1}^{n} \theta_{i_1 i_2} x_{i_1}(k) x_{i_2}(k) \\ &+ \cdots \sum_{i_1=1}^{n} \cdots \sum_{i_l=i_{l-1}}^{n} \theta_{i_1 i_2 \dots i_l} x_{i_1}(k) x_{i_2}(k) \dots x_{i_l}(k) + \xi(k), \end{aligned} \tag{4.8}$$

where l is the order of the polynomial non-linearity, $\theta_{i_1 i_2 \dots i_m}$ are the model parameters, and $x(k) = (x_1(k) \cdots x_{n_y+n_u+n_\xi}(k))^\top$ is a vector consisting of the past input, output and noise regressors as follows

$$x_m(k) = \begin{cases} y(k-m) & 1 \le m \le n_y \\ u(k-(m-n_y-1)) & n_y+1 \le m \le n_y+n_u+1 \\ \xi(k-(m-n_y-n_u-1)) & n_y+n_u+2 \le m \le n_y+n_u+n_\xi+1. \end{cases} \tag{4.9}$$

We will also use the following alternative and equivalent representation for P-NARMAX models:

$$y(k)=\sum_{i=1}^{p}c_i\prod_{j=0}^{n_u}u^{b_{i,j}}(k-j)\prod_{q=1}^{n_\xi}\xi^{d_{i,q}}(k-q)\prod_{m=1}^{n_y}y^{a_{i,m}}(k-m)+\xi(k), \tag{4.10}$$

where p is the number of model terms, c_i are the model parameters, $a_{i,m}, b_{i,j}, d_{i,q} \in \mathbb{Z}_{\geq 0}$ are the exponents for output, input and noise terms, respectively.

TAG Representation

In this section we propose a TAG for the P-NARMAX model class. The proposed TAG captures the structural relationships in (4.10). In the sequel, the time index will be dropped in the context of the proposed TAG, as q^{-1} will be used to denote a backward time shift. For convenience, the following notation is introduced. For a given model in the form of (4.10), define $J_i := \{j \in \mathbb{Z}_{\geq 0} \mid b_{i,j} \neq 0\}$, $Q_i := \{q \in \mathbb{Z}_{\geq 0} \mid d_{i,q} \neq 0\}$ and $M_i := \{m \in \mathbb{Z}_{\geq 0} \mid a_{i,m} \neq 0\}$, the set of indices of input, noise and output factors, respectively, with non-zero exponents in the ith model term. Furthermore, for the ith model term, the sequence of delays in the input, noise and output factors are denoted by $(\bar{j}_n^{(i)})_{n\in J_i}$, $(\bar{q}_n^{(i)})_{n\in Q_i}$, $(\bar{m}_n^{(i)})_{n\in M_i}$ respectively.

Theorem 4.9 ([11]) *Consider the TAG* $G_{\mathrm{N}} = \langle \mathfrak{N}, \mathfrak{T}, S, \mathcal{I}, \mathcal{A}\rangle$ *with*

- $\mathfrak{N} = \{\text{expr0}, \text{expr1}, \text{expr2}, \text{op}, \text{par}\}$,
- $\mathfrak{T} = \{\text{u}, \text{y}, \xi, +, \text{c}, \times, q^{-1}\}$,
- $S = \text{expr0}$,
- $\mathcal{I} = \{\eta_1\}$, *where initial tree* η_1 *is depicted in Fig. 4.6,*
- $\mathcal{A} = \{\alpha_i\}_{i=1}^{7}$, *where the auxiliary trees* α_i*'s are depicted in Fig. 4.7.*

The model set $\mathcal{M}(G_{\mathrm{N}})$ *is equivalent to the set of all models that can be expressed as* (4.10) *with finite values of* p, n_u, n_y *and* n_ξ.

Proof of Theorem 4.9 For the first part of the proof, we show that for any P-NARMAX model in the form of (4.10), there exists a derivation tree such that the resulting derived tree has a yield that is equal to the right hand side (RHS) of (4.10). Algorithm 1 constructs such a derivation tree for a given P-NARMAX model. The algorithm constructs the derivation tree by starting with the introduction of the first factor (u, y or ξ) of each of the p model terms, and subsequently building each of the branches by introducing the remaining factors with the corresponding delays and exponents. The procedure DELAYS (ψ, ν, n) adjoins n auxiliary tree α_7 to the derivation tree ψ at vertex ν.

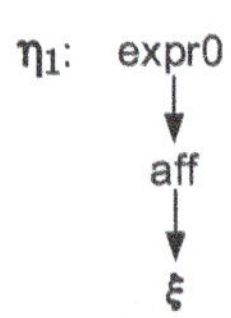

Fig. 4.6 Initial trees $\mathcal{I}$ of TAG G_{N} [11]

Algorithm 1 Parse P-NARMAX model (4.10) to derivation tree [11].

Require: $p, J_i, Q_i, M_i, \left(\bar{j}_n^{(i)}\right)_{n\in J_i}, \left(\bar{q}_n^{(i)}\right)_{n\in Q_i}, \left(\bar{m}_n^{(i)}\right)_{n\in M_i}$

$V \leftarrow \{v_0\};\ \ l(v_0) \leftarrow \eta_1$ ▷ initialize with start tree
$r \leftarrow v_0$
$V \leftarrow \bigcup_{i=1}^{p}\{v_{i,1}\} \cup V$ ▷ Insert p vertices to begin the p summation branches
$E \leftarrow \bigcup_{i=2}^{p}\{\langle v_{i-1,1}, v_{i,1}\rangle\} \cup \{\langle v_0, v_{1,1}\rangle\}$
for $i \leftarrow 1, p$ **do**
  if $J_i \neq \phi$ **then** ▷ If there is an input factor in the ith term
    $l(v_{i,1}) \leftarrow \alpha_1$ ▷ For each summation branch, assign the appropriate label to the first vertex
    $\langle V, E, r\rangle \leftarrow$ DELAYS$(\langle V, E, r\rangle, v_{i,1}, \bar{j}_1^{(i)})$ ▷ Adjoin delay trees
    $b_{i,\bar{j}_1^{(i)}} \leftarrow b_{i,\bar{j}_1^{(i)}} - 1$ ▷ Reduce the corresponding exponent by 1
  else if $Q_i \neq \phi$ **then**
    $l(v_{i,1}) \leftarrow \alpha_3$
    $\langle V, E, r\rangle \leftarrow$ DELAYS$(\langle V, E, r\rangle, v_{i,1}, \bar{q}_1^{(i)})$
    $d_{i,\bar{q}_1^{(i)}} \leftarrow d_{i,\bar{q}_1^{(i)}} - 1$
  else if $M_i \neq \phi$ **then**
    $l(v_{i,1}) \leftarrow \alpha_2$
    $\langle V, E, r\rangle \leftarrow$ DELAYS$(\langle V, E, r\rangle, v_{i,1}, \bar{m}_1^{(i)})$
    $a_{i,\bar{m}_1^{(i)}} \leftarrow a_{i,\bar{m}_1^{(i)}} - 1$
  $cnt_i \leftarrow 1$ ▷ Multiply remaining factors
  for all $j \in J_i$ **do**
    $V \leftarrow \bigcup_{n=1}^{b_{i,j}}\{v_{i,cnt_i+n,1} \cup V\};\ \ l(v_{i,cnt_i+n,1}) \leftarrow \alpha_4$
    $E \leftarrow \bigcup_{n=1}^{b_{i,j}}\{\langle v_{i,cnt_i+n-1,1}, v_{i,cnt_i+n,1}\rangle\} \cup E$
    for $n \leftarrow 1, \bar{b}_j^{(i)}$ **do** ▷ Adjoin corresponding delays for each factor multiplied
      $\langle V, E, r\rangle \leftarrow$ DELAYS$(\langle V, E, r\rangle, v_{i,cnt_i+n,1}, j)$
    $cnt_i \leftarrow cnt_i + \bar{b}_j^{(i)}$
  for all $q \in Q_i$ **do**
    $V \leftarrow \bigcup_{n=1}^{d_{i,q}}\{v_{i,cnt_i+n,1} \cup V\};\ \ l(v_{i,cnt_i+n,1}) \leftarrow \alpha_6$
    $E \leftarrow \bigcup_{n=1}^{d_{i,q}}\{\langle v_{i,cnt_i+n-1,1}, v_{i,cnt_i+n,1}\rangle\} \cup E$
    for $n \leftarrow 1, \bar{d}_q^{(i)}$ **do**
      $\langle V, E, r\rangle \leftarrow$ DELAYS$(\langle V, E, r\rangle, v_{i,cnt_i+n,1}, q)$
    $cnt_i \leftarrow cnt_i + \bar{d}_q^{(i)}$
  for all $m \in M_i$ **do**
    $V \leftarrow \bigcup_{n=1}^{a_{i,m}}\{v_{i,cnt_i+n,1} \cup V\};\ \ l(v_{i,cnt_i+n,1}) \leftarrow \alpha_5$
    $E \leftarrow \bigcup_{n=1}^{a_{i,m}}\{\langle v_{i,cnt_i+n-1,1}, v_{i,cnt_i+n,1}\rangle\} \cup E$
    for $n \leftarrow 1, \bar{a}_m^{(i)}$ **do**
      $\langle V, E, r\rangle \leftarrow$ DELAYS$(\langle V, E, r\rangle, v_{i,cnt_i+n,1}, m)$
    $cnt_i \leftarrow cnt_i + \bar{a}_m^{(i)}$
return $\langle V, E, r\rangle$

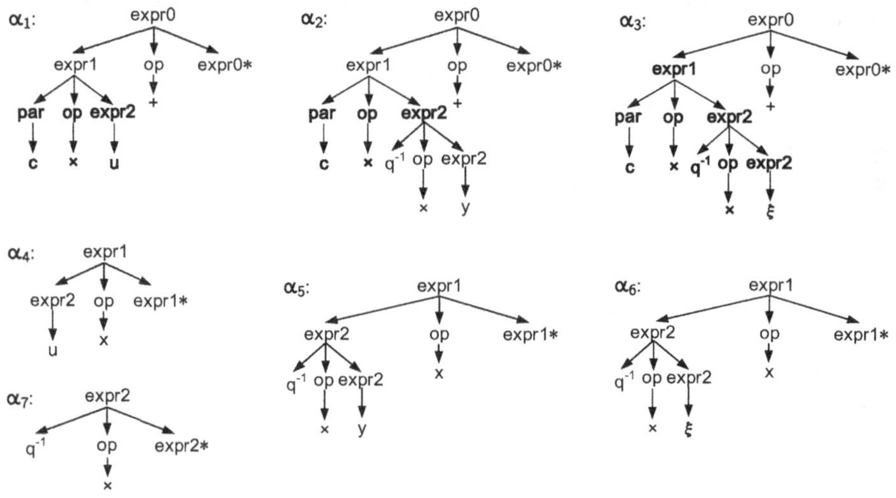

Fig. 4.7 Auxiliary trees $\mathcal{A}$ of TAG G_{N} [11]

For the second part of the proof, it needs to be shown that all expressions in $L(G_{\mathrm{N}})$, i.e., yields of all possible trees generated by G_{N}, are RHS expressions of P-NARMAX models. This is proven by structural induction. We first observe that the simplest tree in $L(G_{\mathrm{N}})$ is the initial tree η_1 with the yield ξ. This corresponds to the model

$$y(k) = \xi(k), \tag{4.11}$$

which belongs to the P-NARMAX class. Now, consider an arbitrary saturated derived tree $\psi \in L_{\mathrm{T}}(G_{\mathrm{N}})$ whose yield is the RHS of a P-NARMAX model. This implies that the yield is a causal polynomial expression in terms of the factors u, y and ξ. To complete the principle of induction, it must be shown that any possible adjunction to ψ results in a new tree in $L_{\mathrm{T}}(G_{\mathrm{N}})$ whose yield is also a causal polynomial model expression in terms of the aforementioned factors.

For convenience, the auxiliary trees are grouped based on the operators involved—$\alpha_1, \alpha_2, \alpha_3$ are called *additive-type* auxiliary trees, $\alpha_4, \alpha_5, \alpha_6$ are called *multiplicative-type*, and α_7 is called *delay-type* auxiliary tree. The following adjunctions be made on ψ:

- Adjunction of an additive-type tree. Such an adjunction introduces an input, output or noise term additively in the expression while respecting the causality of the expression. Hence the resulting expression is also a polynomial.
- adjunction of a multiplicative-type tree. This simply introduces multiplicative factors to an existing model term, and hence, the resulting expression is also a polynomial. Furthermore, since the multiplicative factor introduced is $u(k)$, $y(k-1)$ or $\xi(k-1)$, the expression is also causal.

- adjunction of a delay-type tree. This operation simply adds a positive delay to an existing monomial, and hence preserves the polynomial and causal structure and of the expression.

Since all possible operations yield a causal polynomial expression, it can be concluded that $L(G_{\mathrm{N}})$ consists of only dynamical polynomial expressions in terms of the factors u, y and ξ which corresponds to a P-NARMAX model. This concludes the proof. □

Theorem 4.9 demonstrates that structural properties of a rich class of dynamical models can be captured within a compact set of elementary trees of a TAG. The expansive representational capability of TAG can be used in the context of model structure determination by shaping the search space of GP. This ensures that ill-posed model structures are not generated during the GP algorithm, consequently making the search algorithm more efficient. Furthermore, Algorithm 1 provides a method to compute the derivation tree representation of a given P-NARMAX model in terms of grammar G_{N}. Consequently, available prior information about the model of the system can be translated to TAG representation (or incorporated in tree sets $\mathcal{I}, \mathcal{A}$), thereby making the evolutionary search more efficient. Hence, the use of TAG enables identification within a larger class of dynamical models without requiring user-interaction, while simultaneously allowing the user to restrict the evolutionary search effectively. These ideas will be assembled together in Chap. 6.

Model Generation Using G_{N}

Three illustrative examples are used to demonstrate the generation of models using G_{N}. The models generated belong to the ARX, P-NARX and P-NARMAX model classes. It will be demonstrated that by restricting the elementary trees $\mathcal{I}$ and $\mathcal{A}$ to subsets of the elementary trees in the proposed TAG G_{N}, we can generate models that only belong to model sub-classes that are properly included in the set of P-NARMAX models, such as FIR and truncated Volterra series.

ARX Example

ARX models can be described by the equation

$$y(k) = \sum_{i=0}^{n_u} b_i u(k-i) + \sum_{j=1}^{n_y} a_j y(k-j) + \xi(k), \tag{4.12}$$

where $a_j, b_i \in \mathbb{R}$ are coefficients. The grammar G_{N} can be used to generate ARX models by restricting the auxiliary tree set $\mathcal{A}$ as

$$\mathcal{A}' = \{\alpha_1, \alpha_2, \alpha_7\} \subset \mathcal{A}. \tag{4.13}$$

Consider the example depicted in Fig. 4.8. Tree (A) is a derivation tree with initial tree η_1 at the root node, and auxiliary trees α_1 and α_2 in subsequent vertices. The edges are labelled with Gorn addresses of vertices in the auxiliary trees at which adjunctions take place. Performing the adjunctions results in derived tree (B) in

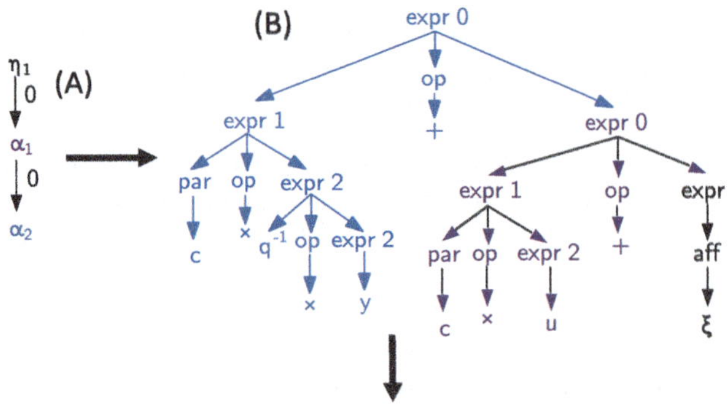

$$y(k) = c_1 y(k-1) + c_2 u(k) + \xi(k)$$

Fig. 4.8 Example 1 [11]—derivation tree (**A**), derived tree (**B**) and the corresponding symbolic model

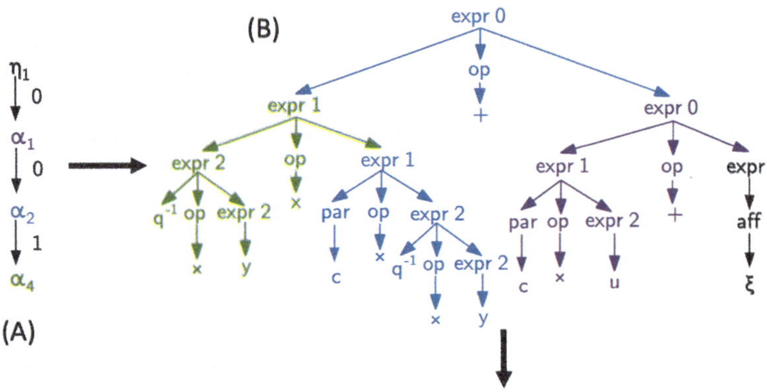

$$y(k) = c_1 y^2(k-1) + c_2 u(k) + \xi(k)$$

Fig. 4.9 Example 2 [11]—derivation tree (**A**), derived tree (**B**) and the corresponding symbolic model

Fig. 4.8. The RHS of the resulting model appears at the leaves of the derived tree, and the corresponding model is

$$y_k = c_1 y(k-1) + c_2 u(k) + \xi(k). \tag{4.14}$$

NARX Example

P-NARX models can be described by the equation

$$y(k) = \sum_{i=1}^{p} c_i \prod_{j=0}^{n_u} u^{b_{i,j}}(k-j) \prod_{m=1}^{n_y} y^{a_{i,m}}(k-m) + \xi(k). \tag{4.15}$$

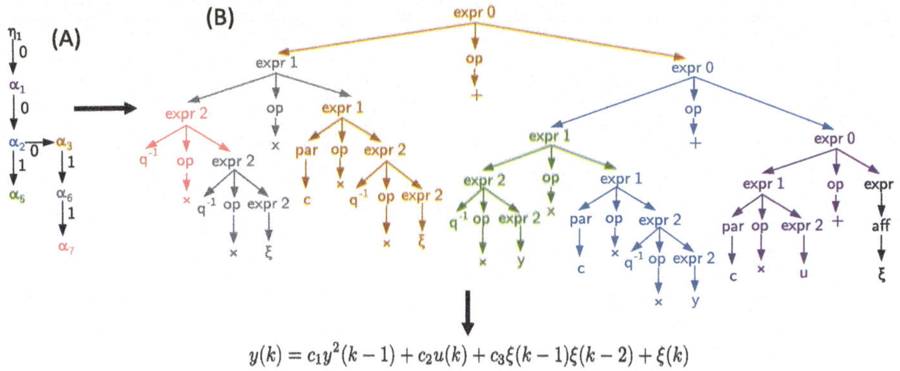

Fig. 4.10 Example 3 [11]—derivation tree (**A**), derived tree (**B**) and symbolic model

By restricting auxiliary trees to the set

$$\mathcal{A}'' = \{\alpha_1, \alpha_2, \alpha_4, \alpha_5, \alpha_7\} \subset \mathcal{A} \tag{4.16}$$

we can restrict the proposed grammar to generate P-NARX models only. Consider the example derivation tree (A) in Fig. 4.9, which is an extension of the previous example. The derivation tree consists of the initial tree η_1, and auxiliary trees α_2, α_3 and α_4. Performing the adjunctions described by the derivation tree results in the derived tree (B) in Fig. 4.9. The corresponding symbolic model is

$$y(k) = c_1 y^2(k-1) + c_2 u(k) + \xi(k). \tag{4.17}$$

P-NARMAX Example

This example builds on the previous example by adjoining trees α_3, α_6 and α_7 to the tree α_2. The new derivation tree and derived tree are depicted in Fig. 4.10. The corresponding model,

$$y(k) = c_1 y^2(k-1) + c_2 u(k) + c_3 \xi(k-1)\xi(k-2)\xi(k) + \xi(k), \tag{4.18}$$

is a P-NARMAX model.

4.3.3 Beyond the P-NARMAX Class

In Sect. 4.3.2, we developed a TAG to generate IO models that belong to the P-NARMAX class, and also demonstrated that the proposed grammar can be scaled down to restrict the generative capacity to more specialized model structures such as

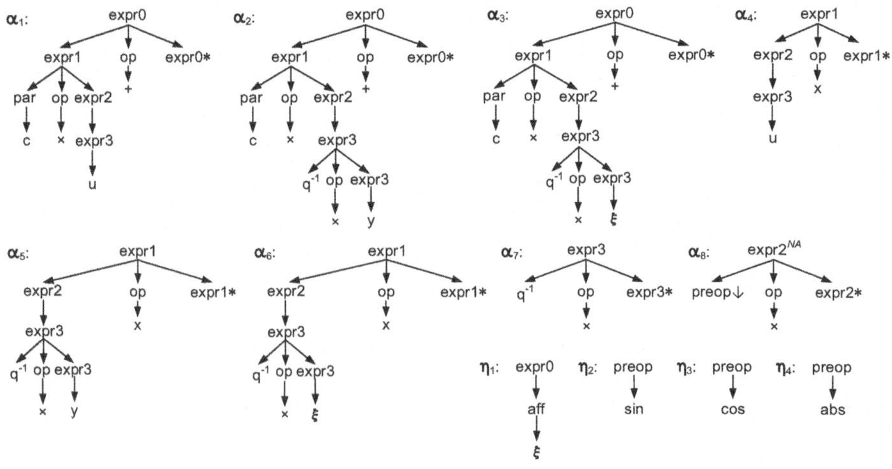

Fig. 4.11 Initial trees $\{\eta_i\}_{i=1}^4$ and auxiliary trees $\{\alpha_j\}_{j=1}^8$ of TAG G_{AT}

linear ARX models. In this section, we demonstrate that the concept of TAG-based specification of model set can be extended beyond the class of P-NARMAX models.

The P-NARMAX class of models can be used to describe a large variety of dynamical systems. However, for certain applications, it is desirable to look beyond the P-NARMAX class. In this section, we proposed two extensions to grammar G_{N}. The first extension is towards general NARMAX models, where the functional relationship between input, output and noise variables are not restricted to polynomials. The second extension is towards an independently parameterized output-error noise structure, leading to the NBJ model structure.

General NARMAX Models

In some applications, a general understanding of the dynamics of the system may provide insights into the nature of functional relationships that exist between the input, output and noise variables of the system. For example, a system that exhibits rotational dynamics should require trigonometric non-linearities to describe the relationship between inputs and outputs. Such knowledge can easily be incorporated into the TAG, thereby extending the model set to include the relevant non-linear models.

A relatively straight-forward extension to TAG G_{N} can be made by including non-linear transformations of the input, output and noise variables in the grammar. Figure 4.11 depicts the initial and auxiliary trees for TAG G_{AT},[1] which extends grammar G_{N} by including trigonometric (sin and cos) and absolute-value transformations of input, output and noise variables. These extensions are encoded in the additional initial trees $\{\eta_i\}_{i=2}^4$ and auxiliary tree α_8.

---

[1] The superscript *NA* in auxiliary tree α_8 refers to a *null adjunction* constraint on the root node, which prohibits the adjunction of any auxiliary tree at that location. See [6] for details.

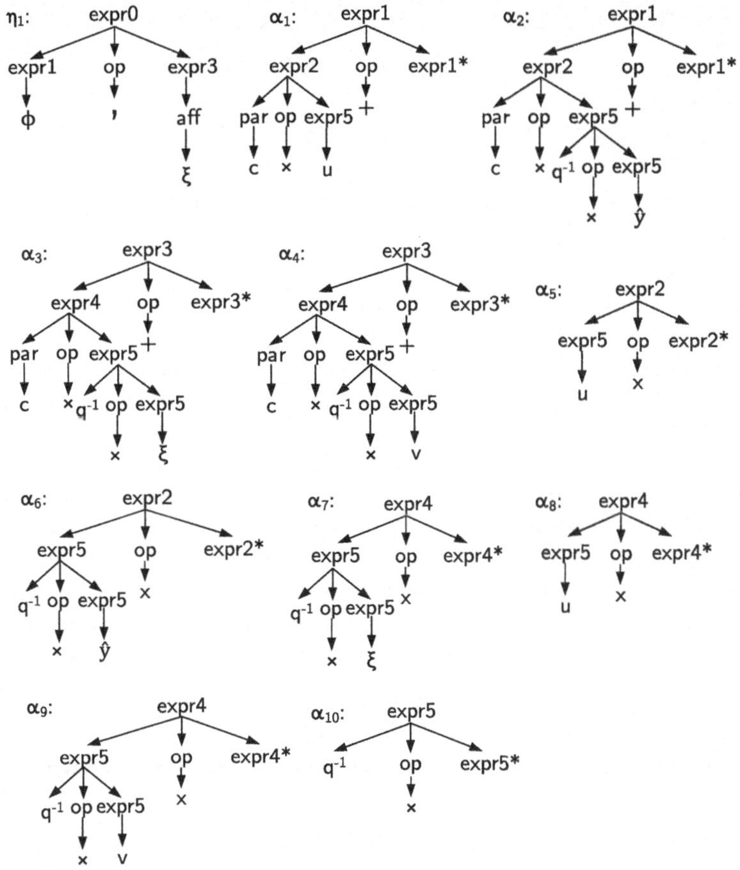

Fig. 4.12 Initial and auxiliary trees ($\mathcal{I}$ and $\mathcal{A}$) of TAG G_{NBJ} [11]

TAG G_{AT} illustrates one of the many ways in which the proposed grammar can be extended to the general NARMAX class. Depending on the insight on the nature of dynamics being modelled, other extensions to the proposed grammar can also be made.

NBJ Models

In the case of linear systems, a BJ model structure is an extension of the OE model structure, where the error is modelled as an ARMA (ARMA) process [13]. The BJ class also includes, as special cases, other linear model structures such as ARMAX and OE. In the same spirit, NBJ model structure can be expressed as a Non-linear Output Error (NOE) model where the noise process is subsequently modelled as a Non-linear Auto-Regressive Moving-Average models (NARMA) process. The NBJ model structure is given by the following equations [11]

$$
\begin{aligned}
y_0(k) &= f\left(y_0(k-1), \ldots, y_0(k-n_y), u(k), \ldots, u(k-n_y)\right), \\
v(k) &= g\left(v(k-1), \ldots, v(k-n_v), u(k), \ldots, u(k-n_u), \xi(k-1), \ldots, \xi(k-n_\xi)\right) \\
&\quad + \xi(k), \\
y(k) &= y_0(k) + v(k),
\end{aligned}
\tag{4.19}
$$

where $f(\cdot)$ and $g(\cdot)$ are polynomial functions in terms of their arguments, and function g satisfies

$$
g(0, \ldots, 0, u(k), \ldots, u(k-n_u), 0, \ldots, 0) = 0 \tag{4.20}
$$

for any input sequence. Notice that the RHS expressions of the equations describing the process and noise dynamics have the same structure that was studied in Sect. 4.3.2 for P-NARMAX models (see (4.10)). Hence, the TAG G_N can be extended to generate NBJ models. Figure 4.12 depicts the initial and auxiliary trees of TAG G_{NBJ} for NBJ model structures. The structure of the initial tree η_1 ensures that all elements in $L(G_{NBJ})$ contain two expressions, separated by a comma, that represent the functions $f(\cdot)$ and $g(\cdot)$ respectively. Each of these expressions can be expanded by adjoining auxiliary trees that ensure that the polynomial structure is maintained. Furthermore, it is ensured that input terms can *only* be introduced multiplicatively (and *not* additively) in noise model $g(\cdot)$, thereby ensuring that the condition in (4.20) is satisfied.

4.4 TAG Representation of Non-linear State-Space Models

In Sect. 4.3, we developed a technique to describe the structure of parametric IO models using TAG. The concept of encoding model structures using the TAG framework is fairly general, and can be extended to parametric model representations other than IO models. To illustrate the generality of this concept, in this section, we propose a TAG for NL SS representations.

Consider a SS models with n states, p inputs and q outputs, described by the following equations

$$
\begin{aligned}
x_1(k+1) &= f_1(x_1(k), \ldots, x_n(k), u_1(k), \ldots, u_p(k)), \\
&\vdots \\
x_n(k+1) &= f_n(x_1(k), \ldots, x_n(k), u_1(k), \ldots, u_p(k)), \\
y_1(k) &= g_1(x_1(k), \ldots, x_n(k), u_1(k), \ldots, u_p(k)), \\
&\vdots \\
y_q(k) &= g_p(x_1(k), \ldots, x_n(k), u_1(k), \ldots, u_p(k)),
\end{aligned}
\tag{4.21}
$$

where $f_1, \ldots, f_n$ are the state equations governing the dynamics of states $x_1, \ldots, x_n$, respectively, and $g_1, \ldots, g_p$ are the output equations that map inputs and states to outputs $y_1, \ldots, y_q$, respectively. For ease of illustration, we do not consider a noise process in the SS model. However, just like the case of IO models, a noise model can be included in the SS model structure. For the purpose of illustration, we consider the state and output equations to be linear or polynomial non-linear functions of the inputs and states. Furthermore, trigonometric transformations (sin and cos) of input and state variables are also considered in this illustration.

TAG Representation

The state-space grammar $G_{SS}(n, p, q)$ is given by the tuple $G_{SS} = \langle \mathfrak{N}, \mathfrak{T}, S, \mathcal{I}, \mathcal{A} \rangle$ with

- $\mathfrak{N} = \bigcup_{i=1}^{n}\{\mathrm{se}_i\} \bigcup_{l=1}^{q}\{\mathrm{oe}_l\} \bigcup \{\mathrm{model, param, op, expr1, st, ip}\}$,
- $\mathfrak{T} = \bigcup_{i=1}^{n} \{\mathrm{x}_i\} \bigcup_{j=1}^{p} \{\mathrm{u}_j\} \bigcup \{\mathrm{c}, +, \times\}$,
- $S = \mathrm{model}$,
- $\mathcal{I} = \{\eta_i\}_{i=1}^{13}$, where initial trees η_i's are depicted in Fig. 4.13,
- $\mathcal{A} = \{\alpha_i\}_{i=1}^{16}$, where the auxiliary trees α_i's are depicted in Fig. 4.13.

Observe that the initial and auxiliary trees in Fig. 4.13 are defined based on variables i, j and l, which serve as index variables for states, inputs and outputs, respectively, and take up values $i \in [1, n], j \in [1, p]$ and $l \in [1, q]$. The role of each of the initial and auxiliary trees in TAG G_{SS} is discussed below.

- The initial tree η_1 sets the structure of the SS model, with n state equations, with labels se_i, and q output equations, with labels oe_l, each separated by commas.
- Initial trees $\eta_2(i)$, $\eta_3(i)$, $\eta_4(i)$ and $\eta_5(i)$ introduce a state, input or non-linear transformations of the state or input terms to the ith state equations. Similarly, initial trees $\eta_6(l)$, $\eta_7(l)$, $\eta_8(l)$ and $\eta_9(l)$ perform similar operations on the lth output equation.
- Initial trees $\eta_{10}(i)$ and $\eta_{11}(j)$ introduce state x_i and input u_j, respectively.
- Initial trees η_{12} and η_{13} introduce a $\sin(\cdot)$ and a $\cos(\cdot)$ transformation, respectively.
- Auxiliary trees $\alpha_1(i)$, $\alpha_2(i)$, $\alpha_3(i)$ and $\alpha_4(i)$ introduce *additive-type* terms (state, input or non-linear transformations of the same) to the existing ith state equation. Similarly, auxiliary trees $\alpha_5(l), \alpha_6(l), \alpha_7(l)$ and $\alpha_8(l)$ introduce additive-type terms to the existing lth output equation.
- Auxiliary trees $\alpha_9(i)$, $\alpha_{10}(i)$, $\alpha_{11}(i)$ and $\alpha_{12}(i)$ introduce *multiplicative-type* terms (state, input or non-linear transformations of the same) to the existing ith state equation. Similarly, auxiliary trees $\alpha_{13}(l)$, $\alpha_{14}(l)$, $\alpha_{15}(l)$ and $\alpha_{16}(l)$ introduce multiplicative-type terms to the existing lth output equation.

Just like in the IO case, TAG G_{SS} can be expanded to include other non-linearities in the model set, based on insights into the nature of dynamics of the system to be modelled. The grammar can also be scaled down to restrict the scope of the model set specification, for example, to generate only linear SS models. However, unlike in the IO case, the order of the SS model, in terms of the number of states, produced by TAG $G_{SS}(n, p, q)$ is fixed to n.

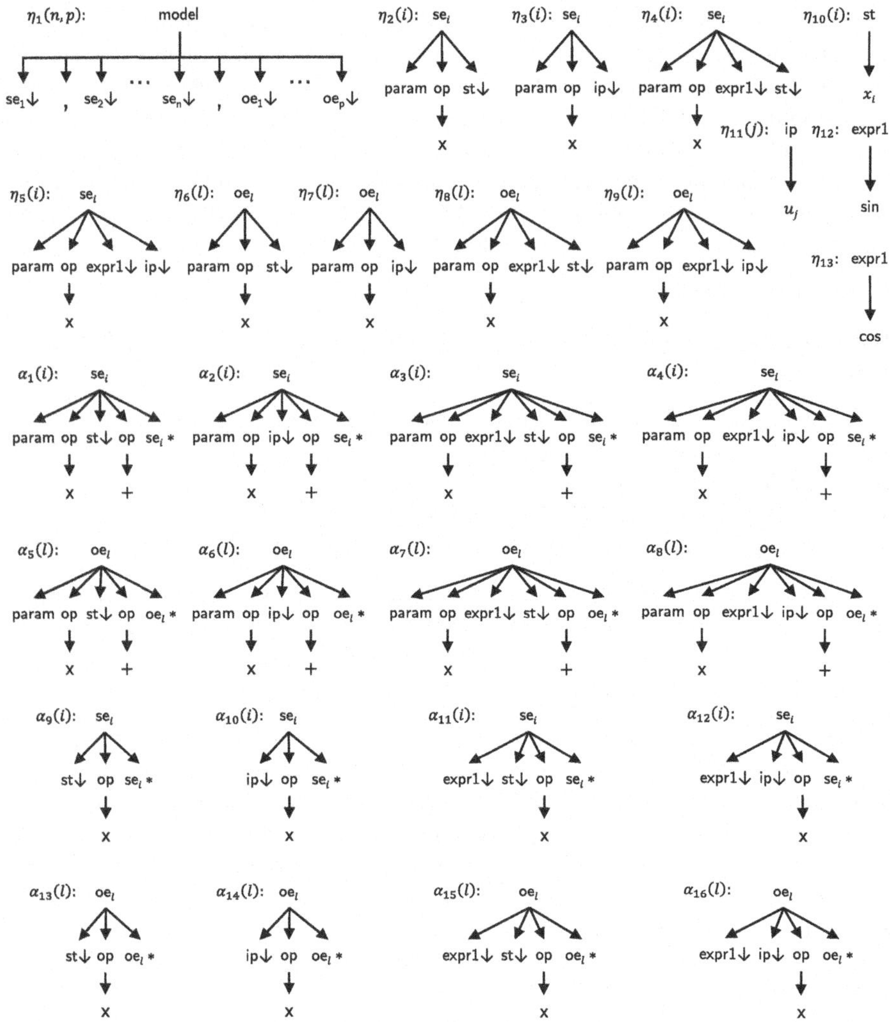

Fig. 4.13 Initial and auxiliary trees ($\mathcal{I}$ and $\mathcal{A}$) of TAG G_{SS}

Model Generations Using TAG G_{SS}

Consider the example depicted in Fig. 4.14. The Figure illustrates the derivation tree representation and the derived tree representation of the following state-space model with $n = 3$ states, $p = 2$ inputs and $q = 2$ outputs:

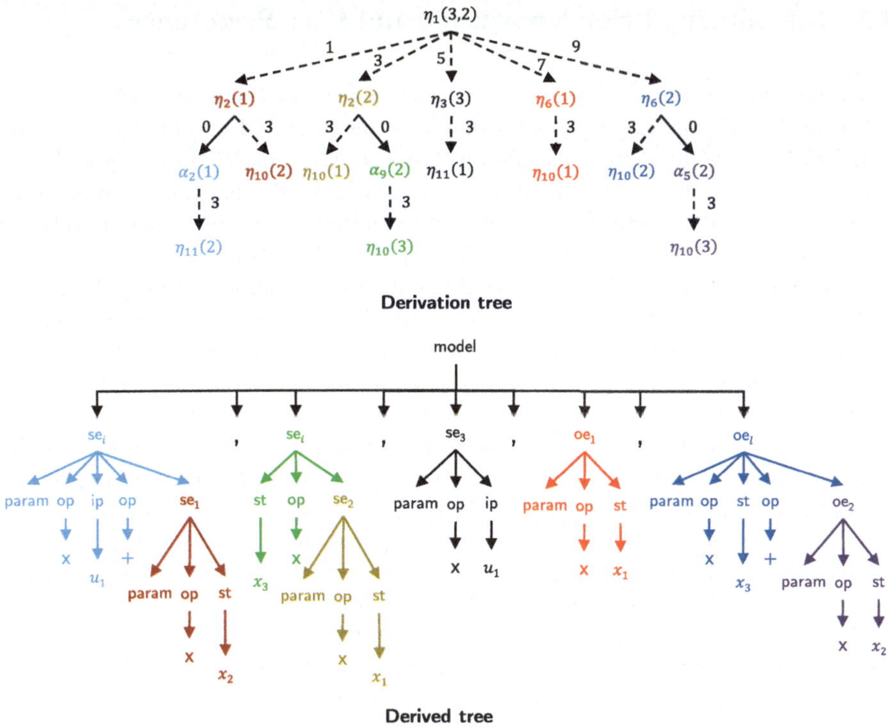

Fig. 4.14 An example of a state-space model structure generated from TAG G_{SS}. The figure depicts the derivation tree and derived tree representation of (4.22)

$$
\begin{aligned}
x_1(k+1) &= c_1 x_2(k) + c_2 u_2(k),\\
x_2(k+1) &= c_3 x_1(k) x_3(k),\\
x_3(k+1) &= c_4 u_1(k),\\
y_1(k) &= c_5 x_1(k),\\
y_2(k) &= c_6 x_2(k) + c_7 x_3
\end{aligned}
\tag{4.22}
$$

The derived tree representation in Fig. 4.14 depicts the structure of (4.22), with three state-expressions and two output-expressions, each separated by commas. The derivation tree representation of the model depicts the initial and auxiliary trees of TAG G_{SS} used to generate the model. In the derivation tree representation, dashed lines denote substitution operations and solid lines denote adjunction operations.

4.5 Introducing Prior Knowledge and User Preferences

It is widely accepted in SI literature (e.g., see [13, part (iii)]) that any prior information regarding the underlying system should be incorporated into the identification framework in order to improve the quality of the estimated model. Similarly, in the EC community, it is generally accepted (e.g., see [3, Chap. 10]) that suitably modifying an EA (e.g., by modifying the solution representation or variation operators) based on problem-specific information typically improves the efficiency of the search for solving a given problem. Thus, it is vital that the identification approach developed in this thesis is capable of incorporating prior domain knowledge or user preferences. In this section, we discuss two mechanisms, based on TAG formulations, for introducing prior knowledge—*model seeding* and *specifications on the genotype*.

Assume an identification scenario in which an initial model estimate for the data generating system (DGS) is known and the identification problem is to infer a model of higher quality compared to the initial model estimate (in terms of user-specified performance criteria) from measured data. In this setting, a derivation tree representation of the low quality model can be computed (for e.g., using Algorithm 1 if the model belongs to the P-NARMAX class), and included in the initial population of GP. This is referred to as model seeding. Model seeding can be used to bias the evolutionary search towards a specific region in the search space. However, since the initial model estimate is bound to be better than the other randomly generated models in the initial population, the GP algorithm may converge to the initial model within a few iterations, leading to degeneracy of the population. Hence, care must be taken to ensure diversity in the population when using model seeding.

We can also consider a second scenario in which the dynamic class of the DGS is known. This information may be known due to domain knowledge of the user, user-preference or informative experiments conducted on the system. Prior knowledge of the dynamic class can be incorporated as specifications on the genotype by suitably modifying the grammar being used. For example, in Sect. 4.3.2, grammar G_{N} was scaled down to restrict the model set to specific sub-classes of the P-NARMAX class. Similarly, the grammar can also be modified to extend the model set, as was demonstrated in Sect. 4.3.3. TAG can also to treat different model representations, e.g. SS models, as illustrated in Sect. 4.4.

The two mechanisms for introducing prior knowledge discussed here do not treat scenarios in which the knowledge is in terms of *specification on the phenotype*. These may include, for e.g., constraints on the model complexity, stability requirements, and so on. Such specifications can be treated as user-specified performance measures, and will be treated in Chap. 5.

4.6 Discussion

In this chapter, we introduced TAG, a tree generating grammar originally used to describe structural relationships between words in a linguistic sentence. Inspired by [5], we demonstrated that TAG can also be used to describe the structural relationship between variables of a dynamical model. Based on this concept, we introduced a TAG-based specification of a model set. In the context of data-driven modelling, a TAG-based model set specification provides a foundation to develop an identification framework in which the user is not required to pre-specify model structure or complexity. Instead, the conceptualized identification framework would be required to explore models of varying structures and complexities that can be generated by the specified TAG. In this sense, the TAG-based model set specification addresses Research Question 1.1 described in Chap. 1, and brings us one step closer to the goal of achieving automated system identification.

We also illustrated, in this chapter, the generality of TAG in terms of representing the structure of dynamical models. We showed that the proposed grammar G_{N} for P-NARMAX models can be (i) used to generate model structures that belong to subclasses of the P-NARMAX class, and (ii) used to generate multiple levels of model complexity within the same model class, and (iii) scaled up to describe models with more complex non-linearities. Furthermore, we also demonstrated that TAG can be used to generate alternate model representations, specifically, SS models. While the scope of this dissertation is restricted to IO models, the generality of TAG-based model generation concept ensures the possibility of extending the identification framework developed in this dissertation beyond the class of IO models.

Besides serving as a tool to generate models with varying structure, complexity and representation, TAG can also be used to introduce prior information or user preference in terms of desired model structure or representation. This can be achieved by scaling down the grammar, as discussed in Sect. 4.5, expanding the grammar, discussed in Sect. 4.3.3, or developing a TAG for the desired alternative representation, e.g., state-space representations, discussed in Sect. 4.4. This addresses Research Question 1.2 proposed in Chap. 1.

Finally, the flexibility of TAG-based model set specification comes at a cost. In order to automate the task of system identification, the conceptualized identification framework must address the problem of model structure determination in a search space that grows combinatorially w.r.t the depth of the derivation tree representation of the model. This is a challenging problem to solve. Nonetheless, EAs such as GP, introduced in Chap. 3, are promising approaches that can be used to tackle the challenging problem of model structure determination. This will be the subject of Chap. 6.

References

1. Billings SA (2013) Nonlinear system identification: NARMAX methods in the time, frequency, and spatio-temporal domains. Wiley (2013)
2. Chomsky N (1956) Three models for the description of language. IRE Trans Inf Theory 2(3):113–124
3. Eiben AE, Smith JE (2003) Introduction to evolutionary computing, vol 53. Springer
4. Gorn S (1965) Explicit definitions and linguistic dominoes. In: Systems and computer science, proceedings of the conference held at University of Western Ontario, pp 77–115
5. Hoai NX, McKay R (2001) A framework for tree adjunct grammar guided genetic programming. In: Proceedings of the post-graduate ADFA conference on computer science (PACCS'01), pp 93–99
6. Joshi AK, Schabes Y (1997) Tree-adjoining grammars. In: Handbook of formal languages. Springer, pp 69–123
7. Joshi, AK, Levy LS, Takahashi M (1975) Tree adjunct grammars. J Comput Syst Sci 10(1):136–163
8. Kallmeyer L (2009). A declarative characterization of different types of multicomponent tree adjoining grammars. Res Lang Comput 7(1), 55–99
9. Kallmeyer L (2010) Parsing beyond context-free grammars. Springer Science & Business Media (2010)
10. Khandelwal D, Schoukens M, Tóth R (2019) Grammar-based representation and identification of dynamical systems. In: Proceedings of the 18th European Control Conference (ECC). Naples, Italy, June 2019, pp 1318–1323
11. Khandelwal D, Schoukens M, Tóth R (2020) A tree adjoining grammar representation for models of stochastic dynamical systems. Automatica 119:109099
12. Leontaritis I, Billings SA (1985) Input-output parametric models for non-linear systems part I: deterministic non-linear systems. Int J Control 41(2):303–328
13. Ljung L (ed) (1999) System identification: theory for the User, 2nd edn. Prentice Hall PTR. ISBN 0-13-656695-2
14. Meduna A (2014) Formal languages and computation: models and their applications. Auerbach Publications
15. Parkes AP (2012) Introduction to languages, machines and logic: computable languages, abstract machines and formal logic. Springer Science & Business Media (2012)
16. Shieber SM (1985) Evidence against the context-freeness of natural language. In: Philosophy, language, and artificial intelligence. Springer, pp 79–89
17. Stone MH (1948) The generalized Weierstrass approximation theorem. Math Mag 21(5):237–254
18. Whigham PA et al (1995) Grammatically-based genetic programming. In: Proceedings of the workshop on genetic programming: from theory to real-world applications, vol 16, pp 33–41

Chapter 5
Performance Measures

> *I didn't have time to write a short letter, so I wrote a long one instead.*
>
> *Mark Twain*

In Chaps. 3 and 4, we introduced two concepts relevant for automating the task of system identification—GP and TAG. GP can be used to explore a generalized model set, spanning across multiple model structures and complexities, specified by the chosen TAG. These ideas help us to address Research Questions 1 and 3 laid out in Chap. 1. This leads us to the next question, i.e. Research Question 2: How can we incorporate multiple user-specified performance measures in the desired identification framework, and what is the appropriate notion of optimality in a multi-objective setting?

In this chapter, we introduce a multi-objective framework for ordering candidate solutions in the solution set based on the notion of *Pareto-dominance*, and discuss some EC-based approaches for multi-objective optimization. Subsequently, we propose and motivate four performance measures for the conceptualized identification framework—mean-squared one-step-prediction and simulation error, parametric complexity and dynamic complexity of the model, and discuss some computation aspects (partially published in [11]) relevant in the context of system identification.

5.1 Introduction

As discussed in Sect. 1.3.3, the choice of an identification criterion should ideally be influenced by the ultimate use-case of the model by user-preferences. For applications such as weather forecasting and fault detection, models must be well-tuned to make accurate short-term predictions, while in applications such as optimal control design for mechatronic systems or stability analysis of structural dynamics, the model must be accurate over longer time-horizons. A typical identification task may also involve multiple modelling objectives for which a model should be optimized. For example,

D. Khandelwal, *Automating Data-Driven Modelling of Dynamical Systems*,
Springer Theses, https://doi.org/10.1007/978-3-030-90343-5_5

for use in a MPC design problem, the desired model should have low run-time complexity while also having sufficiently good multi-step-ahead prediction accuracy. In realistic scenarios, modelling objectives oppose each other, thereby necessitating the need of a trade-off between objectives. For example, a more complex model may achieve high prediction accuracy, however a simpler model with slightly worse accuracy might be more suitable for systems analysis and control design. Again, the "right choice of trade-off" typically depends on the application and on user-preferences.

When modelling objectives contradict each other, there may not exist a single solution that is optimal for all performance measures. This makes solving multi-objective problems more involved than single-objective problems. Many optimization techniques have been developed to solve multi-objective problems, see [6, 15]. A commonly used approach involves converting a multi-objective problem to a single-objective problem, for example, by taking a linear combination of the criterion. The coefficients of the linear combination play the role of *trade-off parameters* between the various performance measures, and are typically chosen a priori by the user or treated as a hyper-parameter to be optimized, for example using grid-search. A second approach relies on a more direct approach of optimizing models simultaneously for multiple objectives. Such methods rely on a new notion of optimality based on partial ordering w.r.t the feasible set of decision variables and the performance measures. Multi-Objective Optimization (MOO) may yield a (possibly infinite) set of solutions that are *Pareto optimal*, i.e., solutions for which no single performance measure can be further improved without worsening any of the other performance measures. The Pareto optimal set of solutions reveal trade-offs that can be made by the user a posteriori, thereby enabling the user to make informed multi-criteria-based selection.

SI approaches discussed in Sect. 2.4 typically estimate models for a single performance measure. Hence, optimizing for multiple performance measures with these approaches is typically handled in the following ways:

- Using the first solution approach discussed in the prequel, i.e., via trade-off parameters forming linear combinations of the performance measures.
- Reformulating all but one of the performance measures as constraints to be satisfied in a single-objective optimization problem. The user is required to formulate reasonable constraints for the other performance measures.
- Using prior-knowledge and user-preferences to restrict the model set based on some of the modelling objectives. For example, in PPEM typical modelling objectives include the one-step-ahead prediction error and model complexity. User experience and prior knowledge is used to restrict the model set to a specific model complexity such that the identification problem can be formulated in terms of prediction error alone.

In contrast, EC-based methods provide an opportunity to estimate a set of models that lie along the Pareto-front of a MOO problem without having to convert the given problem to a single-objective problem. This is due to the fact that:

(i) By default, EAs work with a set of candidate models, making it possible to estimate a set of models that lie on the Pareto front in a single run of the algorithm.
(ii) In each iteration of EC, candidate solution structures are first generated, based on the solutions obtained in the previous iteration, and subsequently evaluated. This *generate-and-validate* feature of EC makes it possible to evaluate candidate models on multiple performance measures.

In the field of SI, the idea of EC-based MOO was first explored by Fonseca and Fleming [8]. The authors demonstrated that the estimate of the Pareto front, obtained using GA, clearly illustrated the effect of optimizing for conflicting objectives. This makes the Pareto front a valuable tool for the user to make an informed selection of the identified model, such that it is better suited for the eventual use-case.

In Sect. 5.2, we introduce the notions of partial ordering, non-dominance of solutions and the Pareto front, and discuss how notions can be used to design MOEAs. For system identification, when the model set is specified using TAG (as introduced in Chap. 4), the candidate models may vary in terms of their dynamical properties and their complexities. To meaningfully evaluate models in such a general setting, we propose several performance measures in Sect. 5.3, along with related computational aspects.

5.2 The Multiobjective Framework

5.2.1 *Ordering and Dominance in Sets*

In the following, we introduce the notions of ordering and dominance of solutions. This requires the introduction of some relevant mathematical preliminaries, which can be found in detail in textbooks and tutorials such as [6, 7]. For completeness, we include the relevant notions here.

Definition 5.1 (*Binary relations and its properties*) A binary relation on a given set $\mathcal{M}$ is a set $R \subseteq \mathcal{M} \times \mathcal{M}$. The binary relation R is said to be

- reflexive, if and only if $\forall x \in \mathcal{M} : (x, x) \in R$,
- irreflexive, if and only if $\forall x \in \mathcal{M} : (x, x) \notin R$,
- symmetric, if and only if $\forall x \in \mathcal{M} : \forall y \in \mathcal{M} : (x, y) \in R \Leftrightarrow (y, x) \in R$,
- asymmetric, if and only if $\forall x \in \mathcal{M} : \forall y \in \mathcal{M} : (x, y) \in R \Leftrightarrow (y, x) \notin R$,
- antisymmetric, if and only if $\forall x \in \mathcal{M} : \forall y \in \mathcal{M} : (x, y) \in R \wedge (y, x) \in R \Rightarrow x = y$,
- transitive, if and only if $\forall x \in \mathcal{M} : \forall y \in \mathcal{M} : \forall z \in \mathcal{M} : (x, y) \in R \wedge (y, z) \in R \Rightarrow (x, z) \in R$.

A pair $(x, y) \in R$ can also be denoted as xRy.

Definition 5.2 (*Pre-order, partial order, strict partial order*) A binary relation R is said to be a

- pre-order, if and only if it is transitive and reflexive,
- partial order, if and only if it is an antisymmetric pre-order,
- strict partial order, if and only if it is irreflexive and transitive.

Definition 5.3 (*Partially ordered set*) Let $\mathcal{M}$ be a set and let R be a partial order on $\mathcal{M}$. The pair $(\mathcal{M}, R)$ is a partially ordered set.

Denote the set $\Delta = \{(x, x) \mid x \in \mathcal{M}\}$ as the diagonal of $\mathcal{M}$. Note that if R is a partial order on $\mathcal{M}$, then $(R \setminus \Delta)$ is a strict partial order on $\mathcal{M}$.

Definition 5.4 (*Minimal element*) An element $x \in \mathcal{M}$ is said to be a minimal element of the partially ordered set $(\mathcal{M}, R)$ if there exists no $x' \in \mathcal{M}$ such that $x'Rx$ and $x' \neq x$.

Based on these notions, we introduce the concepts of Pareto-dominance in the context of multi-criteria system identification. Let the set of models be denoted by $\mathcal{M}(G)$, where G is the specified TAG. It is assumed that any model structure $M \in \mathcal{M}(G)$ can be parameterized as $M(\theta)$, where the dimension of the parameter vector θ depends on the structure of M. Let the user-specified multiple performance measures be denoted by $\{J_i(M, \theta, \mathcal{D}_N)\}_{i=1}^{m}$, and let the vector-valued function $J(M, \theta, \mathcal{D}_N) = (J_1(M, \theta, \mathcal{D}_N), \ldots, J_m(M, \theta, \mathcal{D}_N))^\top$ map from $\mathcal{M}(G)$ to $\mathbb{R}^m$. In MOO literature, the set of models $\mathcal{M}(G)$ and the space of vector-valued performance vectors are referred to as the *decision space* and *objective space*, respectively.

Definition 5.5 (*Pareto-dominance in objective space*) Let $z, z' \in \mathbb{R}^m$ be two vectors in the objective space. The objective vector z is said to Pareto-dominate the objective vector z' if and only if

1. $z_i \leq z'_i$ for all $i \in \{1, \ldots, m\}$, and
2. $\exists j \in \{1, \ldots, m\}$ such that $z_j < z'_j$.

The Pareto order on the objective space is a strict partial operator, and is denoted as $\prec$. Hence, an objective vector $z \in \mathbb{R}^m$ that Pareto dominates (in short, dominates) an objective vector $z' \in \mathbb{R}^m$ is denoted as $z \prec z'$. Based on the strict partial ordering $\prec$, we can define the binary relation $\preceq = (\prec \cup \Delta)$ which is a partial order allowing for equality between vectors in the objective space that do not Pareto-dominate each other.

We have so far defined the notion of Pareto dominance in terms of vectors in the objective space. This idea can be extended to solutions in the decision space, i.e., the model set $\mathcal{M}(G)$, as follows.

Definition 5.6 (*Pareto dominance in decision space*) Let $M(\theta) \in \mathcal{M}(G)$ and $M'(\theta') \in \mathcal{M}(G)$ be two candidate solutions in the model set, where θ and θ' are the corresponding parameter vectors. The solution $M(\theta)$ Pareto dominates the solution $M'(\theta')$, denoted by $M(\theta) \prec_J M'(\theta')$, if and only if $J(M, \theta, \mathcal{D}_N) \prec J(M', \theta', \mathcal{D}_N)$.

The binary relation $\prec_J$ on $\mathcal{M}(G)$ is a strict partial order.

Definition 5.7 (*Efficient set and Pareto front*) The minimal elements of the strict partial order $\prec_J$ on $\mathcal{M}(G)$ are called efficient points. The set of all efficient points in $\mathcal{M}(G)$ is called efficient set, denoted by $\mathcal{E}(G, \prec_J)$. The corresponding set of objective vectors $\mathcal{P}_J(G, \prec) := \{J(M, \theta, \mathcal{D}_N) \mid M(\theta) \in \mathcal{E}(G)\}$ is called the non-dominated or Pareto-optimal objective vectors, a.k.a the Pareto front.

Definitions 5.5–5.7 give the relevant notions of optimality in a multiobjective problem setting. In this setting, we are interested in estimating the set of candidate models that lie in the efficient set $\mathcal{E}(\mathcal{M}(G), \prec_J)$ that achieve objective values that lie on the Pareto front $\mathcal{P}_J(\mathcal{M}(G), \prec)$. Note that since the model set $\mathcal{M}(G)$ spans a continuous space (the parameters θ lie in a continuous domain), the efficient set $\mathcal{E}(\mathcal{M}(G), \prec_J)$ may possibly contain infinitely many efficient solutions. Thus, in reality, we can only estimate a subset of the efficient set $\mathcal{E}(\mathcal{M}(G), \prec_J)$.

5.2.2 Handling Multiple Criteria in EC

Multiple objectives can be incorporated in an EA is several ways. Two common paradigms include Pareto-based algorithms and decomposition based algorithms, described below.

1. Pareto-based algorithms—These algorithms directly utilize the notion of Pareto dominance in order to perform selection in each iteration of the EA. The basic idea is as follows. Once a new population $\tilde{X}^{(j)}$ is proposed using variation operators in iteration j, the population can be evaluated based on m performance measures specified by the user. The candidate solutions can then be combined with other solutions evolved by the EA, for example, solutions from the previous iterations $X^{(j-1)}$, or an archive of efficient solutions $\mathcal{E}(\{X^{(l)}\}_{l=1}^{j}, \prec_J)$ found up to iteration j (see [12]). Subsequently, the efficient subset of the combined set of solutions can be selected and propagated to the next iteration. In accordance with the heuristics of EA, by introducing variations to candidate solutions closer to the current Pareto front estimate, we expect to obtain new solutions that are also close to the estimated Pareto front with some probability. As the number of iterations increases, more models in the model set are explored, and as a consequence, the estimated Pareto-front approaches the true Pareto-front $\mathcal{P}_J(G, \prec)$.
2. Decomposition-based algorithms—These algorithms decompose the objective space into several regions using a scalarization technique. Subsequently, individuals in the population are partitioned and each subset of individuals are evolved for a specific scalarization of the objective function. As the EA iterations proceed, the various subsets of individuals approach the corresponding regions of the Pareto front, ensuring a good distribution of solutions along the front.

A number of MOEAs have been proposed in the literature. Some of the popular algorithms include Non-dominated Sorting Genetic Algorithm II (NSGA2) [5], Strength

Pareto Evolutionary Algorithm 2 (SPEA2) [20] and MOEA based on Decomposition (MOEAD) [18]. A comparison of some of the MOEAs can be found in [19]. In the identification methodology developed in this dissertation, we will make use of the non-dominated sorting algorithm proposed in [5]. However, in principle, any of the other approaches can also be used.

5.3 Proposed Multi-objective Criteria

For the proposed identification framework, a user can specify any desired performance measures. However, in this dissertation we will be making use of the following identification criteria: one-step-ahead prediction error, simulation error, parametric complexity and dynamic complexity. The proposed criteria, and the underlying motivation is described in the sequel. To assist the discussion, we assume that the model set is generated by grammar G_{N}. In other words, the model set is restricted the class of models that can be represented as a P-NARMAX model, see (4.8) and (4.10). For convenience, we recall the equivalent representations as follows:

$$\begin{aligned} y(k) = f(&u(k), \ldots, u(k-n_u), y(k-1), \ldots, y(k-n_y), \xi(k-1), \ldots, \xi(k-n_\xi)) \\ &+ \xi(k), \end{aligned} \tag{5.1}$$

where $f(\cdot)$ is a polynomial non-linear function, and

$$y(k) = \sum_{i=1}^{p} c_i \prod_{j=0}^{n_u} u^{b_{i,j}}(k-j) \prod_{l=1}^{n_\xi} \xi^{d_{i,l}}(k-l) \prod_{m=1}^{n_y} y^{a_{i,m}}(k-m) + \xi(k). \tag{5.2}$$

5.3.1 *One-Step-Ahead Prediction Error*

The one-step-ahead predicted output at time instant k, denoted by $\hat{y}(k \mid k-1)$, of a P-NARMAX model is given by the conditional expectation

$$\hat{y}(k \mid k-1) := E_\xi[y(k) \mid \mathcal{D}_{k-1}, u_m(k)], \tag{5.3}$$

where, $E_\xi[\cdot]$ is the expectation operator with respect to the probability distribution of noise ξ. For a P-NARMAX model in the form of (5.1), the one-step-ahead predictor can be formulated as the following non-linear filter (see [4, Chap. 2])

$$\begin{aligned} \hat{y}(k \mid k-1) = f(&u_m(k), \ldots, u_m(k-n_u), y_m(k-1), \ldots, y_m(k-n_y), \\ &\varepsilon_p(k-1), \ldots, \varepsilon_p(k-n_\xi)), \end{aligned} \tag{5.4}$$

where $\varepsilon_p(k)$ is the one-step-ahead prediction error defined as

$$\varepsilon_p(k) := y_m(k) - \hat{y}(k \mid k-1). \tag{5.5}$$

Hence, if the initial conditions of the one-step-ahead predictor in (5.4) are known, the predicted output $\{\hat{y}(k \mid k-1)\}_{i=n_1+1}^{N}$, where $n_1 := \max\{n_u, n_y, n_\xi\}$, can be computed recursively. If the initial conditions must be estimated and if the predictor model (5.4) is asymptotically stable, then any transient errors introduced will asymptotically reduce to 0.

We propose the first performance measure to be the sum-of-squares of the one-step-ahead prediction error. For a given model $M(\theta)$, this can be computed as

$$J_1(M, \theta, \mathcal{D}_N) = \frac{1}{N - n_1} \sum_{k=n_1+1}^{N} \left(y_m(k) - \hat{y}(k \mid k-1)\right)^2. \tag{5.6}$$

In principle, any other signal norm can be used in (5.6) to quantify the one-step-ahead prediction error. The identification criterion J_1 based on the 2-norm has been widely used in the identification literature for many decades. its popularity is primarily due to the ease of use and the maximum-likelihood interpretation of the identification criterion under certain conditions [2] (see Sect. 2.4.2 for discussion).

5.3.2 *Simulation Error*

Motivation

The use of grammars, introduced in Chap. 4, allows us to describe general model sets that may encompass multiple classes of dynamical system. However, such general model sets present a challenge to the use of one-step-ahead prediction error as the sole identification criterion to measure the performance of a given model. One of the assumptions under which the statistical interpretation of identification criterion J_1 in (5.6) holds true, is that the true data-generating system can be represented by the chosen model set. However, model structures generated from grammars provide no such guarantee, even when using a grammar that may generate model structures suitable to describe the DGS. This is due to two factors:

- A grammar capable of generating model structures that are complex enough to represent the DGS can also generate simpler model structures.
- The complexity of a model generated by the grammar is not pre-specified. Hence, it is possible to obtain model structures that are significantly under-parameterized (or over-parameterized) in comparison with the DGS.

It has been observed (for example, in [14, 17]) that the presence of an auto-regressive component in the prediction model can allow a relatively simpler model to achieve optimistic prediction results for an otherwise complex dynamical system. This can be problematic when the ultimate use-case of the model is anything but one-step-ahead prediction. This issue is also illustrated in a simple academic example in Appendix 3. This motivates our second performance criterion—simulation error. Simulation error is typically more sensitive to mismatch between the model and the system structure than prediction error (see [1, 17]). Moreover, simulation models can be defined for models that describe different classes of system dynamics. Simulation error has been used in the past for non-linear structure selection, for example in [17].

The Simulation Model

A simulation model can be defined as the expectation of the output with respect to the distribution of the noise conditioned on the past inputs, i.e.,

$$y_s(k) := E_\xi[y(k) \mid \{u(\tau)\}_{\tau=1}^{k}]. \tag{5.7}$$

The simulation response $y_s(k)$ in (5.7) represents the deterministic response of model (5.2). By comparing (5.7) and (5.3), we observe that the simulation response can also be interpreted as an infinite-step-ahead prediction model. This effectively negates the auto-regressive components of a prediction model.

We propose the second performance measure to be the sum-of-squares of the simulation error $\varepsilon_s(k) := y_m(k) - y_s(k)$. For a given model $M(\theta)$, this can be computed as

$$J_2(M, \theta, \mathcal{D}_N) = \frac{1}{N - n_1} \sum_{k=n_1+1}^{N} (y_m(k) - y_s(k))^2 . \tag{5.8}$$

Just as in the case of quantifying the one-step-ahead prediction error, any other signal norm can be used in (5.8) to quantify the simulation error incurred by a model. The 2-norm-based measure used in (5.8), however, provides a physical interpretation as it quantifies the power in the simulation error of the model.

For a given stochastic model in the form of (5.2), the computation of simulated output as per (5.7) is not trivial. Details on the necessary computations are provided in Sect. 5.4.

5.3.3 Parameter Complexity

When measuring the accuracy of models based on a finite length dataset $\mathcal{D}_N$, the size of the models generated by a given TAG may grow arbitrarily large.[1] More complex models can, in general, capture a relationship that fits the measured data better. However, complex models may capture certain features of the dataset $\mathcal{D}_N$ that are specific to the noise realization in that dataset. Such models do not generalize well to data that was not used to estimate the model parameters. This is known as overfitting. In an EA, when candidate solutions in the population are allowed to grow freely,

[1] Limited only by the maximum tree-depth m_{d}, see Sect. 3.2.1.

the population as a whole may become increasingly complex and overfitting. Consequently, the solution space corresponding to low complexity models may remain relatively unexplored by the EA. This is known as bloating of the population.

In a multi-objective setting, to avoid bloating of the population, we introduce a third performance measure that describes the parametric complexity of the model. For a given model $M(\theta)$, the parametric complexity can be measured as

$$J_3(M, \theta, \mathcal{D}_N) = \dim(\theta), \tag{5.9}$$

where $\dim(\theta)$ is the dimension of vector θ.

The motivation for using performance measure J_3 is akin to the motivation for using information theoretic statistics such as AIC or BIC to perform model selection. Such methods typically estimate the generalization error, i.e. expectation of the error achieved by the model on an independent dataset, by penalizing the performance achieved on data $\mathcal{D}_N$ by a factor that depends on the number of parameters $\dim(\theta)$ and the length of the dataset N. Unlike AIC or BIC, for which the scaling factor of the penalizing term is fixed, in a multi-objective setting parametric complexity can be treated as a separate objective, and an accuracy-complexity trade-off can be made a posteriori.

Remark 5.1 The notion of parametric complexity introduced in this section can be further refined by separating the complexity of the deterministic and stochastic part of the model. Although, in the P-NARMAX case, this leads to ad-hoc notions of model complexity.

5.3.4 Dynamic Complexity

In system identification, Occam's Razor principle is often used as a guiding heuristic for model selection. A prerequisite for using this principle is that one should be able to rank various models in terms of some complexity measure, for example using J_3. However, a model set generated by a grammar, for example G_{N}, may include models belonging to multiple dynamic classes. When measuring the complexity of a parametric model $M(\theta) \in \mathcal{M}(G_{\mathrm{N}})$, parametric complexity $J_3(M, \theta, \mathcal{D}_N)$ only provides partial information since it does not take into account the complexity of the dynamics.

TAG-based representations provide an opportunity to rank models that differ in terms of the dynamics they represent. In order to rank a model in terms of the underlying dynamics, one must take into account the specific auxiliary trees used to generate the model. Various auxiliary trees add to the complexity of a model differently. For example, for grammar G_{N}, adjunction of auxiliary tree α_3 (see Fig. 4.7) introduces a multiplicative non-linearity, while adjunction of auxiliary tree α_1 introduces a new linear term to a model. For grammar G_{N}, the dynamic complexity of the models can be coarsely divided into two categories—linear and non-linear dynamics. A model $M(\theta) \in \mathcal{M}(G_{\mathrm{N}})$ is linear if the corresponding derived tree is generated exclusively from the subset of auxiliary trees $\{\alpha_1, \alpha_2, \alpha_3, \alpha_7\}$. If the derived tree of model $M(\theta)$ contains at least one auxiliary tree from the subset $\{\alpha_4, \alpha_5, \alpha_6\}$, then the model is non-linear. Hence, based on the auxiliary trees used to generate the model, the dynamic complexity of the model can be computed as a ranking of the model as follows

$$J_4(M, \theta, \mathcal{D}_N) = \begin{cases} 0 & \text{if } M \text{ is linear,} \\ 1 & \text{if } M \text{ is non-linear.} \end{cases} \tag{5.10}$$

The classification of dynamics in (5.10) is fairly coarse, and can be easily customized based on user-preferences. For example, if an affine input model structure is preferable, for instance for linearizing feedback control design application, then that preference can be incorporated in (5.10). Similarly, model structures such as ARX, ARMAX and Non-linear Finite Impulse Response (NFIR) can also be considered. The notion of dynamic complexity can also be used to quantize the level of non-linearity, for example, based on the degree of the polynomial expression in (5.2).

5.4 Computation of a Simulation Model

5.4.1 Motivation

In the literature, simulation response of models, for example (5.2), are often computed with noise terms ξ set to 0 [13, ch. 5]. In the case of linear systems, it is always possible to lump the noise sources as an additive term on the output of the deterministic response of the system. If f in (5.1) is linear, it can be re-written as

$$\begin{aligned} y(k) = {} & g(y(k-1), \ldots, y(k-n_y), u(k), \ldots, u(k-n_u)) \\ & + h(\xi(k-1), \ldots, \xi(k-n_\xi) + \xi(k), \end{aligned} \tag{5.11}$$

where g and h are linear functions. Using the definition in (5.7), the simulation response of a model of the form (5.11) can indeed be computed by setting noise contributions to 0, since noise ξ is assumed to be zero-mean (without loss of generality). However, in the non-linear setting, neglecting the process noise terms leads to a biased simulation response (see [9, 10]). For the case of P-NARMAX models, the bias can be attributed to two factors:

(i) the higher order moments of the noise process, and
(ii) recursive non-linear dependencies of the output terms y on noise terms ξ.

Hence, in order to avoid computation of a biased simulation response of (5.2), we need to formulate the effect of these factors carefully and express it in terms of a so-called "simulation model" or deterministic part of (5.2). Here we will call the resulting model as a simulation model of (5.2), but in fact it is the process part of (5.2).

The two following examples from [11] demonstrate how the contributing factors influence the computation of a simulation model of a non-linear model.

Example 5.2 (*High order moments of noise* [11]) Consider a system described by the following equation

$$y(k) = f(u, \xi) = u(k) + \xi^2(k-1) + \xi(k), \tag{5.12}$$

where $\{u(k)\}$ and $\{\xi(k)\}$ are i.i.d sequences distributed as $\mathcal{N}(0, 1)$. The simulation model can be computed by taking the expectation of (5.12) with respect to the noise process ξ,

$$\begin{aligned} y_s(k) &= E_\xi[y(k)] = u(k) + E_\xi[\xi^2(k-1)] + E[\xi(k)], \\ &= u(k) + 1. && (5.13) \\ &= f_s(u(k)) && (5.14) \end{aligned}$$

Observe that if the simulation model was computed by setting the contributions of the noise terms to 0, one would obtain the following simulation model

$$y_s'(k) = u(k). \tag{5.15}$$

The two simulation models are offset by a scalar factor 1.

Example 5.3 (*recursive non-linear dependencies of the output on noise* [11]) Consider a system governed by the following equation

$$y(k) = u(k) - 0.1y^2(k-1) + \xi(k), \tag{5.16}$$

where, $u(k) \sim \mathcal{N}(\mu_u, \sigma_u)$ and $\xi(k) \sim \mathcal{N}(0, \sigma_\xi)$. By taking the expectation and recursively replacing the $y(k-1)$ term in the model, we get the following set of equations

$$\begin{aligned} E_\xi[(y(k)] &= E_\xi\left[u(k) - 0.1y^2(k-1) + \xi(k)\right], \\ &= u(k) - 0.1E_\xi[y^2(k-1)], \\ &= u(k) - 0.1E_\xi\left[(u(k-1) - 0.1y^2(k-2) + \xi(k-1))^2\right], \\ &= u(k) - 0.1u^2(k-1) - 0.1\sigma_\xi^2 - 0.001E_\xi[y^4(k-2)] \\ &\quad + 0.02u(k-1)E_\xi[(y^2(k-2)], \\ &= u(k) - 0.1u^2(k-1) - 0.1 - 0.001E_\xi[y^4(k-2)] \\ &\quad + 0.02u(k-1)E_\xi[(u(k-2) - 0.1y^2(k-3) + \xi(k-2))^2], \\ &= u(k) - 0.1u^2(k-1) - 0.1 - 0.001E_\xi[y^4(k-2)] \\ &\quad + 0.02u(k-1)u^2(k-2) + 0.0002u(k-1)E_\xi[y^4(k-3)] \\ &\quad + 0.02u(k-1)\sigma_\xi^2 - 0.004u(k-2)E_\xi[y^2(k-3)], \\ &= \dots \end{aligned}$$

From these equations, it can be seen that the simulation model of a relatively simple finite order P-NARMAX model may be described by an infinitely long recursive filter. In other words, the simulation model of a finite order P-NARMAX model may be an Non-linear Infinite Impulse Response (NIIR) filter, under the assumption that the applied recursive substitution is convergent.

Examples 5.2 and 5.3 demonstrate that one must take care when computing the simulation model of a non-linear dynamical model. In the next section, we introduce techniques to compute the simulation model of a P-NARMAX model.

5.4.2 Computational Aspects

In order to compute a simulation model of a P-NARMAX model, we propose an alternate representation of the model computed via a change of basis. The noise terms $\xi[\cdot]$ in the polynomial function in (5.1) can be reformulated as multi-linear functions in terms of Hermite polynomial basis. Hermite polynomials, denoted as $\{He_n(\cdot)\}_{n=0}^{\infty}$, form an orthogonal basis on the real line with respect to the standard normal distribution. Some relevant mathematical background on Hermite polynomials can be found in Appendix 1.

It is important to note that in order to compute simulation models as in (5.7), higher order moments of the noise distribution must be known. Hence, in the sequel, we assume that the noise process $\xi(k)$ is i.i.d with $\xi(k) \sim \mathcal{N}(0, 1)$. Note that a noise process $v(k) \sim \mathcal{N}(\mu, \sigma)$, can be equivalently expressed as

$$v(k) = \sigma\xi(k) + \mu, \tag{5.17}$$

and hence, can be included in the model formulation in (5.2).

The Simplified Case

We first consider the simpler problem of computing a simulation model for a subclass of P-NARMAX models. Consider models of the form

$$\begin{aligned} y(k) = f(&u(k-1), \ldots, u(k-n_u), \xi(k-1), \ldots, \xi(k-n_\xi)) \\ &+ g(y(k-1), \ldots, y(k-n_y)) + \xi(k), \end{aligned} \tag{5.18}$$

where the function f is a *polynomial* function of the regressor variables and g is a *linear* function of the past outputs. This is a specific class of models that can be represented as (5.1). The model in Eq. (5.18) can be reformulated by expressing all noise polynomial terms in the model in terms of Hermite polynomials (see (5.43) in the Appendix):

$$\begin{aligned}
y(k) &= f(y(k-1), \ldots, u(k-n_u), \xi(k-1), \ldots, \xi(k-n_\xi)) \\
&\quad + g(y(k-1), \ldots, y(k-n_y)) + \xi(k), \\
&= \tilde{f}(u(k-1), \ldots, u(k-n_u), He_0(\xi(k-1)), \ldots, He(\bar{d}_1)(\xi(k-1)), \ldots, \\
&\quad He_0(\xi(k-n_\xi)), \ldots, He(\bar{d}_{n_\xi})(\xi(k-n_\xi))) + g(y(k-1), \ldots, y(k-n_y)) \\
&\quad + \xi_k, \qquad (5.19)
\end{aligned}$$

where $\bar{d}_i$ is the maximum exponent of the $\xi(k-i)$ term, and the function $\tilde{f}$ is multi-linear in terms of the Hermite polynomials $He_n(\cdot)$. The function $\tilde{f}$ will be later derived in Lemma 5.8. This results in an equivalent model that can be easily transformed to a simulation model (see relations (5.44), (5.45) and (5.47) in the Appendix). These derivations yield a simulation model represented as

$$\begin{aligned}
y_s(k) &= E_\xi\big(f(u(k-1), \ldots, u(k-n_u), \xi(k-1), \ldots, \xi(k-n_\xi))) \\
&\quad + g(y(k-1), \ldots, y(k-n_y)) + \xi(k)\big), \\
&= E_\xi\Big(\tilde{f}(u(k-1), \ldots, u(k-n_u), He(0)(\xi(k-1)), \ldots, He(\bar{d}_1)(\xi(k-1)), \ldots, \\
&\quad He(0)(\xi(k-n_\xi)), \ldots, He(\bar{d}_{n_\xi})(\xi(k-n_\xi))) + g(y(k-1), \ldots, y(k-n_y))\Big), \qquad (5.20) \\
&= f_s(u(k-1), \ldots, u(k-n_u), y_s(k-1), \ldots, y_s(k-n_y)), \qquad (5.21)
\end{aligned}$$

where $f_s(\cdot)$ denotes the simulation model (i.e., the deterministic part) of the stochastic model. Simulation model (5.21) is computed from (5.20) by making use of properties of Hermite polynomials described in Eqs. (5.44), (5.45) and (5.47) in Appendix 1.

Remark 5.4 It should be noted that an equation of form (5.18) can be easily transformed to the equivalent simulation model as in (5.21) without the use of Hermite polynomials, since the moments of a Gaussian distribution are known. However, the use of Hermite polynomials yields an equivalent prediction model, as in (5.19) that offers a convenient representation. It turns out that, in order to obtain a simulation model like (5.21), one must only "switch off" a number of Hermite polynomial terms in (5.19), as per Eqs. (5.44), (5.45) and (5.47).

To summarize the results of our derivation and give a computational form that can be used readily in an EC algorithm, the following equivalent representation of (5.18) is used

$$y(k) = \sum_{i=1}^{p}\left(c_i \prod_{j=0}^{n_u} u^{b_{i,j}}(k-j) \prod_{l=1}^{n_\xi} \xi^{d_{i,q}}(k-q)\right) + \sum_{r=1}^{n_y} a_{ry}(k-r) + \xi(k), \quad (5.22)$$

where $p \in \mathbb{Z}_{>0}$ is the number of monomial terms of u and ξ in the model, $c_i \in \mathbb{R}$ are the coefficients, $b_{i,j}$ and $d_{i,q}$ are the exponents of the $u(k-j)$ and $\xi(k-q)$ factors in the ith monomial term respectively and a_r is the linear co-efficient of the rth output term $y(k-r)$.

Lemma 5.8 [11] *For a model of form* (5.22), *and under the assumption that* $\xi(k) \sim \mathcal{N}(0, 1)$ *is independent of the input* $u(l)$ *for all* $l \in [1, N]$, *the simulation model, computed in the sense of* (5.7), *is given by*

$$y_s(k) = \sum_{i \in P_e} c_i \prod_{j=1}^{n_u} u^{b_{i,j}}(k-j) \prod_{l=1}^{n_\xi} (d_{i,q} - 1)!! + \sum_{r=1}^{n_y} a_r y_s(k-r), \tag{5.23}$$

where $P_e := \left\{ i \in [1, p] \mid d_{i,q} \text{ is even } \forall q \in [1, n_\xi] \right\}$ *and* $(n-1)!! := \frac{n!}{\frac{n}{2}! 2^{\frac{n}{2}}}$.

Proof See Appendix 2. □

Note that the result in (5.23) is exact, as per the definition of the simulation model in (5.7).

The General Case

We now consider the full polynomial NARMAX model as shown in (5.2). Lemma 5.8 does not deal with the non-linear dependence of the output signal y on the noise process ξ. Although $\{u(k)\}$ is a known sequence and $\xi(k)$ is distributed normally, the random variable $y(k)$ is typically not distributed normally. Hence, model terms involving random variables $y(k)$ cannot be equivalently represented in terms of Hermite polynomials of $y(k)$ (the inner-product in (5.41) no longer corresponds to the expectation with respect to the distribution of $y(k)$). Consequently, in order to compute the expectation of (5.2), one must recursively eliminate all y terms, which yields a NIIR in terms of the input, as demonstrated in Example 5.3. Computing an NIIR is intractable because next to infinite time lags, it also contains infinite order polynomial exponents. As a result, just like in the LTI case, an approximation must be made in order to keep the simulation model tractable. In the sequel, we develop an approximation scheme to compute the simulation model of a P-NARMAX model.

Let $U_a^b(k) := \{u(k-i)\}_{i=a}^b$ be the sequence of delayed inputs with the delays ranging from a to b with $a, b \in \mathbb{Z}_{\geq 0}$ and $a < b$. Similarly, define the sequence notation $Y_a^b(k) := \{y(k-i)\}_{i=a}^b$ and $\Xi_a^b(k) := \{\xi(k-i)\}_{i=a}^b$. Additionally, define a set-valued discrete-time shift operator $U_a^b(k) \diamond 1 := U_{a+1}^{b+1}(k) = \{u(k-i)\}_{i=a+1}^{b+1}$. With a slight abuse of notation, the model in (5.1) will be re-written as

$$y(k) = f(U_0^{n_u}, Y_1^{n_y}, \Xi_1^{n_\xi}) + \xi(k). \tag{5.24}$$

For the proposed approximation scheme, we perform l recursive substitutions of the output terms y in the model expression in (5.24), as follows:

$$\begin{aligned} y(k) &= f(U_0^{n_u}(k), \{y(k-1), \ldots, y(k-n_y)\}, \Xi_1^{n_\xi}(k)) + \xi(k), \\ &= f\big(U_0^{n_u}(k), \{f(U_0^{n_u}(k) \diamond 1, Y_1^{n_u}(k) \diamond 1, \Xi_1^{n_\xi}(k) \diamond 1), \ldots, \\ &\quad f(U_0^{n_u}(k) \diamond n_y, Y_1^{n_u}(k) \diamond n_y, \Xi_1^{n_\xi}(k) \diamond n_y)\}, \Xi_1^{n_\xi}(k)\big) + \xi(k), \end{aligned}$$

$$= f_1(U_0^{n_u+1}(k), Y_2^{n_y+1}(k), \Xi_1^{n_\xi+1}(k)) + \xi(k),$$
$$\vdots$$
$$= f_1(U_0^{n_u+l}(k), Y_{l+1}^{n_u+l}(k), \Xi_1^{n_\xi+l}(k)) + \xi(k). \quad (5.25)$$

It should be noted that the functions $f, f_1, \ldots, f_l$ are not identical. However, they are equivalent, in the sense that for given sequences $\{u(k); k \in [0, N]\}$ and $\{\xi(k); k \in [0, N]\}$, these functions produce the same output sequence $\{y_k; k \in [n_1 + 1, N]\}$. Moreover, since f is a polynomial function, so are $f_1, \ldots, f_l$. The parameterized approximation can now be defined as follows.

Definition 5.9 (*l-approximate simulation* [11]) For the P-NARMAX model in (5.24), the l-approximate simulation model is defined as

$$y_{sl}(k) := E_\xi\left[(f_l\left(U_0^{n_u+l}(k), Y_{l+1}^{n_y+l}(k), \Xi_1^{n_\xi+l}(k)\right) + \xi(k)\right], \quad (5.26)$$

where $Y_{sl_a}^b(k) := \{y_{sl}(k-a), \ldots, y_{sl}(k-b)\}$.

The parameter l achieves a trade-off between the exact NIIR realization of the simulation model and the naïve approximation of ignoring the correlation between output y and noise ξ. As l approaches ∞, the approximation $y_{sl}(k)$ approaches the true simulated output $y_s(k) = E_\xi[y(k)]$. To compute the l-approximate simulation model, the following alternate representation of (5.25) is used

$$y(k) = \sum_{i=1}^{p}\left(c_i \prod_{j=0}^{n_u+l} u^{b_{i,j}}(k-j) \prod_{r=1}^{n_y} y^{a_{i,r}}(k-l-r) \prod_{q=1}^{n_\xi+l} \xi^{d_{i,q}}(k-q)\right) + \xi(k)), \quad (5.27)$$

where $a_{i,r}$ is the exponent of $y(k-r)$ in the ith term.

Theorem 5.10 [11] *The l-approximate simulation model for the P-NARMAX model in* (5.2), *under the assumption that ξ is independent of u and that $\xi(k) \sim \mathcal{N}(0, 1)$, is given by*

$$y_{sl}(k) = \sum_{i \in P_e} c_i \prod_{j=0}^{n_u+l} u^{b_{i,j}}(k-j) \prod_{r=1}^{n_y} y_{sl}^{a_{i,r}}(k-l-r) \prod_{q=1}^{n_\xi+l} (d_{i,q}-1)!!. \quad (5.28)$$

Proof The proof makes use of the representation in (5.27), the definition of l-approximate simulation, and the use of Hermite polynomials of the noise process as in the proof of Lemma 5.8. The proof follows the same line of reasoning as in the proof of Lemma 5.8, and is thus omitted here. □

Remark 5.5 It should be noted that the proof of Lemma 5.8 rests on the assumption that the noise terms in (5.2) are polynomial. While Lemma 5.8 and Theorem 5.10 are derived for models of the form (5.22) and (5.27), respectively, the input terms u and output terms y were not explicitly required to be in a polynomial form. Hence, the derived results can easily be extrapolated to compute simulation models from prediction models with arbitrary non-linear functions on the input and output terms, as long as the noise terms appear in a polynomial form.

Remark 5.6 While we assumed that $\xi(k) \sim \mathcal{N}(0, 1)$, the results can be extended to a non-standard Gaussian distribution (see (5.17)). For example, a white Gaussian noise process $\{v(k)\}_{k=1}^{N}$ with $v(k) \sim \mathcal{N}(0, \sigma_\xi)$ can be expressed as $v(k) = \sigma_\xi \xi(k)$. Propagating the constant standard deviation factor σ_ξ through the expressions in the proof of Lemma 5.8 (and, by extension, to the proof of Theorem 5.10), we get the l-approximate simulation response with respect to the more general noise process as

$$y_{sl}(k) = \sum_{i \in P_e} c_i \prod_{j=0}^{n_u+l} u^{b_{i,j}}(k-j) \prod_{r=1}^{n_y} y_{sl}^{a_{i,r}}(k-l-r) \prod_{q=1}^{n_\xi+l} \sigma_\xi^{d_{i,q}}(d_{i,q}-1)!!. \tag{5.29}$$

Furthermore, the results can also be extended to any exponential family distribution by suitably changing the probability measure and the Hilbert space of random functions in Appendix 1.

Lemma 5.8 and Theorem 5.10 can be used to compute the exact or approximate simulation model of a P-NARMAX model. When the simulation output must be approximated, the second performance measure can be computed using the l-approximate simulation output $y_{sl}(k)$. For a given model $M(\theta)$ and a given approximation level l, this can be computed as

$$J_2(M, \theta, \mathcal{D}_N) = \sum_{k=n_1}^{N} \left(y_m(k) - y_{s,l}(k)\right)^2. \tag{5.30}$$

5.4.3 *Simulation Examples*

In this section, we illustrate Lemma 5.8 and Theorem 5.10 using Examples 5.2 and 5.3, respectively.

A simulation study was carried out to illustrate the result in Lemma 5.8 based on the system (5.12) in Example 5.2. The system in (5.12) was simulated with an input $u \sim 3 + \mathcal{N}(0, 1)$ with length $N = 3000$ samples, and with noise disturbance $\xi \sim \mathcal{N}(0, 1)$. In order to compute an empirical estimate of the simulation response of the system, the input excitation was repeated periodically with number of periods $n_\text{p} = 1024$. Data from the first 5 experiments was discarded to remove any errors due to transients. Assuming ergodicity, the ideal simulation response can be computed empirically as the ensemble average

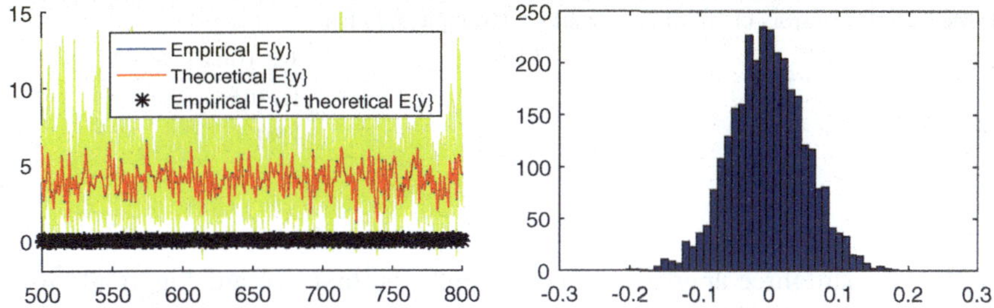

(a) Comparison of y_s and $\bar{y}_s$. The yellow lines correspond to several realizations of the measured response y and indicate the severity of the process noise.

(b) Distribution of $\bar{y}_s - y_s$

Fig. 5.1 Example 5.2—simulation results

$$\bar{y}_s = \frac{1}{n_p - 5} \sum_{i=6}^{n_p} y^{(i)}, \tag{5.31}$$

where $y^{(i)}$ is the output corresponding to the ith period of the input. The theoretical simulation of the model is computed using (5.14). The empirical and the theoretical simulation response is plotted in Fig. 5.1a. Several realizations of the noisy output are plotted in yellow. The unscaled histogram of $\bar{y}_s - y_s$ is depicted in Fig. 5.1b. The difference $\bar{y}_s - y_s$ appears to be centered at 0. This implies that the simulation model computed by neglecting the noise contributions (see (5.15)) would be biased by a scalar factor of 1 (compare with (5.14)). The difference between the two simulation models is also quantified in terms of the root mean squared (RMS) error in Table 5.1, where the RMS of error $e(k)$ is computed as

$$\text{rms}(e) = \sqrt{\frac{1}{N} \sum_{k=1}^{N} e(k)^2}. \tag{5.32}$$

The first column of numbers in Table 5.1 indicate the RMS error between the empirical mean and the proposed simulation models. The second column indicates ensemble average of the RMS errors between the noisy response of the system and the simulation model approximations. It can be observed that the proposed simulation model performs significantly better in approximating the noisy output than the conventional simulation model that sets the noise contributions to 0.

A similar simulation study was carried out to illustrate the result in Theorem 5.10 using the system in (5.16) in Example 5.3. The system in (5.16) was simulated with input $u(k) \sim \mathcal{N}(3, 1)$ of length $N = 3000$ samples, and noise $\xi(k) \sim \mathcal{N}(0, 1)$. Again, the input excitation was repeated periodically with number of periods $n_p = 1024$ and the simulation response was computed empirically using $p - 5$ periods of excitation.

Table 5.1 RMS errors of simulation results—Example 5.2 [11]

	$\mathrm{rms}[\bar{y}_s - E[y]]$	$\mathrm{avg}_i(\mathrm{rms}(y^{(i)} - E[y]))$
$E[y] = y_s$	0.0544	1.7305
$E[y] \approx y_s'$	1.0001	1.9976

We compare three approximation concepts, the first two are commonly used in literature, while the third is the proposed l-approximate simulation.

(i) Ignore the correlation between the output y and noise ξ in (5.16). This yields

$$y_s^{(1)}(k) = u(k) - 0.1(y_s^{(1)}(k-1))^2, \tag{5.33}$$

(ii) Truncate the NIIR after a certain number of recursive substitutions of the $y(k)$ terms. We get the simulation responses

$$y_s^{(2)}(k) = u(k) - 0.1u^2(k-1) - 0.1, \tag{5.34}$$

$$y_s^{(3)}(k) = u(k) - 0.1u^2(k-1) - 0.1 + 0.02u(k-1)(u^2(k-2)+1) \tag{5.35}$$

for one and two recursive substitutions, respectively, followed by truncation of the NIIR.

(iii) The l-approximate simulation model for $l = 1$,

$$\begin{aligned} y_{sl}(k) = u(k) - 0.1u^2(k-1) - 0.1 - 0.001(y_{s1}(k-2))^4 \\ + 0.02u(k-1)(y_{s1}(k-2))^2, \end{aligned} \tag{5.36}$$

and for $l = 2$,

$$\begin{aligned} y_{s2}(k) = u(k) - 0.1u^2(k-1) - 0.1 \\ - 0.001\big(u(k-2) - 0.1(y_{s2}(k-3))^2 + \xi(k-2)\big)^4 \\ + 0.02u(k-1)\big(u(k-2) - 0.1(y_{s2}(k-3))^2 + \xi(k-2)\big)^2. \end{aligned} \tag{5.37}$$

The distribution of the approximation errors achieved by the simulation models in (5.33)–(5.37) w.r.t the empirical simulation $\bar{y}_s$ are depicted in Fig. 5.2. Additionally, the RMS errors are presented in Table 5.2. It can be verified, both visually and numerically, that the truncation-based approximation concepts achieve the worst approximation errors. In comparison, the naïve approximation $y_s^{(1)}$, which is also equal to 0-approximate simulation $y_{s,0}$, performs surprisingly well. The approximation error further reduces for increasing values of parameter l. This is to be expected, since for increasing values of l. the approximation $y_{s,l}$ approaches the true simulation output y_s. This illustrates the result in Theorem 5.10.

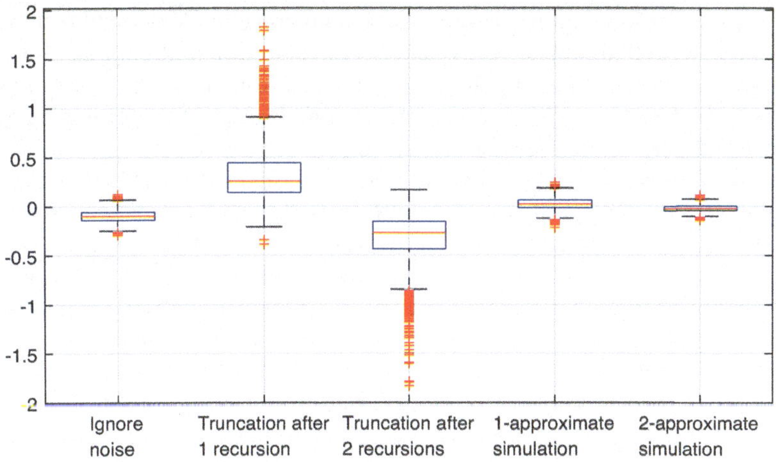

Fig. 5.2 Distribution of error between $\bar{y}_s$ and the various approximate simulation responses

Table 5.2 RMS errors of simulation results—Example 5.3

	$\text{rms}(\bar{y}_s - E[y])$	$\text{avg}_i(\text{rms}(y^{(i)} - E[y]))$
$E[y] \approx y_s^{(1)}(k)$	0.1071	1.1566
$E[y] \approx y_s^{(2)}(k)$	0.4223	1.2203
$E[y] \approx y_s^{(3)}(k)$	0.3999	1.2147
$E[y] \approx y_{s,1}(k)$	0.0683	1.1534
$E[y] \approx y_{s,2}(k)$	0.0369	1.1521

5.5 Conclusions

In this chapter, we introduced the basic notions of partial ordering, non-dominated solutions and the Pareto front. These notions allow us to evaluate and rank-order candidate models based on multiple objectives. The fact that EC applies a *generate-and-validate* policy on a population of candidate solutions can be exploited to focus the evolutionary search along the estimated Pareto front.

In Chap. 4, we demonstrated that TAGs can be used to defined model sets that span across multiple classes of dynamical systems with varying degrees of complexity. Such a general model set specification makes it challenging to evaluate candidate models in a meaningful way, since the nature of dynamics and degrees of complexity of models should be taken into account when comparing any two models belonging to the model set.

In this dissertation, we propose to use four performance measures that capture the following information related to a dynamical model—(i) one-step-ahead prediction accuracy, (ii) simulation accuracy, (iii) parametric complexity and (iv) dynamic complexity. Modelling objectives (i) and (ii) relate to different notions of accuracy of the model with respect to measured data obtained from the system, and objectives (iii) and (iv) relate to two different notions of complexities of the dynamical model, and are not dependent on the measured data.

The one-step-prediction error can be used to measure short-term prediction capabilities, while long-term prediction capability is measured by simulation error. Including these two somewhat complimentary measures for model accuracy ensures that the estimated models are suitable for a broad range of applications. While parametric complexity ensures that models achieving different bias-variance trade-offs along the Pareto front can be explored, dynamic complexity allows models with various dynamic properties to be explored. A Pareto front estimated based on these four dimensions allows a user to select a model a posteriori based on a better understanding of the trade-offs that can achieved by the models belonging to the chosen model set and with respect to the measured data. This enables the user to easily align the selection of identified model with the ultimate use-case of the model.

Appendix 1: Mathematical Preliminaries

Hermite Polynomials

Let $(\mathbb{R}, \mathcal{B}_{\mathbb{R}}, p_x)$ be a probability space on the real line $\mathbb{R}$, where $\mathcal{B}_{\mathbb{R}}$ is the Borel σ-algebra on $\mathbb{R}$, and p_x is the probability measure of the Gaussian distribution. Consider the Hilbert space $L_2(\mathbb{R}, \mathcal{B}_{\mathbb{R}}, p_x)$ of random functions f that satisfy,

$$E[f] = 0 \text{ and } E\left[f^2\right] < \infty. \tag{5.38}$$

where E is associated with the inner product

$$\langle f, g\rangle = E[fg] = \frac{1}{\sqrt{2\pi}} \int_{-\infty}^{\infty} f(x)g(x)e^{-\frac{x^2}{2}}\mathrm{d}x, \tag{5.39}$$

for any $f, g \in L_2(\mathbb{R}, \mathcal{B}_{\mathbb{R}}, p_x)$.

Definition 5.11 Hermite polynomials $He_n(x)$ of degree n are defined as

$$He_n(x) = (-1)^n e^{\frac{x^2}{2}} \frac{\mathrm{d}^n}{\mathrm{d}x^n} e^{-\frac{x^2}{2}}. \tag{5.40}$$

Hermite polynomials form a closed and complete orthogonal basis in the L_2 space with the inner product defined in (5.39) (see [16]), i.e.,

$$\langle He_n, He_m \rangle = \frac{1}{\sqrt{2\pi}} \int_{-\infty}^{\infty} He_n(x) He_m(x) e^{-\frac{x^2}{2}} \mathrm{dx}$$
$$= n!\delta_{n,m}, \tag{5.41}$$

where $\delta_{n,m} = 1$ when $n = m$ and 0 otherwise. Hermite polynomials possess a number of properties that will be used in order to derive simulation models automatically from a given stochastic model of form (5.2).

Since Hermite polynomials form a complete orthogonal basis in $L_2(\mathbb{R}, \mathcal{B}_{\mathbb{R}}, p_x)$, any function $f \in L_2(\mathbb{R}, \mathcal{B}_{\mathbb{R}}, p_x)$ can be expressed as a convergent linear combination of basis functions

$$f = \sum_{n=0}^{\infty} \frac{1}{n!} \langle f, He_n \rangle He_n. \tag{5.42}$$

Given a monomial in $x \in \mathbb{R}$, one can compute an equivalent representation in terms of Hermite polynomials $He(x)$ using the following inverse relationship

$$x^n = n! \sum_{m=0}^{\lfloor \frac{n}{2} \rfloor} \frac{1}{2^m m!(n-2m)!} He_{n-2m}(x), \tag{5.43}$$

where $\lfloor . \rfloor$ is defined as the *floor* operator, i.e., $\lfloor n \rfloor = a$, where $a \in \mathbb{Z}$ is the largest integer such that $a \leq n$. Hermite polynomials of normally distributed random variables have useful properties. Let X be a random variable distributed as $\mathcal{N}(0, 1)$. It can be shown that

$$E[He_n(X)] = \begin{cases} 1 & \text{for } n = 0, \\ 0 & \text{for } n \neq 0. \end{cases} \tag{5.44}$$

The following result can be obtained by using (5.41)

$$E[He_n(X)\ He_m(X)] = \int_{-\infty}^{\infty} He_n(x) He_m(x) \frac{1}{\sqrt{2\pi}} e^{-\frac{x^2}{2}} dx,$$
$$= \langle He_n, He_m \rangle,$$
$$= n!\delta_{n,m}. \tag{5.45}$$

Based on these properties, it has been shown in [3] that for $X_1, X_2 \sim \mathcal{N}(0, 1)$,

$$E[He_n(X_1)\ He_m(X_2)] = n!\delta_{n,m}(E[X_1 X_2])^n. \tag{5.46}$$

In particular, if X_1, X_2 are independent, we get

$$E[He_n(X_1)\, He_m(X_2)] = \begin{cases} 1 & \text{if } n = 0, m = 0, \\ 0 & \text{otherwise.} \end{cases} \tag{5.47}$$

Appendix 2: Mathematical Proofs

Proof of Lemma 5.8 [11] The proof relies on the use of the proposed Hermite polynomial based representation in (5.19). Let $\gamma_i := c_i \prod_{j=1}^{n_u} u(k-j)^{b_{i,j}}$. Using (5.43), Eq. (5.22) can be written as

$$y(k) = \sum_{i=1}^{p} \left(\gamma_i \prod_{q=1}^{n_\xi} \left(d_{i,q}! \sum_{m=0}^{\lfloor \frac{d_{i,q}}{2} \rfloor} \frac{He_{d_{i,q}-2m}(\xi(k-q))}{2^m m!(d_{i,q}-2m)!} \right) \right) + \sum_{r=1}^{n_y} g_r y_{k-r}. \tag{5.48}$$

Note that (5.48) is of the form (5.19). Define partitions $P_{\mathrm{e}} := \{i \in [1, p] \mid d_{i,q} \text{ is even } \forall q \in [1, n_\xi]\}$ and $P_{\mathrm{o}} := \{i \in [1, p] \mid i \notin P_{\mathrm{e}}\}$. This yields

$$\begin{aligned} y(k) = &\sum_{i \in P_{\mathrm{e}}} \gamma_i \prod_{q=1}^{n_\xi} \left(d_{i,q}! \sum_{m=0}^{\lfloor \frac{d_{i,q}}{2} \rfloor} \frac{He_{d_{i,q}-2m}(\xi(k-q))}{2^m m!(d_{i,q}-2m)!} \right) \\ &+ \sum_{i \in P_{\mathrm{o}}} \gamma_i \prod_{q=1}^{n_\xi} \left(d_{i,q}! \sum_{m=0}^{\lfloor \frac{d_{i,q}}{2} \rfloor} \frac{He_{d_{i,q}-2m}(\xi(k-q))}{2^m m!(d_{i,q}-2m)!} \right) + \sum_{r=1}^{n_y} a_r y(k-r). \end{aligned} \tag{5.49}$$

Taking the expectation with respect to the noise process ξ, and recognizing that the second term on the right hand side drops out (due to (5.44)), we get that

$$\begin{aligned} y_s(k) = &\sum_{i \in P_{\mathrm{e}}} \gamma_i \prod_{q=1}^{n_\xi} E_\xi \Bigg(d_{i,q}! \Bigg(\frac{He_0(\xi(k-q))}{2^{\frac{d_{i,q}}{2}} \frac{d_{i,q}}{2}!} \\ &+ \sum_{m=0}^{\lfloor \frac{d_{i,q}}{2} - 1 \rfloor} \frac{He_{d_{i,q}-2m}\xi(k-q)}{2^m m!(d_{i,q}-2m)!} \Bigg) \Bigg) + E_\xi \left[\sum_{r=1}^{n_y} a_r y(k-r) \right], \\ = &\sum_{i \in P_{\mathrm{e}}} \gamma_i \prod_{q=1}^{n_\xi} (d_{i,q}-1)!! + \sum_{r=1}^{n_y} a_r y_s(k-r), \end{aligned} \tag{5.50}$$

Appendix 3: Experimental Verification—Simulation Examples

Motivating Example for Simulation Error

Example 5.7 Consider the example of a linear connection of 15 mass spring damper systems with mass $m_i = 60 + 10\cos(\frac{2\pi i}{15} + \frac{\pi}{4})$ for $i \in [1, 15]$, spring coefficients $k_i = 500 - 60\cos(\frac{2\pi i}{16})$ for $i \in [1, 15]$, and $k_{16} = 100$, and damping coefficients $d_i = 150 + 25\sin(\frac{2\pi i}{16})$ for $i \in [1, 16]$. An input force is applied to mass m_{15}, and the position of mass m_1 is measured as the output of the system. A dataset consisting of 2000 data-points was generated using a white noise excitation signal u with $\sigma_u = 5$. The output signal y is corrupted by additive white noise with $\sigma_e = 1$. This results in an approximate signal-to-noise ratio of 16.5 dB.

The dataset was used to estimate, using a PPEM method, the parameters of two parameterized model structures—(a) when the system is in the model set, i.e., when the model structure corresponds to the structure of the true data-generating system (DGS), and (b) when the system does not belong in the chosen model set. More specifically, in both scenarios, a BJ structure is chosen, which is represented as:

$$y(k) = G(q^{-1})u(k) + H(q^{-1})\xi(k) = \frac{B(q^{-1})}{F(q^{-1})}u(k) + \frac{C(q^{-1})}{D(q^{-1})}\xi(k), \tag{5.51}$$

where $\xi \sim \mathcal{N}(0, \sigma_\xi^2)$, and B, F, C and D are polynomials of order n_b, n_f, n_c and n_d respectively. Define $P := \{n_b, n_c, n_d, n_f\}$ as the set of orders that specify the structure of the model to-be-identified. For scenario (a), the polynomial orders are given by $P_a = \{31, 0, 0, 30\}$ which corresponds to the structure of the DGS. For scenario (b), we choose $P_b = \{2, 2, 2, 3\}$. PPEM was used to estimate the parameters based on one-step-ahead prediction error in the two scenarios.

To quantify the performance of the estimated model, we use the 1-step ahead prediction error and simulation error. The simulation error of a BJ model is defined as

$$\varepsilon_s(k) := y_m(k) - y_s(k), \tag{5.52}$$

where, $y_s(k)$ is the simulated output for a measured excitation signal $\{u_m(k)\}_{k=1}^N$ can be computed as

$$y_s(k) = \frac{B(q^{-1})}{F(q^{-1})}u_m(k). \tag{5.53}$$

Note that, unlike the one-step-ahead-prediction output, the simulated output $y_s(k)$ does not depend on the past measured output $y_s(k-1)$ but its own past values $y_s(k-1)$.

Figure 5.3 depicts the 1-step ahead prediction and the simulation output achieved by the models estimated using PPEM in scenario (a). This sets a reference for the best prediction that can be achieved by an unbiased estimator. Figure 5.4 depicts the 1-step ahead prediction and the simulation output achieved by PPEM in scenario (b),

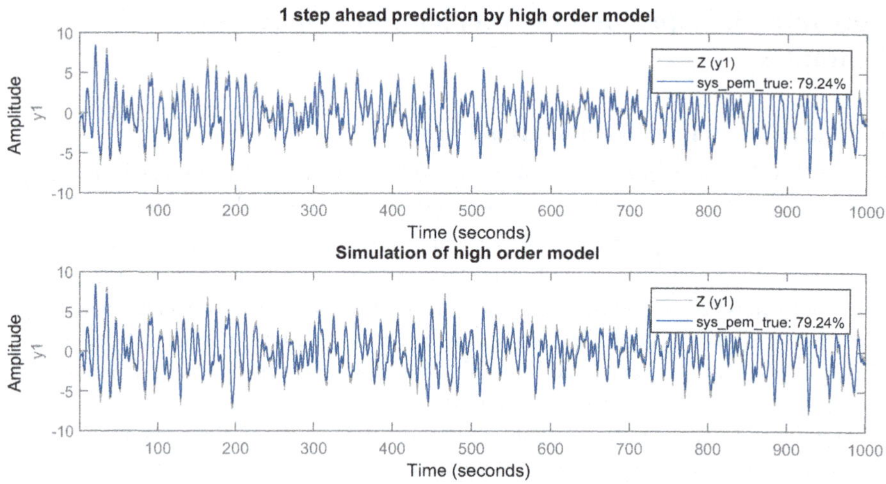

Fig. 5.3 PPEM result for "system in model class" setting—1-step ahead prediction and simulation

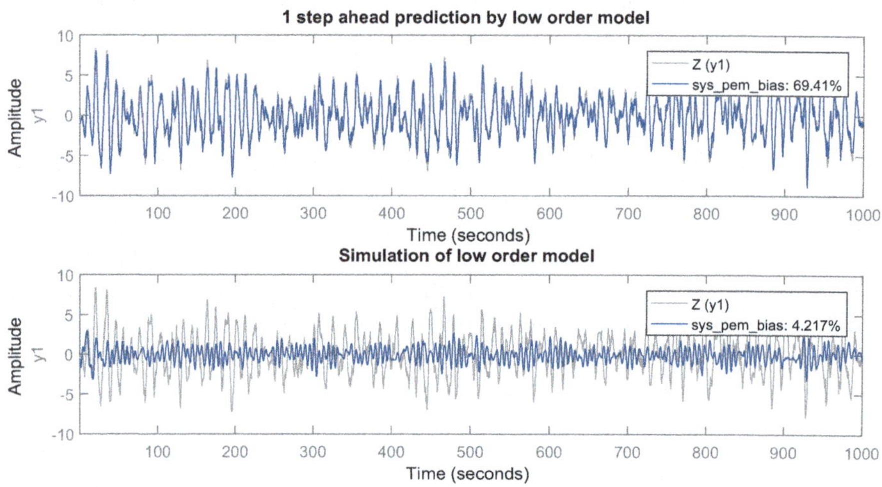

Fig. 5.4 PPEM result for "system *not* in model class" setting—1-step ahead prediction and simulation

i.e. in the case where the system is not in the model class. It can be observed here that the 1-step ahead prediction error performs worse compared to scenario (a), but not by a great margin. More specifically, the best fit ratio (BFR) drops from 79.2 to 69.4%. However, the simulation of the same model with the given input captures none of the dynamics observed in the output measurements, and achieves a BFR of only 4.2%. This illustrates that when the system does not belong to the model class, the auto-regressive component of the model may contribute in re-aligning the predictions with the measurements, whereas the estimated model of the system does not capture any of the dynamics of the underlying DGS. In this case, the low order identified model cannot be trusted to perform well for other objectives than prediction, like optimal control design or system analysis.

Remark 5.8 It should be highlighted that the results depicted here are subject to high variances, and that the results are not always bleak. The purpose of this example is not to provide a complete empirical analysis, but to merely show that, when the system is *not* in the model set, a model identified solely based on prediction error may achieve deceptively good performance.

Remark 5.9 Note that a cross-correlation test would reveal that the residual error ε_p is correlated to the input excitation u_m, implying that the model structure in scenario (b) was not rich enough to capture the dynamics present in the data. However, performing such tests requires human intervention and goes against the research goals of this dissertation. Moreover, correlation tests do not generalize to non-linear dynamics.

References

1. Aguirre LA, Barbosa BH, Braga AP (2010) Prediction and simulation errors in parameter estimation for nonlinear systems. Mech Syst Signal Process 24(8):2855–2867
2. Astrom K (1979) Maximum likelihood and prediction error methods. IFAC Proc 12(8):551–574
3. Bershad NJ, Celka P, McLaughlin S (2001) Analysis of stochastic gradient identification of Wiener-Hammerstein systems for nonlinearities with Hermite polynomial expansions. IEEE Trans Signal Process 49(5):1060–1072
4. Billings SA (2013) Nonlinear system identification: NARMAX methods in the time, frequency, and spatio-temporal domains. Wiley, Chichester
5. Deb K, Pratap A, Agarwal S, Meyarivan T (2002) A fast and elitist multiobjective genetic algorithm: NSGA-II. IEEE Trans Evol Comput 6(2):182–197
6. Ehrgott M (2005) Multicriteria optimization, vol 491. Springer, Berlin
7. Emmerich MT, Deutz AH (2018) A tutorial on multiobjective optimization: fundamentals and evolutionary methods. Nat Comput 17(3):585–609
8. Fonseca CM, Fleming PJ (1993) Genetic algorithms for multiobjective optimization: formulation, discussion and generalization. In: Proceedings of the 5th international conference on genetic algorithms. Morgan Kaufmann Publishers Inc., pp 416–423
9. Giordano G, Sjöberg J (2016) Consistency aspects of Wiener-Hammerstein model identification in presence of process noise. In: Proceedings of IEEE conference on decision and control, pp 3042–3047
10. Hagenblad A, Ljung L, Wills A (2008) Maximum likelihood identification of wiener models. Automatica 44(11):2697–2705
11. Khandelwal D, Schoukens M, Tóth R (2018) On the simulation of polynomial NARMAX models. In: Proceedings of the IEEE conference on decision and control. IEEE, pp 1445–1450
12. Laumanns M, Thiele L, Zitzler E, Deb K (2002) Archiving with guaranteed convergence and diversity in multi-objective optimization. In: Proceedings of the 4th annual conference on genetic and evolutionary computation. Morgan Kaufmann Publishers Inc., pp 439–447
13. Ljung L (ed) (1999) System identification. Theory for the user, 2nd edn. Prentice Hall PTR. ISBN 0-13-656695-2
14. Ljung L (2001) Estimating linear time-invariant models of nonlinear time-varying systems. Eur J Control 7(2–3):203–219

15. Miettinen K (2012) Nonlinear multiobjective optimization, vol 12. Springer, Boston
16. Olver FW, Lozier DW, Boisvert RF, Clark CW (2010) NIST handbook of mathematical functions. Cambridge University Press, Cambridge
17. Piroddi L, Spinelli W (2003) An identification algorithm for polynomial NARX models based on simulation error minimization. Int J Control 76(17):1767–1781
18. Zhang Q, Li H (2007) Moea/d: a multiobjective evolutionary algorithm based on decomposition. IEEE Trans Evol Comput 11(6):712–731
19. Zitzler E, Deb K, Thiele L (2000) Comparison of multiobjective evolutionary algorithms: empirical results. Evol Comput 8(2):173–195
20. Zitzler E, Laumanns M, Thiele L (2001) Spea2: improving the strength Pareto evolutionary algorithm. TIK-report 103

Chapter 6
Evolutionary Multi-criteria System Identification

Black
then
white are
all I see,
in my infancy.
Red and yellow then came to be, reaching out to me,
lets me see.

James Maynard Keenan, Tool

In order to realize the concept of automated identification, we introduced three key ingredients: Genetic Programming (GP), Tree Adjoining Grammar (TAG)- and Pareto-dominance-based multi-objective framework. In this chapter, we fuse these ingredients to develop an approach for automated system identification. The proposed methodology makes use of a modified GP algorithm to automatically search for Pareto-optimal models in the model set defined by a user-specified TAG, based on multiple identification criteria that can be aligned to the ultimate use-case of the model. From the user's perspective, the identification task becomes easier, since (i) selecting a suitable TAG is a far easier task than selecting a suitable model structure and complexity, and (ii) the set of Pareto-efficient models estimated by GP makes it possible for the user to select, *a posteriori*, a model that achieves a desirable trade-off between the various performance measures used.

After introducing the proposed identification approach in detail, we analyse the asymptotic behaviour and computational complexity of the proposed algorithm. Much of the discussion and analysis presented in this chapter relies on concepts and notations introduced in Chaps. 3, 4 and 5.

D. Khandelwal, *Automating Data-Driven Modelling of Dynamical Systems*,
Springer Theses, https://doi.org/10.1007/978-3-030-90343-5_6

6.1 Setup of the Identification Problem

6.1.1 A Brief Recap—Tree Adjoining Grammar and Pareto-optimality

Recall that the primary research goal of the present dissertation is to automate the task of system identification from a user's perspective. In Chap. 2, based on a review of a number of identification methodologies, we proposed the following set of desired features that the conceptualized identification methodology should possess:

- The identification methodology should be applicable for multiple classes of dynamical systems, with minimal changes to the overall approach.
- Instead of requiring intervention from the user, the identification methodology should be able to infer the appropriate model structure and complexity based on measured data obtained from the system-to-be-modelled.
- In the presence of prior information related to the dynamic properties of the system or desired structure of the model, it should be possible to suitably restrict the scope of the identification algorithm.
- In order to estimate models that are aligned to the ultimate utilization scenario, it should be possible to incorporate multiple user-specified identification criterion.

In order to develop the fundamentals of an identification methodology that incorporate some of these desired features, the following notions have been introduced:

- In Chap. 4, TAG was used to specify model sets that may span across multiple classes of dynamical systems and models with varying structures and complexity. We also demonstrated that prior information or preference can be utilized to restrict the model set, thereby restricting the scope of the methodology. In the absence of prior knowledge, a general TAG can be used in order to cast a wide net across dynamic classes and model structures, at the cost of making the underlying optimization problem more challenging.
- In Chap. 5, we introduced the notion of partial ordering of solutions based on multiple performance measures, leading to the notion of Pareto-optimality. The notion of Pareto-optimality can be used to develop a multi-criterion identification methodology.

The chapter is structured as follows. Based on the concepts of TAG-based model set and Pareto-optimality, we formalize the multi-criterion identification problem in Sect. 6.1.2. In Sect. 6.2, we proposed a novel identification method, based on GP (introduced in Chap. 3), that brings together the notions of TAG-based model set specification and Pareto-optimality, in order to solve the multi-criterion system identification problem. The proposed identification method is formulated on a fairly general level: TAG can be used to describe a variety of model structures and representations, and arbitrary user-specified performance measures can be incorporated. For specific choices of TAG G_{N} (proposed in Chap. 4) and the performance measures $\{J_i(M, \theta, \mathcal{D}_N)\}_{i=1}^4$ (suggested in Chap. 5), details for parameter estimation in the proposed identification framework are discussed in Sect. 6.3. The asymptotic behaviour and computational complexity of the proposed algorithm are studied in Sect. 6.4. Finally, the main aspects of the proposed identification approach are discussed in Sect. 6.5.

6.1.2 The Multi-criterion System Identification Problem

Let G be the chosen TAG with the corresponding tree language $L_{\mathrm{T}}(G)$ and model set $\mathcal{M}(G)$ (see Definition 4.8). Recall that for any saturated derived tree $\psi \in L_{\mathrm{T}}(G)$, the corresponding model structure M can be obtained by the genotype-phenotype mapping $\Phi : L_{\mathrm{T}}(G) \to \mathcal{M}(G)$, i.e., $M = \Phi(\psi)$. Furthermore, it is assumed that any model $M \in \mathcal{M}(G)$ can be parameterized by a parameter vector $\theta \in \Theta(M)$, where the domain $\Theta(M)$ is determined by the structure of M (or alternatively, by the structure of saturated derived tree ψ). We also assume that model M is parameterized such that the parameterized model $M(\theta)$ does not contain any duplicate model terms. Let $J(M, \theta, \mathcal{D}_N)$ be the set-valued performance measure with number of objectives n_{obj}.

The general problem

The general multi-criteria system identification problem can be formalized as the following optimization problem,

$$\min_{M \in \mathcal{M}(G), \theta \in \Theta(M)} \quad J(M, \theta, \mathcal{D}_N), \tag{6.1}$$

where the minimization of the set-valued performance measures is based on the strict partial ordering $\prec_J$ on the set $\mathcal{M}(G)$. Notice that the feasible set $\Theta(M)$ for decision variable θ is only determined once a model structure $M \in \mathcal{M}(G)$ is fixed. Hence, there exists a hierarchical structure built into the problem. In order to suitably refine the optimization problem, we divide the set of objective functions into two subsets—the set of continuous objective functions $J_{\mathrm{cont}}(M, \theta, \mathcal{D}_N)$ that explicitly depend on model parameters θ, and the set of categorical objective functions $J_{\mathrm{cat}}(M, \theta, \mathcal{D}_N)$ that do not explicitly depend on the values of parameters θ or on measured data $\mathcal{D}_N$, but only depend on the structure of the model. Continuous objective functions take values in a continuous domain, while categorical objective functions typically take values in a discrete domain. For example, for the performance measures proposed in Sect. 5.3, we have $J_{\mathrm{cont}}(M, \theta, \mathcal{D}_N) = \{J_i(M, \theta, \mathcal{D}_N)\}_{i=1}^{2}$ and $J_{\mathrm{cat}}(M, \theta, \mathcal{D}_N) = \{J_i(M, \theta, \mathcal{D}_N)\}_{i=3}^{4}$. Consequently, a more natural representation of the multi-criteria identification problem would be the following:

$$\min_{M \in \mathcal{M}(G)} \quad J(M, \theta, \mathcal{D}_N), \tag{6.2a}$$

$$\text{subject to} \quad \theta \in \arg\min_{\theta \in \Theta(M)} J_{\mathrm{cont}}(M, \theta, \mathcal{D}_N), \tag{6.2b}$$

where, again, the minimization of the set-valued performance measures is based on the strict partial ordering $\prec_J$ on the set $\mathcal{M}(G)$. The optimization problem in (6.2a) and (6.2b) is a *bilevel multi-objective* optimization problem; it consists of two levels, the upper level in (6.2a) and the lower level in (6.2b), both of which are multi-objective. The upper level optimization problem is formulated in terms of decision variable M, and represents the model-structure-determination problem that is to be solved. The lower level optimization problem is formulated in terms of decision variable $\theta \in \Theta(M)$ and is defined only for a given model structure M, as it determines the domain $\Theta(M)$ of the lower level decision variable. Observe that the lower-level problem is formulated for performance measures in $J_{\text{cont}}(M, \theta, \mathcal{D}_N)$ since, for a fixed model structure M, the performance measures in $J_{\text{cat}}(M, \theta, \mathcal{D}_N)$ are also fixed. The Pareto-efficient set of the lower level optimization problem is the set of all parameters θ that lie in the Pareto-front corresponding to the set-valued objective $J_{\text{cont}}(M, \theta, \mathcal{D}_N)$. This sets up the following hierarchical structure in the optimization problem: for any model structure $M \in \mathcal{M}(G)$, the feasible set of parameters θ is determined by the Pareto-efficient set of the lower level optimization problem in (6.2b).

Due to the interplay of decision variables between the two levels of the optimization problem, bilevel optimization problems can be difficult to solve. Hansen et al. [14] demonstrated that even a single-objective bilevel problem with linear objectives and linear constraints in both levels is an NP hard problem to solve. Nevertheless, bilevel optimization problems are relevant for many decision-making problems in supply chain management [29], power grid management [30], transportation systems [12], control theory [2] and network security [33]. In the field of game theory, bilevel programs are studied in terms of the dynamics between a leader and a follower in a Stackelberg game [31]. Due to the relevance of bilevel programs, several solution techniques have been proposed in the literature, see Sinha et al. [32] and Colson et al. [6] for a review.

The optimization problem in (6.2) encapsulates the general identification problem as it combines the problems of model structure selection (in the upper-level) and parameter estimation (in the lower-level). Furthermore, the set valued identification criteria can be adapted to incorporate performance measures suitable for the use-case of the model. For the performance measures proposed in Sect. 5.3, the optimization problem remains the same as in (6.2) with $J(M, \theta, \mathcal{D}_N) = \{J_i(M, \theta, \mathcal{D}_N)\}_{i=1}^4$ and $J_{\text{cont}}(M, \theta, \mathcal{D}_N) = \{J_i(M, \theta, \mathcal{D}_N)\}_{i=1}^2$.

Simplifications to the general problem

In the optimization problem (6.2a) and (6.2b), since both levels are multi-objective, for any given model structure $M \in \mathcal{M}(G)$, the Pareto-efficient set of the lower level problem can be infinitely large, achieving a continuous trade-off between objectives in J_{cont}, for example, between one-step-ahead-prediction error on one end and simulation error on the other.[1] This leads to the problem of storing a rather large set

[1] In the event that the data generating system (DGS) can be represented exactly by model structure M, it is possible that the Pareto-efficient set of the lower level problem consists of a single point, i.e.,

of parameter values θ for a single model structure M, which in itself is a decision variable in a discrete combinatorial space $\mathcal{M}(G)$ in the upper level.

To avoid such issues, we simplify the general problem in (6.2a, 6.2b) by replacing the multi-objective problem in lower level 6.2b with multiple single-objective optimization problems using a scalarization technique, e.g., convex combination or Chebyshev scalarization (for details, see Ishibuchi et al. [17] or Emmerich and Deutz [8]). This approach effectively discretizes the Pareto-front of the lower-level optimization problem. Hence, the bi-level optimization problem can be simplified to

$$\min_{M \in \mathcal{M}(G), \theta \in \Theta(M)} \quad J(M, \theta, \mathcal{D}_N), \tag{6.3a}$$

$$\text{subject to} \quad \theta \in \bigcup_{i=1}^{n_\beta} \left\{ \arg\min_{\theta \in \Theta(M)} \sum_{j=1}^{|J_{\text{cont}}|} \beta_{i,j} J_j(M, \theta, \mathcal{D}_N) \right\}, \tag{6.3b}$$

where n_β is the number of combinations considered in the scalarization of the lower-level multi-objective function, $\beta_{i,j}$ are trade-off parameters pre-selected by the user based on the scalarization technique used, and $J_j \in J_{\text{cont}}$ for all $j \in [1, |J_{\text{cont}}|]$.

For the performance measures proposed in Sect. 5.3, and using convex combinations of the proposed identification criteria, the simplified problem can be written as

$$\min_{M \in \mathcal{M}(G), \theta \in \Theta(M)} \quad \{J_i(M, \theta, \mathcal{D}_N)\}_{i=1}^{4}, \tag{6.4a}$$

$$\text{subject to} \quad \theta \in \bigcup_{i=1}^{n_\beta} \left\{ \arg\min_{\theta \in \Theta(M)} \left(\beta_i J_1(M, \theta, \mathcal{D}_N) + (1 - \beta_i) J_2(M, \theta, \mathcal{D}_N)\right) \right\}, \tag{6.4b}$$

As a special case of the optimization problem in (6.4), for $\beta_1 = 1$ and $\beta_2 = 0$, the lower level multi-objective problem can be replaced with two single-objective problems of minimizing the one-step-prediction error and simulation error, i.e.,

$$\min_{M \in \mathcal{M}(G), \theta \in \Theta(M)} \quad \{J_i(M, \theta, \mathcal{D}_N)\}_{i=1}^{4}, \tag{6.5a}$$

$$\text{subject to} \quad \theta \in \left\{ \arg\min_{\theta \in \Theta(\psi)} J_1(M, \theta, \mathcal{D}_N) \right\} \bigcup \left\{ \arg\min_{\theta \in \Theta(M)} J_2(M, \theta, \mathcal{D}_N) \right\}. \tag{6.5b}$$

It should be stressed that the Pareto-efficient set of the simplified problems in (6.3), (6.4) or (6.5) is not the same as that of (6.2), since, for a given M, the set of optimal

---

the optimal parameter vectors for objectives J_1 and J_2 coincide. However, this is a rare scenario, specially in realistic identification problems and also due to the generality of the model set described by TAG.

parameters for (6.3b), (6.4b) and (6.5b) form subsets of the Pareto-efficient set of (6.2b). Nonetheless, since objectives in J_{cont} are also included in the upper-level problems in (6.3a), (6.3a) and(6.5a), we expect the resulting Pareto-fronts of the simpler bi-level problems to be a fair (albeit sparse) approximation of the Pareto-front of (6.2).

6.2 The Proposed Algorithm

In this section, we fuse the ingredients of GP, TAG and Pareto-based ordering in order to develop an algorithm to solve the multi-criteria system identification problem. We first give a general overview of the proposed identification approach in Sect. 6.2.1. This is followed by detailed discussions of each of the steps in the proposed methodology.

6.2.1 A Bird's Eye View

In order to solve the bilevel optimization problem in (6.5) (or, in general, in 6.4) we propose a GP-based algorithm constructed in the following ways.

- The proposed GP-based algorithm is adapted to be compatible with derivation tree representations of models described by a user-specified TAG G. This requires suitable modifications to
  - The initialization scheme, such that the GP population can be initialized by derivation tree representations of candidate solutions.
  - The variation operators, such that the candidate solutions obtained after the crossover and mutation operator also satisfy the grammar rules encoded in G.

  The modifications made to the standard GP are inspired by the first TAG-based GP algorithm proposed in the literature by [15], and are discussed in Sect. 6.2.2.
- The algorithm is extended to a multi-objective setting using Pareto-based selection. In this dissertation, we use the non-dominated sorting algorithm proposed in Deb et al. [7], however, in principle, relevant concepts from other MOEAs such as SPEA2 or MOEA/D can also be used, see discussion in Sect. 5.2.2 or more extensive reviews in Emmerich and Deutz [8] and Ishibuchi et al. [17].
- A local search step is nested within the algorithm to estimate model parameters solving the lower-level optimization problem. This is an example of a memetic algorithm, see discussion in Sect. 3.3.

Each candidate solution is represented using (i) a derivation tree representation (introduced in Sect. 4.2.3), (ii) a derived tree representation (introduced in Sect. 4.2.3), and (iii) a dynamic IO representation (for example, as in (4.10)). The role of these representations is discussed in Sect. 6.2.2.

An overview of the proposed multi-objective grammar-based memetic algorithm is given in Algorithm 1 and illustrated in Fig. 6.1. In each iteration of GP, modified

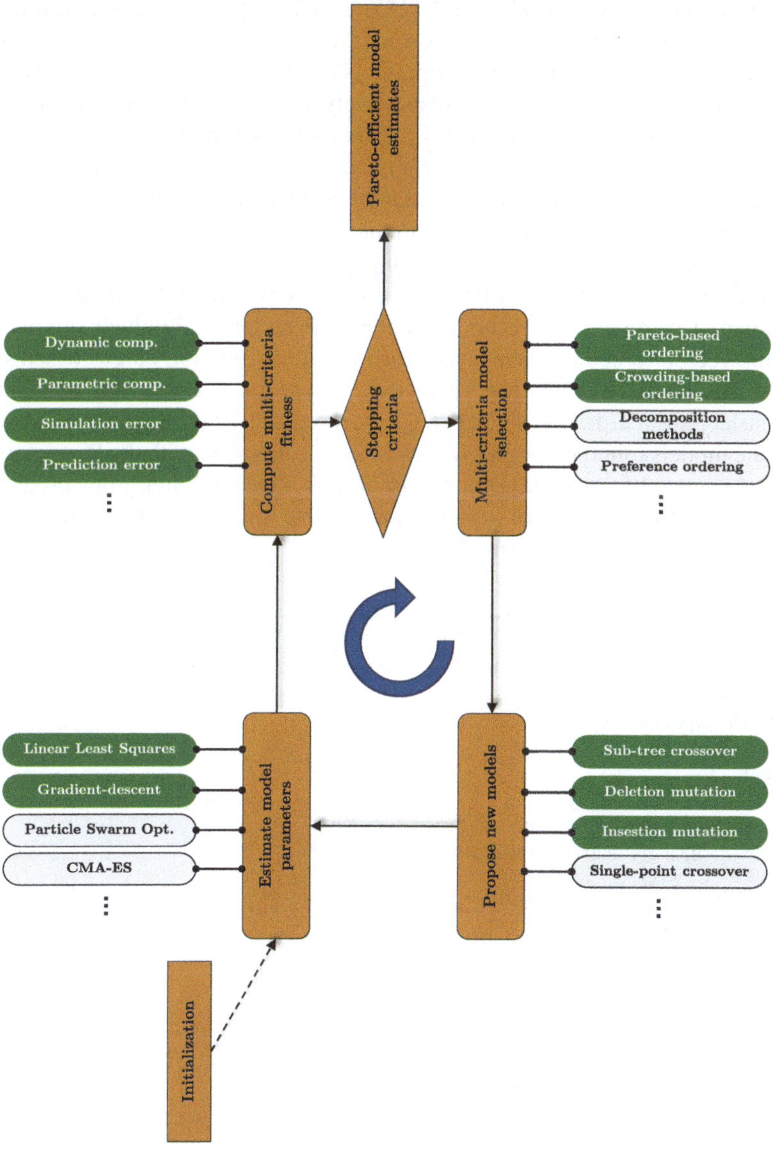

Fig. 6.1 A schematic illustration of the proposed identification algorithm, where the operations described for each step act upon a population of models. The crucial ingredients that we used to carefully design each phase of the proposed algorithm (described in Sect. 6.2.2) are indicated with green. However, the modular nature of the method allows to take other choices for more extensive grammars or performance measures, some of which are indicated in gray boxes

variation operators (based on Hoai et al. [16]) are used to propose candidate models, in terms of derivation trees and derived trees, of the population (see Step 8). By virtue of the nature of crossover and mutation operators used, the derivation and derived trees belong to the sets $L_D G$ and $L_T(G)$, respectively, defined by the user-specified TAG G. These individuals are subsequently translated to their phenotype representation, i.e., parametric dynamical IO model structures. Before these model structures can be evaluated based on user-specified performance measures, the parameter vectors of each of the model structures must be estimated. This corresponds to the local search step of the algorithm in Step 4. The result of the local search step is the solution to the lower level problem in (6.5b) for the corresponding model structure. Based on the estimated parameters, in Step 5, the newly proposed individuals are evaluated using the vector-valued performance measure J. In Step 6, the Pareto-based sorting algorithm proposed in Deb et al. [7] is used to order and select candidate models based on the strict partial ordering $\prec_J$ on $\mathcal{M}(G)$. An archive, consisting of Pareto-efficient candidate models found in the previous iterations, is updated with the Pareto-efficient subset of the newly proposed solutions in Step 7. As the GP iterations proceed, the set of Pareto-efficient models stored in the archive approaches the Pareto-efficient set of the upper level problem in (6.5a).

Algorithm 1 The multi-objective grammar-based memetic algorithm

Require: population size $n_s > 0$, number of iterations $L > 0$, TAG G, crossover rate p_c, mutation rate p_m, maximum tree-depth m_d

1: *Initialize* population $X^{(0)}, X^{(-1)} = \{\}, j = 0, \mathfrak{A}^{(0)} = \{\}$ ▷ See Algorithm 2
2: $\tilde{X}^{(0)} \leftarrow X^{(0)}$
3: **repeat**
4: Estimate model parameters in $\tilde{X}^{(j)}$ ▷ See discussion of parameter estimation in Sect. 6.2.2
5: Compute multi-objective fitness of models in $\tilde{X}^{(j)}$
6: $X^{(j)} \leftarrow n_s$ individuals selected from $(\tilde{X}^{(j)} \cup X^{(j-1)})$ ▷ See Algorithms 5 and 6
7: $\mathfrak{A}^{(j+1)} \leftarrow \mathcal{E}(\mathfrak{A}^{(j)} \cup \tilde{X}^{(j)}, \prec_J)$
8: Propose new population $\tilde{X}^{(j+1)}$ using variation operations ▷ See discussion on variation operators in Sect. 6.2.2
9: $j \leftarrow j + 1$
10: **until** $j \leq L + 1$
11: **return** $X^{(L)}, \mathfrak{A}^{(l)}$

6.2.2 *Step-by-Step Procedure*

In the following section, we discuss details of each of the steps in the proposed identification procedure.

Data collection and pre-processing

The measured dataset can be pre-processed in a number of ways in order to make the modelling task easier and to obtain models that are suitable for their use-case. Data pre-processing is commonly performed to

- remove trends, such as constant or linear offset, from data,
- pre-filter data in order to focus the modelling task on a particular range of frequencies,
- treat outliers or missing data, for example, by estimating the true values or removing them altogether.

Pre-processing of data is often specific to the application, and in this thesis we assume that suitable pre-processing has been performed on the measured data. Some commonly used techniques for pre-processing of data can be found in Ljung [22] and Pintelon and Schoukens [24].

The measured dataset $\mathcal{D}_N$ plays a crucial role in Steps 4 and 5 of Algorithm 1. In Step 4, data is used to obtain parameter estimates for a given model structure, and in Step 5, data is used to evaluate each model (structure and parameters) in the population. When TAG G is used to generate complex model structures that are over-parameterized w.r.t the true system, the use of finite-length datasets may result in over-fitting of the model and high variance in the parameter estimates. In order to prevent optimistic training error estimates and overfitting, we use two independent datasets—an estimation dataset $\mathcal{D}_{\text{est}}$, of length N_{est}, for estimating model parameters in Step 4 and a validation dataset $\mathcal{D}_{\text{val}}$, of length N_{val}, to evaluate the performance of proposed model structures. Whenever possible, we also use a third independent dataset $\mathcal{D}_{\text{test}}$, of length N_{test}, to estimate the general performance of the Pareto-efficient set returned at the end of Algorithm 1. The three independent datasets can be obtained either from three independent experiments or by splitting a sufficiently long experiments into three parts. Splitting a dataset into three subsets may introduce transient errors at the beginning of each of the subsets. Transient errors can be reduced by estimating the initial conditions, or by provisioning a burn-in period (i.e., by ignoring the first few samples of the dataset where the transient errors are more prominent) that allows the transient errors to decay to an acceptable level (assuming that the estimated model is exponentially stable).

Representations and encoding

In the proposed TAG-based GP, the population $X^{(j)}$ in iteration j consists of n_s candidate solutions encoded as derivation trees, denoted as $\{\rho_i^j\}_{i=1}^{n_s}$. Derivation tree representations serve as the genotype for the GP algorithm, replacing the commonly-used expression trees. Hence, variation operators are used on the derivation tree representation of the candidate solutions. Each candidate solution can also be represented as a derived tree δ_i^j that can be computed by performing the adjunctions and substitutions captured in the derivation tree ρ_i^j (see Sect. 4.2.3). This operation is denoted as $\delta_i^j = \Delta G[\rho_i^j]$. Finally, the derivation tree representation ρ_i^j of individual i in iteration j can be mapped to a IO model $M_i^j \in \mathcal{M}(G)$, which forms the phenotype of the GP algorithm. The mapping from genotype ρ_i^j to phenotype M_i^j is denoted by $M_i^j = \Phi_G(\rho_i^j)$. In practice, the derived tree δ_i^j serves as an intermediate step in the conversion from genotype (i.e., the derivation tree) ρ_i^j to phenotype (i.e., the IO model) M_i^j.

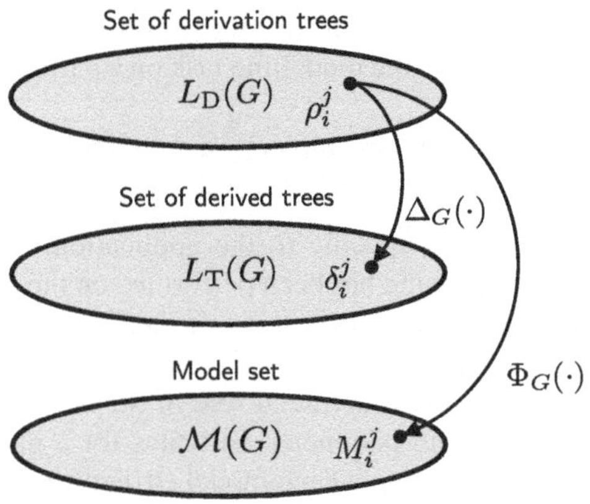

Fig. 6.2 Model representations used in the proposed algorithm

The candidate model M_i^j can be parameterized as $M_i^j(\theta_i^j)$ by replacing the placeholder parameters in δ_i^j with a parameter vector θ_i^j, whose dimensionality depends on the structure of M_i^j. The various model representations permitted by grammar G, the corresponding sets they belong to, and mappings between these representations are illustrated in Fig. 6.2.

Initialization

In Step 1 of Algorithm 1, the algorithm is initialized with population $X^{(0)}$ that consists of candidate solutions generated from grammar G. The initialization schemes for GP presented in Sect. 3.2.2 must be modified in order to generate individuals using a user-specified TAG G. Hoai and McKay [15] developed an approach for initializing derivation trees based on a TAG for use in GP. We adopt and formalize the same initialization scheme for our method.

Before introducing the initialization scheme, we recall the relevant notations introduced in Chaps. 3 and 4. A TAG G is given by the tuple $\langle \mathfrak{N}, \mathfrak{T}, \mathcal{I}, \mathcal{A}, S \rangle$, where $\mathfrak{N}$ is the set of non-terminal labels, $\mathfrak{T}$ is the set of terminal labels, $\mathcal{I}$ is the set of initial trees, $\mathcal{A}$ is the set of auxiliary trees, and S is the label of the root node. An initial tree η is described by the tuple V, E, r, where V is the set of vertices, E is the set of edges and r is the root-node. Similarly, an auxiliary tree α is described by the tuple V, E, r, f, where f is the foot-node of the auxiliary tree. Finally, ϕ denotes the null set.

The algorithms used for initialization are presented in Algorithms 2, 3 and 4.

In the proposed initialization, each derivation tree ρ_i^0 in the initial population $X^{(0)}$ is grown upto a depth d that is randomly chosen from the range $[2, m_{\mathrm{id}}]$, where m_{id} is the maximum tree depth in the initial population, a hyper-parameter chosen by the user. The tree is initialized with an initial tree whose root node is labelled with start symbol S of grammar G, see Step 3 of Algorithm 2. Throughout the initialization process, we maintain lists Π_{s} and Π_{a} to track leaves and internal vertices in the derived tree that can participate in a substitution and adjunction operation, respectively. If the chosen initial tree contains leaves available for substitution (listed in Π_{s} in Step 6), compatible[2] initial trees, chosen randomly with uniform distribution, are substituted in Step 7 of Algorithm 2 and in Algorithm 3. After any substitution or adjunction operation, list Π_a, consisting of all vertices in the derived tree that are available for adjunction, is updated. Each vertex listed in Π_s and Π_a is also paired with the corresponding node in the derivation tree that generates the vertex in the derived tree.

Algorithm 2 The initialization scheme

Require: Grammar $G = \langle \mathfrak{N}, \mathfrak{T}, \mathcal{I}, \mathcal{A}, S \rangle$, maximum initial tree-depth m_{id}, population size n_s

1: Initialize $X[0] \leftarrow \{\}$, $D^{(0)} \leftarrow \{\}$

2: **while** $i \leq n_s$ **do** ▷ Iterate through each individual in the population

3: Select $\eta_i = \langle V, E, r \rangle \in \mathcal{I}$ randomly, with uniform distribution, such that $l(r) = S$

4: $\rho_i^0 \leftarrow \langle V_i, E_i, r_i \rangle$, where $V_i = \{r_i\}$, $E_i = \phi$, and $l(r_i) \leftarrow \eta_i$ ▷ Initialize the derivation tree

5: $\delta_i^0 \leftarrow \eta_i$ ▷ Initialize the derived tree

6: $\Pi_s \leftarrow \{(v_k, v) \mid (v \in \Lambda(\delta_i^0)) \wedge (l(v) \in \mathfrak{N}) \wedge (\exists \eta = \langle V'', E'', r'' \rangle : l(v) = l(r''))\}$ ▷ List vertices in the derived tree that can be substituted to, and the corresponding vertex in the derivation tree

7: $(\rho_i^0, \delta_i^0) \leftarrow$ SUBSTITUTE$(\rho_i^0, \delta_i^0, r_i, \Pi_s, G)$

8: $l \leftarrow 1 + |\Pi_s|$

9: $k \leftarrow 1$ ▷ A counter for derivation tree vertices

10: $v_k \leftarrow r_i$

11: Select $d \in [2, m_{id}]$ randomly with uniform distribution

12: **while** $d_t(\rho_i^0) \leq d$ **do** ▷ Grow derivation tree until chosen depth d

13: $\Pi_a \leftarrow \{(v_k, v) \mid (v \in V) \wedge (v \notin \Lambda(\delta_i^0)) \wedge (\exists \alpha = \langle V', E', r', f' \rangle : l(v) = l(f'))\}$ ▷ List vertices in the derived tree that can be adjoined to, and the corresponding vertex in the derivation tree

14: $(\rho_i^0, \delta_i^0, \Pi_a, V) \leftarrow$ ADJOIN$(\rho_i^0, \delta_i^0, k + l, \Pi_a, G)$

15: $k \leftarrow k + l$

16: $\Pi_s \leftarrow \{(v_k, v) \mid (v \in \Lambda(\delta_i^0)) \wedge (l(v) \in \mathfrak{N}) \wedge (\exists \eta = \langle V'', E'', r'' \rangle \in \mathcal{I} : l(v) = l(r''))\}$ ▷ List vertices in the derived tree that can be substituted to, and the corresponding vertex in the derivation tree

17: $(\rho_i^0, \delta_i^0) \leftarrow$ SUBSTITUTE$(\rho_i^0, \delta_i^0, v_k, \Pi_s, G)$

18: $l \leftarrow 1 + |\Pi_s|$

19: $\rho_i^0 \leftarrow \langle V_i, E_i, r_i \rangle$

20: $k \leftarrow k + 1$

return $X[0] = \{\rho_i^0\}_{i=1}^{n_s}$, $D^{(0)} = \{\delta_i^0\}_{i=1}^{n_s}$

After initializing the derivation tree with an initial tree, along with the corresponding substituted trees, the derivation tree can be grown iteratively by adjoining auxiliary trees. In each iteration, a vertex available for adjunction is chosen randomly, with uniform distribution, from list Π_a (see Step 13 in Algorithm 2). Subsequently, a compatible auxiliary tree is chosen randomly with uniform distribution and adjoined to the individual at the chosen location, see Step 14 in Algorithms 2 and 4. Again, if required, suitable initial trees substituted to the newly adjoined tree in Steps 16 and 17 in Algorithm 2. This process is repeated until the derivation tree has reached the required depth.

---

[2] See Sect. 4.2.3 for conditions under which a substitution operation is well-defined.

Parameter estimation and multi-criteria evaluation

In each iteration of the proposed identification method, i.e., Algorithm 1, a new population $\tilde{X}^{(j)}$ is proposed. Let the corresponding models $\tilde{M}_i^j$ be parameterized by parameter vectors $\tilde{\theta}_i^j$. In Step 4 of Algorithm 1, parameter estimates of $\tilde{\theta}_i^j$ that optimize the lower level criterion (or criteria) are computed.

Consider a single candidate model structure $\tilde{M}_i^j$ in iteration j of the algorithm. For the general multi-criterion identification problem in (6.2), the Pareto-efficient set of parameters that optimize the lower-level optimization problem for model structure $\tilde{M}_i^j$ alone may be infinitely large. This is due to the fact that performance measures in J_{cont} may contradict each other, leading to a continuum of possible trade-offs. For example, among the performance measures proposed in Chap. 5, prediction and simulation error minimization may lead to non-identical parameter estimates, leading to infinite trade-off possibilities. Hence, a more computationally feasible approach is to simplify the lower-level objective by considering a finite union of single-objective problems, as proposed in in (6.4) and (6.5). For each lower-level objective and each candidate model structure $\tilde{M}_i^j$, the estimation dataset $\mathcal{D}_{\text{est}}$ is used to obtain a parameter estimate denoted by $\tilde{\theta}_{i,r}^j$, where r is the index of the lower-level objective. Hence, in iteration j, the set of candidate parametric models is given by $\{\cup_{r=1}^{n_\beta} \tilde{M}_i^j(\tilde{\theta}_{i,r}^j)\}_{i=1}^{n_s}$.

Algorithm 3 Substitute

Require: Derivation tree $\rho_i^j = \langle V, E, r\rangle$, derived tree δ_i^j, vertex $v_k \in V$, list of vertex pairs available for substitution Π_s, grammar G

$l \leftarrow 1$
while $\Pi_s \neq \phi$ **do**
  Select any pair $(v_\rho, v_\delta) \in \Pi_s$
  Select $\eta = \langle V', E', r'\rangle \in \mathcal{I}$ such that $l(v_\delta) = l(r')$
  $V \leftarrow V \cup \{v_{k+l}\}$ with $l(v_{k+l}) = \eta$ ▷ Update the derivation tree
  $E \leftarrow E \cup \{e\}$ where $e = \langle v_\rho, v_{k+l}\rangle$ and $g(e) = p_G(v_\rho, \rho_i^j)$
  $\delta_i^j \leftarrow \delta_i^j[v_\delta, \eta]$ ▷ Update the derived tree
  $l \leftarrow l + 1$
  $\Pi_s \leftarrow \Pi_s \setminus \{(v_\rho, v_\delta)\}$
return ρ_i^j, δ_i^j

Algorithm 4 Adjoin

Require: Derivation tree $\rho_i^j = \langle V, E, r\rangle$, derived tree δ_i^j, index k for next vertex position in derivation tree, list of vertices pair available for adjunction Π_s, grammar G

Select a pair $(v_\rho, v_\delta) \in \Pi_a$ randomly with uniform distribution ▷ Select a vertex that can be adjoined to
Select $\alpha = \langle V', E', r', f'\rangle \in \mathcal{A}$ such that $l(v_\delta) = l(f')$. In case of multiple compatible auxiliary trees, select one randomly with uniform distribution.
$V \leftarrow V \cup \{v_k\}$ with $l(v_k) = \alpha$ ▷ Update the derivation tree
$E \leftarrow E \cup \{e\}$ where $e = \langle v_\rho, v_k\rangle$ and $g(e) = p_G(v_\rho, \rho_i^j)$
$\delta_i^j \leftarrow \delta_i^j[\![v_\delta, \alpha]\!]$ ▷ Update the derived tree
$\Pi_a \leftarrow \Pi_a \setminus \{(v_\rho, v_\delta)\}$ ▷ Update the list of vertices available for adjunction
return $\rho_i^j, \delta_i^j, \Pi_a, V'$

Table 6.1 List of variation operators and their corresponding types

Variation operator	Hyper-parameter	Operator ypes	Illustration
Crossover	p_c	Sub-tree	Fig. 6.3
Mutation	p_m	Node insertion	Fig. 6.4
		Branch insertion	Fig. 6.5
		Node deletion	Fig. 6.6
		Branch deletion	Fig. 6.7

Optimization techniques that can be used for the parameter estimation step in Algorithm 1 are dependent on the grammar G used to generate models and the performance measures J_{cont} that depend on the value of the parameters. In general, when the chosen grammar G and objective function J_{cont} lead to formulation of non-linear optimization problems, one must use global optimization techniques such as Covariance matrix adaptation evolutionary strategies (CMA-ES) or PSO. Due to the computation-intense nature of such algorithms, they can be challenging to implement in practice, especially for large datasets. Alternatively, local optimization techniques such as gradient-descent-based algorithms may also be used. While local techniques may be more efficient computationally, they are susceptible to local equilibria. When the chosen grammar G and objective function J_{cont} lead to simpler (for example, convex) optimization problems, more efficient and specialized optimization techniques can be utilized. In Sect. 6.3, we discuss the numerical aspects of minimizing prediction and simulation error for model structures generated by TAG G_{N}.

After the estimation of parameters, in Step 5 of Algorithm 1, proposed model structures $\tilde{M}_i^j$ and the corresponding parameter estimates $\tilde{\theta}_{i,r}^j$ are used to compute the vector-valued performance measure $J(M, \theta, \mathcal{D}_{\text{val}})$ for each candidate model in the population. For the P-NARMAX model class generated by TAG G_{N} and the performance measures proposed in Sect. 5.3, the necessary computational details can be found in Sects. 5.3 and 5.4.

Variation operators—Crossover

In each iteration of Algorithm 1, crossover and mutation operators are used to propose a new population. In the proposed algorithm, variation operators are performed on the derivation tree representation of each candidate solution. Recall that one of the main motivations for using TAG is to ensure that any model generated using a given TAG belongs the desired model set. In order to maintain this property during the evolutionary search, the crossover and mutation operators must be modified. The variation operators used in the proposed method are based on those used in [15], where the authors proposed crossover and mutation operators that ensure the resulting model structures also belong to the desired model set encoded in the TAG.

We employ standard sub-tree crossover operator, introduced in Sect. 3.2.2, with additional constraints that ensure validity of the newly proposed adjunction operators. The crossover operator is applied to a pair of individuals in the population with probability p_c. Let ρ_1 and ρ_2 be two derivation trees selected for crossover and denote the corresponding derived trees as δ_1 and δ_2. The crossover operation consists of the following steps:

1. Select a vertex ν_1 in derivation tree ρ_1 randomly[3] and let ν'_1 denote the parent of vertex ν_1. The label of vertex ν_1 denotes the auxiliary tree α_1 that is to be adjoined, in the derived tree representation δ_1, to the auxiliary tree[4] $\alpha'_1 = \langle V_1, E_1, r_1, f_1\rangle$ represented by the label of the parent vertex ν'_1.
2. Select a vertex ν_2 and its parent vertex as ν'_2 in derivation tree ρ_2 with corresponding auxiliary trees α_2 and $\alpha'_2 = \langle V_2, E_2, r_2, f_2\rangle$, respectively, such that the following conditions are satisfied

$$\exists\, \nu_3 \in V_1 : \alpha'_1[\![\nu_3, \alpha_2]\!] \text{ is defined,} \tag{6.6a}$$

$$\exists\, \nu_4 \in V_2 : \alpha'_2[\![\nu_4, \alpha_1]\!] \text{ is defined.} \tag{6.6b}$$

   In case of multiple possibilities, select one pair of vertices (ν_3, ν_4) randomly with uniform distribution. If there does not exist a pair (ν_3, ν_4) that satisfy (6.6) for the chosen vertex ν_1 in derivation tree ρ_i, select another vertex ν_1 randomly and without replacement and repeat item 2.
3. Swap the sub-tree rooted at ν_1 in ρ_1 with the sub-tree rooted at ν_2 in ρ_2, and label the newly formed edges with the corresponding Gorn addresses

$$g(\langle \nu'_1, \nu_2\rangle) = p_{\mathrm{G}}(\nu_3, \alpha'_1), \tag{6.7a}$$

$$g(\langle \nu'_2, \nu_1\rangle) = p_{\mathrm{G}}(\nu_4, \alpha'_2). \tag{6.7b}$$

If the conditions in (6.6) are not satisfied for any vertex ν_1 in ρ_1, then crossover cannot be performed for the two chosen individuals.

In Fig. 6.3 we illustrate the crossover operator on two individuals generated by TAG G_{N}. The models corresponding to the original individuals are the following.

$$y(k) = c_1 y^2(k-1) + c_2 u(k) + c_3 \xi(k-1)\xi(k-2) + \xi(k), \tag{6.8}$$
$$y(k) = c_1 y(k-2) + c_2 \xi(k-1) + \xi(k). \tag{6.9}$$

Subsequent to the crossover operator, the models corresponding to the new individuals are the following

$$y(k) = c_1 y(k-3) + c_2 u(k) + c_3 \xi(k-1)\xi(k-2) + \xi(k), \tag{6.10}$$
$$y(k) = c_1 y^2(k-1) + c_2 \xi(k-1) + \xi(k). \tag{6.11}$$

Remark 6.1 Observe that, as illustrated in Fig. 6.3, the standard sub-tree crossover operator acting on the derivation tree representation is able to exchange components

[3] The vertex may be selected randomly with uniform distribution, however it is a convention to bias the random selection towards internal vertices, i.e., vertices that are not leaves. See [20].

[4] Note that the parent node label may correspond to an initial tree rather than auxiliary tree. For convenience, we assume that the parent node is an auxiliary tree, however the steps described here also apply for the case when the parent is an initial tree..

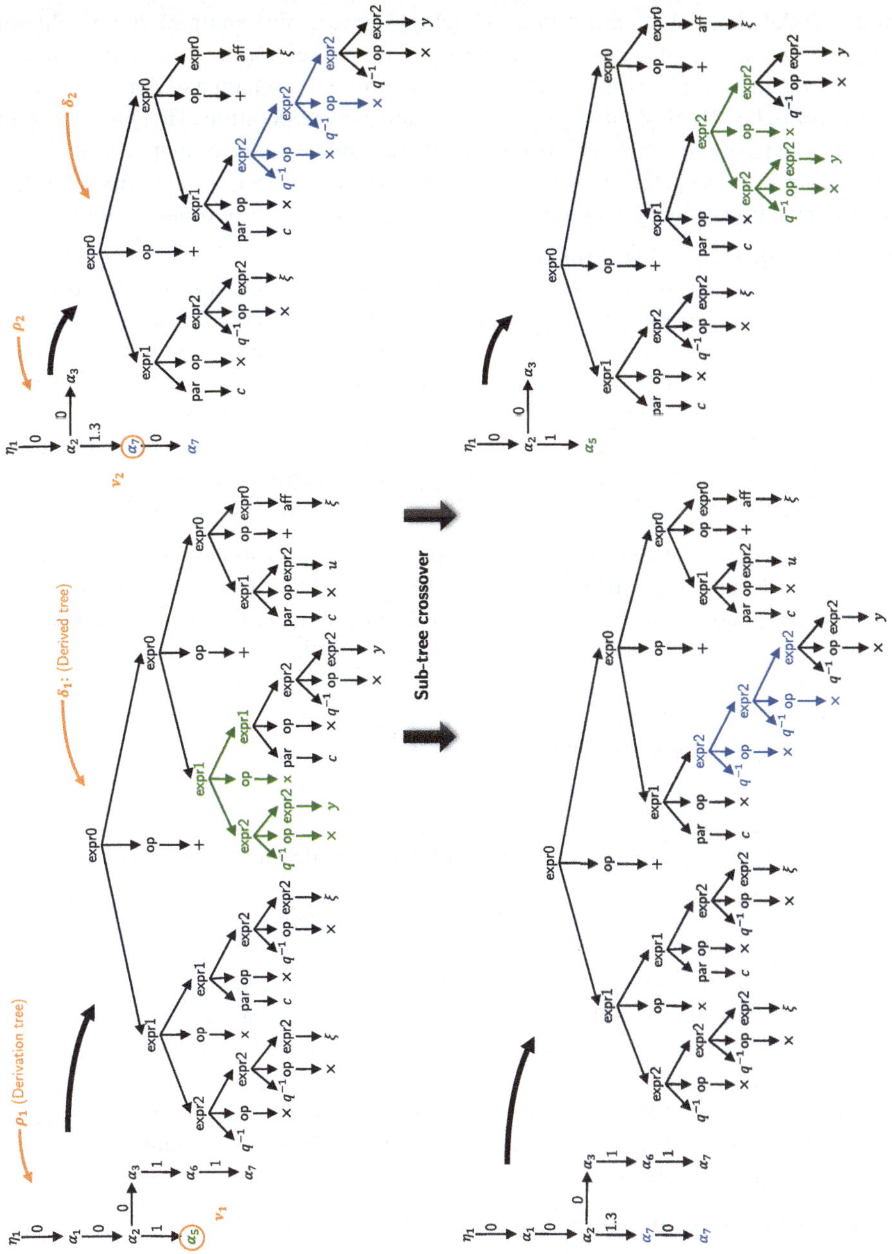

Fig. 6.3 The crossover operator illustrated on two individuals generated using G_N. The illustration depicts the two derivation trees ρ_1, ρ_2, the corresponding derived trees δ_1, δ_2 and vertices ν_1, ν_2 chosen for crossover. The coloured sections in the derivation trees and derived trees track the changes made in each individual

of the model expression (exponents and delays in the given example) that would not be possible for the same operator in the standard expression tree encoding used in GP. In fact, in an expression tree encoding of the model, exchange of delays or exponents would be achieved through a two-point crossover operation. This highlights an interplay between the representation of the model and the range of mapping achieved by the variation operator. In Ashlock et al. [1], the authors argue that this interplay determines, in part, the effectiveness of an EA in finding the optimal solution.

Variation operator—Mutation
In addition to the crossover operator, we also use the mutation operator to add variations to the proposed population. The mutation operator is applied to an individual in the population with a probability of p_m. The mutation operator used in the proposed algorithm are of two types—insertion and deletion. Furthermore, there are two variations for each type of mutation operator—node or branch mutation. See overview in Table 6.1. For any chosen individual selected for mutation, the type of mutation operator used is determined randomly with uniform distribution.[5]

The two variations of **insertion mutation** operator are described below.

1. Let the derivation tree of the individual selected for insertion mutation be denoted by ρ_1. Select a vertex ν in the derivation tree randomly with uniform distribution on all vertices except the root node, and denote the parent node of ν as ν'. Denote the auxiliary trees[4] corresponding to vertices ν and ν' in the derived tree representation as α and $\alpha' = \langle V', E', r', f' \rangle$, respectively. If node-type insertion is selected, go to Step 2, else if branch-type insertion is selected, go to Step 3.
2. **For node-type insertion operator**
   (a) select an auxiliary tree $\alpha_m = \langle V, E, r, f \rangle \in \mathcal{A}$ that satisfies the following conditions
$$\exists\ \nu_1 \in V' : \alpha'[\![\nu_1, \alpha_m]\!] \text{ is defined,} \tag{6.12a}$$
$$\exists\ \nu_2 \in V : \alpha_m[\![\nu_2, \alpha]\!] \text{ is defined.} \tag{6.12b}$$
   For multiple choices of vertices pair (ν_1, ν_2), select one randomly with uniform distribution. If there exists no pair of vertices that satisfy the conditions in (6.12), repeat Step 1 to choose a new vertex ν for mutation (without replacement).
   (b) In ρ_1, delete edge $\langle \nu', \nu \rangle$ and insert a new vertex ν_m with label $l(\nu_m) = \alpha_m$. Make new edges $\langle \nu', \nu_m \rangle$ and $\langle \nu_m, \nu \rangle$ with the corresponding edge labels $g(\langle \nu', \nu_m \rangle) = p_{\mathrm{G}}(\nu_1, \alpha')$ and $g(\langle \nu_m, \nu \rangle) = p_{\mathrm{G}}(\nu_2, \alpha_m)$.
3. **For branch-type insertion operator**
   (a) Select an auxiliary tree $\alpha_m = \langle V, E, r, f \rangle \in \mathcal{A}$ that satisfies
$$\exists\ \nu_1 \in V : \alpha[\![\nu_1, \alpha_m]\!] \text{ is defined.} \tag{6.13}$$

[5] Note that deletion operator cannot be used if the chosen derivation tree has tree depth less than 2. Similarly, insertion operator cannot be used if the derivation tree has depth equal to m_{d}.

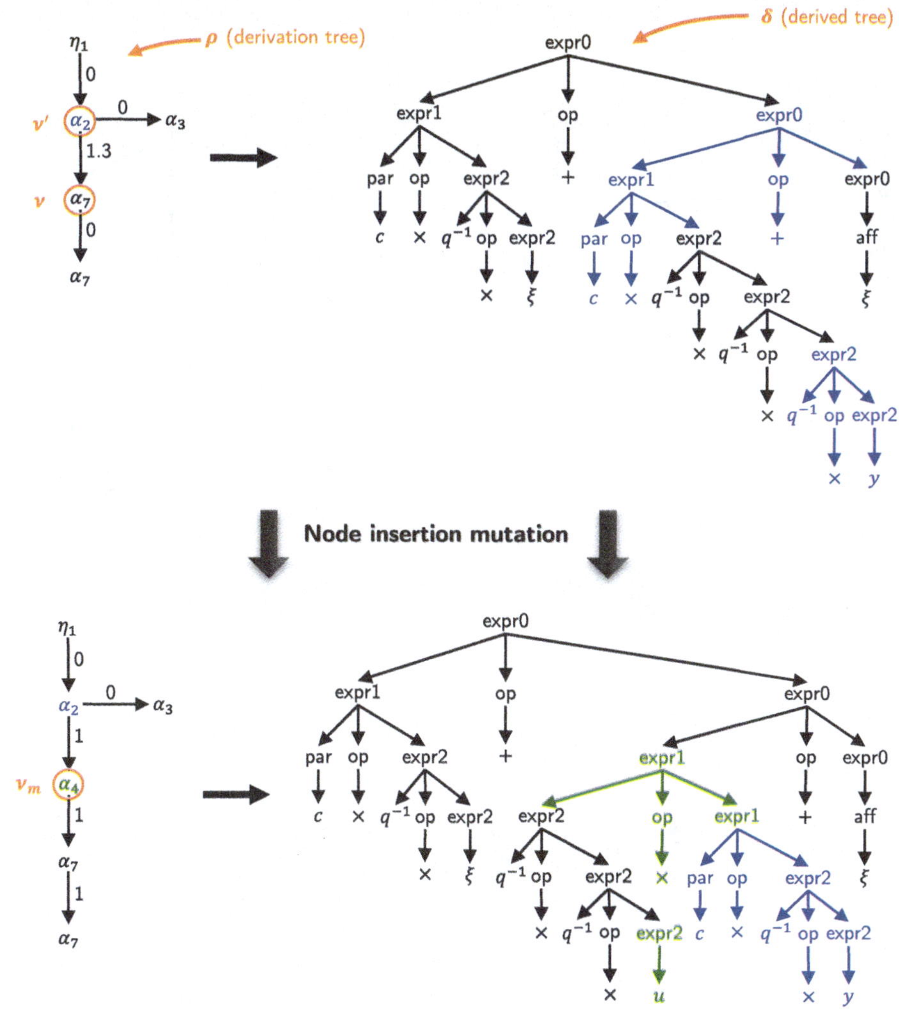

Fig. 6.4 The node insertion mutation operator illustrated on an individual generated using G_N. The illustration depicts the derivation tree ρ, the corresponding derived trees δ and vertex ν chosen for mutation. The coloured sections in the derivation trees and derived trees track the changes made in the individual. The new auxiliary tree (shaded in green) introduced into the individual is $\alpha_m = \alpha_4$

In the case of multiple choices for the pair (ν_1, α_m), select one pair randomly with uniform distribution. If there exists no pair (ν_1, α_m) that satisfies the conditions in (6.13), repeat Step 1 to choose a new vertex ν for mutation (without replacement).

(b) In ρ_1, insert a new vertex ν_m with label $l(\nu_m) = \alpha_m$. Make new edge $\langle \nu, \nu_m \rangle$ with edge label $g(\langle \nu, \nu_m \rangle) = p_G(\nu_1, \alpha)$.

The two variations of the insertion mutation operator are illustrated in Figs. 6.4 and 6.5. In both examples, the model corresponding to the parent individual ρ is

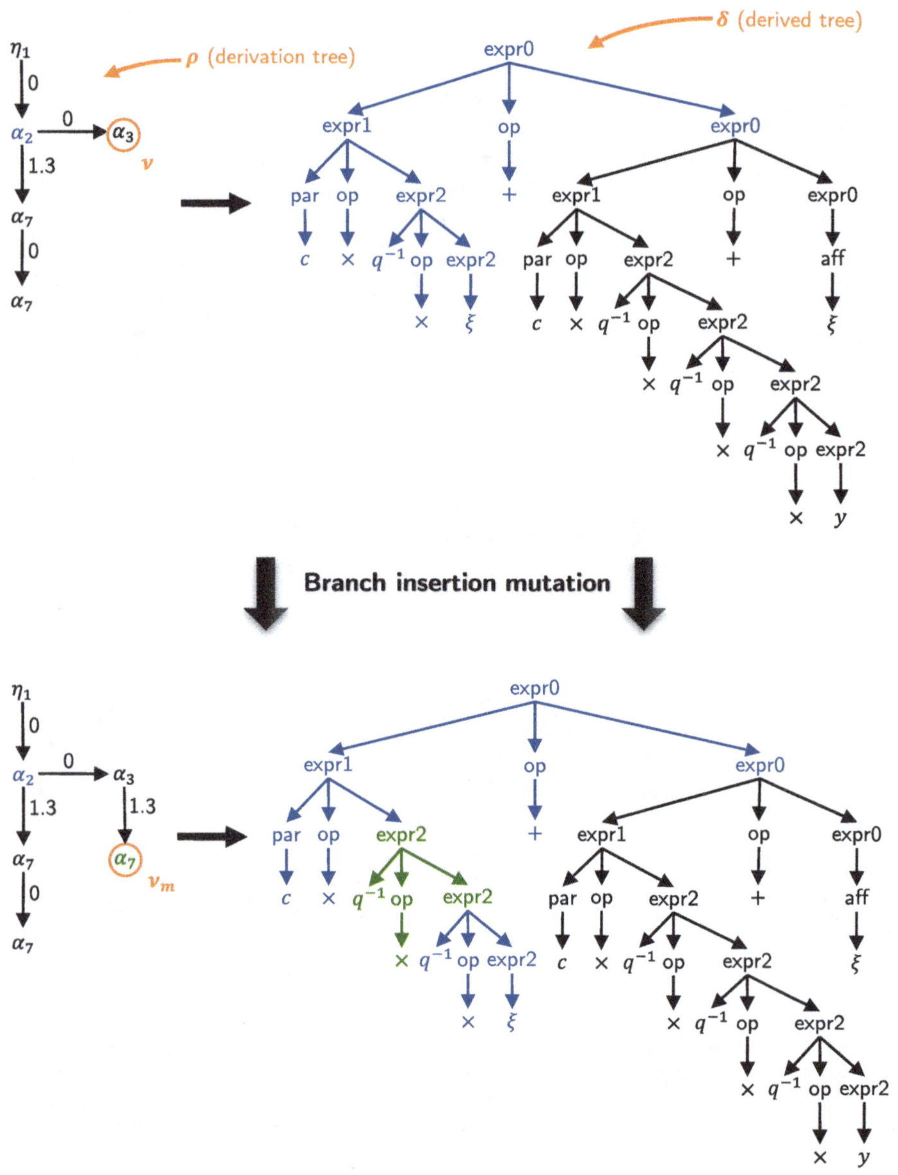

Fig. 6.5 The branch insertion mutation operator illustrated on an individual generated using G_N. The illustration depicts the derivation tree ρ, the corresponding derived trees δ and vertex ν chosen for mutation. The coloured sections in the derivation trees and derived trees track the changes made in the individual. The new auxiliary tree (shaded in green) introduced into the individual is $\alpha_m = \alpha_7$

$$y(k) = c_1 y(k-3) + c_2 \xi(k-1) + \xi(k). \tag{6.14}$$

The model obtained as a result of node insertion mutation in Fig. 6.5 is

$$y(k) = c_1 u(k-2) y(k-1) + c_2 \xi(k-1) + \xi(k). \tag{6.15}$$

In this example, the node insertion mutation introduces an input term $u(k)$ in the model expression, and moves the two delay-type auxiliary trees from the existing output term $y(k-3)$ to the newly introduced input-term.

The branch-insertion mutation operation in Fig. 6.5 results in the following model

$$y(k) = c_1 y(k-3) + c_2 \xi(k-2) + \xi(k). \tag{6.16}$$

In this example, the mutation operator simply introduces a delay-type auxiliary tree to the existing noise-term $\xi(k-1)$.

The two variations of the **deletion operator** are described below.

1. Let ρ be the derivation tree chosen for the deletion operation. Select a vertex ν in the derivation tree randomly with uniform distribution on all vertices except the root node. Denote the parent node of ν as ν' and the auxiliary tree[4] generated by ν' in the derived tree representation as $\alpha' = \langle V', E', r', f' \rangle$. If the chosen variation of deletion operation is node-type, go to Step 2, else if the chosen variation is sub-tree type, go to Step 3.
2. **For node-type deletion operator**

   (a) Let the chosen vertex ν have n children vertices $\nu''_1, \ldots, \nu''_n$. Let the auxiliary trees generated by the children vertices be $\alpha''_1, \ldots, \alpha''_n$. The chosen vertex ν can be deleted if the auxiliary trees $\alpha''_1, \ldots, \alpha''_n$ can be adjoined to the auxiliary tree α'. This can be formalized as the following condition

   $$\exists \{v_1, \ldots, v_n\} \subset V' : \alpha' [\![' v_1, \alpha''_1]\!] \ldots [\![v_n, \alpha''_n]\!] \text{ is defined.} \tag{6.17}$$

   If there exist many choices for subset $\{v_1, \ldots, v_n\} \subset V'$ that satisfy (6.17), select one randomly with uniform distribution. If there exists none, then vertex ν cannot be deleted. In that case, go to Step 1 to select another vertex ν in ρ without replacement.

   (b) Delete vertex ν and edges connected to $\langle \nu, \nu''_1 \rangle, \ldots, \langle \nu, \nu''_n \rangle$. Make new edges $\langle \nu', \nu''_1 \rangle, \ldots, \langle \nu', \nu''_n \rangle$ with labels $g(\langle \nu', \nu''_1 \rangle) = p_G(v_1, \alpha'), \ldots, g(\langle \nu', \nu''_n \rangle) = p_G(v_n, \alpha')$.

3. **For sub-tree deletion operator**

   (a) Select a vertex ν in derivation tree ρ randomly with uniform distribution on all vertices except the root node.

   (b) Delete the sub-tree rooted at ν.

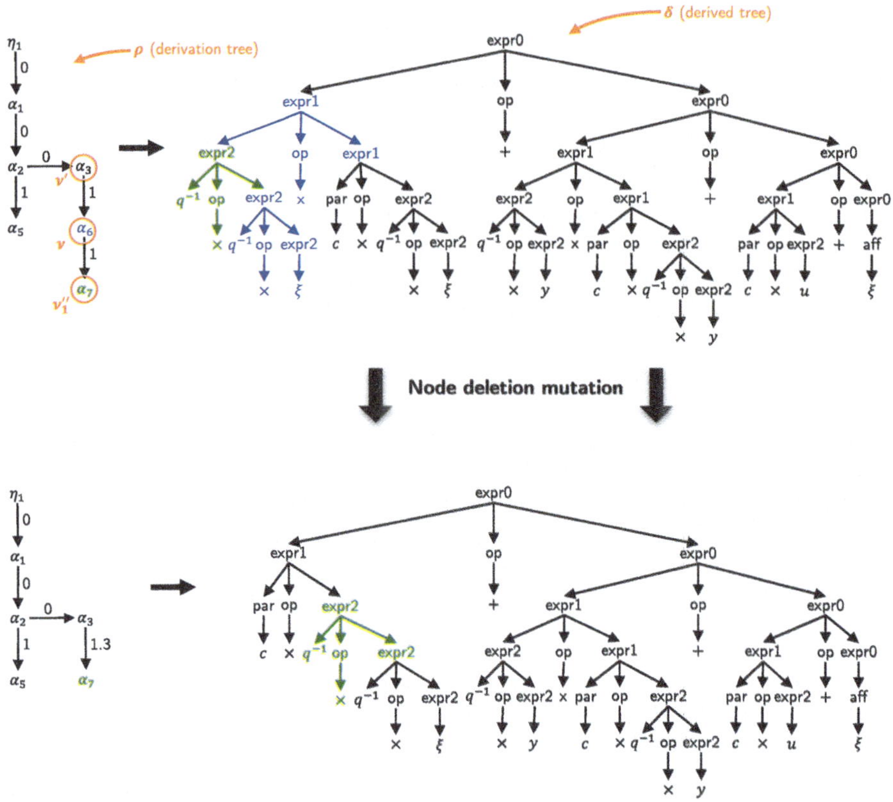

Fig. 6.6 The node deletion mutation operator illustrated on an individual generated using G_N. The illustration depicts the derivation tree ρ, the corresponding derived trees δ and vertex ν chosen for mutation. The coloured sections in the derivation trees and derived trees track the changes made in the individual

The two operations of deletion are illustrated in Figs. 6.6 and 6.7. The model corresponding to the parent individual in both examples is

$$y(k) = c_1 u(k) + c_2 y^2(k-1) + c_3 \xi(k-1)\xi(k-2) + \xi(k). \tag{6.18}$$

The result of the node deletion operator in Fig. 6.6 is

$$y(k) = c_1 u(k) + c_2 y^2(k-1) + c_3 \xi(k-2) + \xi(k). \tag{6.19}$$

In this example, the mutation operator deletes one of the noise factors $\xi(k-2)$ and adjoins the delay-type tree to the other noise term $\xi(k-1)$ in the model expression. The model obtained due to branch deletion operator in Fig. 6.5 is

$$y(k) = c_1 u(k) + c_2 y^2(k-1) + c_3 \xi(k-1) + \xi(k). \tag{6.20}$$

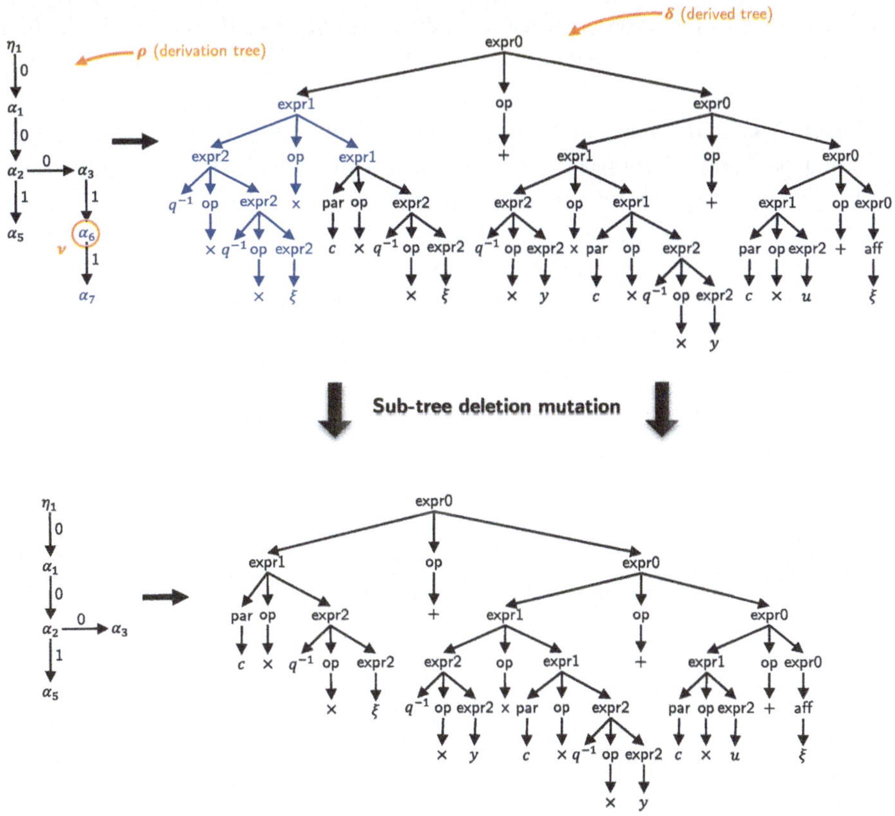

Fig. 6.7 The branch deletion mutation operator illustrated on an individual generated using G_N. The illustration depicts the derivation tree ρ, the corresponding derived trees δ and vertex ν chosen for mutation. The coloured sections in the derivation trees and derived trees track the changes made in the individual

In this example, the sub-tree corresponding to $\xi(k-2)$ is completely deleted (i.e., including the adjoining delay-type tree).

Selection and archiving

The selection mechanism adopted in the proposed algorithm plays a crucial role in the search for appropriate model structures in a multi-objective setting. The importance of the selection operator can be attributed to two factors: (i) the selection operator guides the evolutionary search, in a heuristic way, towards regions of the solution space that may contain "good" solutions, and (ii) the selection operator determines the goodness of a model structure based on the notion of Pareto-dominance allowing the proposed algorithm to address the upper-level optimization problems in (6.2), (6.4) or (6.5).

In the proposed algorithm, we employ the Pareto-based sorting and selection algorithm proposed in Deb et al. [7]. The selection mechanism relies on two ordering relations—Pareto-based ordering and crowding-distance-based ordering. In the first step, the individuals of a population are sorted based on Pareto-dominance relationships.

1. **Pareto-based ordering**: In this step, candidate solutions are ranked and grouped based on Pareto-dominance ordering. For an individual $M_i^j(\theta_i^j)$ in population $X^{(j)}$, the rank $r_{\mathrm{ND}}(M_i^j(\theta_i^j))$, based on non-dominated ordering is determined by the number of solutions in $X^{(j)}$ that Pareto-dominate $M_i^j(\theta_i^j)$. A model $M_i^j(\theta_i^j)$ that is not Pareto-dominated by any other model in $X^{(j)}$ is said to have rank $r_{\mathrm{ND}}(M_i^j(\theta_i^j)) = 1$. Similarly, a model $M_i^j(\theta_i^j)$ that is Pareto-dominated by only one other model in $X^{(j)}$ has rank 2, and so on. In iteration j, all solutions $M_i^j(\theta_i^j)$ in $X^{(j)}$ with $r_{\mathrm{ND}}(M_i^j(\theta_i^j)) = 1$ are grouped together as rank one solutions (denoted as $\mathcal{F}_1^{(j)}$), all solutions $M_i^j(\theta_i^j)$ with $r_{\mathrm{ND}}(M_i^j(\theta_i^j)) = 2$ are grouped as rank two solutions (denoted as $\mathcal{F}_2^{(j)}$), and so on. This can be expressed as follows[6]:

$$\begin{aligned}
\mathcal{F}_1^{(j)} &= \{M_i^j(\theta_i^j) \mid r_{\mathrm{ND}}(M_i^j(\theta_i^j)) = 1\} := \mathcal{E}(X^{(j)}, \prec_J),\\
\mathcal{F}_2^{(j)} &= \{M_i^j(\theta_i^j) \mid r_{\mathrm{ND}}(M_i^j(\theta_i^j)) = 2\} := \mathcal{E}(X^{(j)} \setminus \mathcal{F}_1^{(j)}, \prec_J),\\
&\vdots\\
\mathcal{F}_{n_s}^{(j)} &= \{M_i^j(\theta_i^j) \mid r_{\mathrm{ND}}(M_i^j(\theta_i^j)) = n_s\} := \mathcal{E}(X^{(j)} \setminus \mathcal{F}_1^{(j)} \setminus \cdots \setminus \mathcal{F}_{n_s-1}^{(j)}, \prec_J),
\end{aligned} \tag{6.21}$$

   Computation of the solution fronts described in (6.21) can be performed using the Pareto-dominance-based sorting algorithm presented in Algorithm 5.
2. **Crowding-distance-based ordering**: Since Pareto-based ranking allows multiple candidate solutions to occupy any given rank, the solutions within each rank group are further sorted using an estimate of the crowding distance with respect to other models in the same rank group. Crowding distance is a measure of the density of solutions surrounding a given candidate model. In Deb et al. [7], the authors estimate the crowding distance of a model by measuring the largest cuboid that includes the given model in objective space such that no other model in the rank group is included in the cuboid. This idea in illustrated in Fig. 6.8. The crowding-based sorting algorithm is described in Algorithm 6. The rank of a model $M_i^j(\theta_i^j)$ in a solution front $\mathcal{F}_l^{(j)}$ based on descending order of crowding-distance is denoted as $r_c(M_i^j(\theta_i^j))$. Hence, a model $M_{i_1}^j(\theta_{i_1}^j)$ in solution front $\mathcal{F}_l^{(j)}$ with the largest crowding distance is said to have rank $r_c(M_i^j(\theta_i^j)) = 1$, a model $M_{i_2}^j(\theta_{i_2}^j)$ in solution front $\mathcal{F}_l^{(j)}$ with the second-largest crowding distance is said to have rank $r_c(M_i^j(\theta_i^j)) = 2$, and so on.

Based on the two ordering procedures described in the prequel, the combined selection ordering $\prec_{\mathrm{sel}}$ on the set $X^{(j)}$ is obtained as the following

6 Note that, since the population $X^{(j)}$ contains n_s candidate solutions, there can maximally exist n_s solution fronts in each iteration.

Algorithm 5 Non-dominated sorting [7]

Require: Candidate solutions $\mathcal{M}^{(j)} = \{M_i^j(\theta_i^j)\}_{i=1}^{n_s}$
  Initialize $l \leftarrow 1$
  while $\mathcal{M}^{(j)} \neq \phi$ **do**
    $\mathcal{F}_l^{(j)} \leftarrow \phi$ ▷ Rank l solution front
    for each $M_i^j(\theta_i^j) \in \mathcal{M}^{(j)}$ **do**
      $\mathcal{F}_l^{(j)} \leftarrow \mathcal{F}_l^{(j)} \cup \{M_i^j(\theta_i^j)\}$ ▷ Include the individual temporarily in the solution front
      for each $M_q^j(\theta_q^j) \in \mathcal{F}_l \wedge M_q^j(\theta_q^j) \neq M_i^j(\theta_i^j)$ **do**
        if $M_i^j(\theta_i^j) \prec_J M_q^j(\theta_q^j)$ **then** ▷ If the newly added model dominates any other model in the solution front
          $\mathcal{F}_l^{(j)} \leftarrow \mathcal{F}_l^{(j)} \setminus \{M_q^j(\theta_q^j)\}$ ▷ remove the dominated model
        else if $M_q^j(\theta_q^j) \prec_J M_i^j(\theta_i^j)$ **then** ▷ If the newly added model is dominated by any other model in the solution front
          $\mathcal{F}_l^{(j)} \leftarrow \mathcal{F}_l^{(j)} \setminus \{M_i^j(\theta_i^j)\}$ ▷ Remove the newly added model
    $\mathcal{M}^{(j)} \leftarrow \mathcal{M}^{(j)} \setminus \mathcal{F}_l^{(j)}$ ▷ Remove all rank l models from the set of models yet to be sorted
    $l \leftarrow l + 1$
  return $\mathcal{F}_1^{(j)}, \ldots, \mathcal{F}_{l-1}^{(j)}$

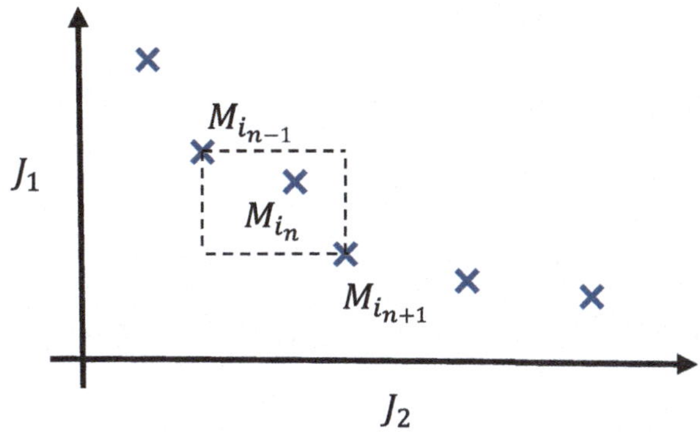

Fig. 6.8 Illustration of estimation of the crowding-distance based on the largest cuboid that can be drawn around individual M_{i_n} that does not include any other individual in the same rank

$$M_{i_1}^j(\theta_{i_1}^j) \prec_{\text{sel}} M_{i_2}^j(\theta_{i_2}^j) \Rightarrow \left(r_{\text{ND}}(M_{i_1}^j(\theta_{i_1}^j)) < r_{\text{ND}}(M_{i_2}^j(\theta_{i_2}^j))\right) \vee \left(\left(r_{\text{ND}}(M_{i_1}^j(\theta_{i_1}^j)) = r_{\text{ND}}(M_{i_2}^j(\theta_{i_2}^j))\right) \wedge \left(r_{\text{c}}(M_{i_1}^j(\theta_{i_1}^j)) < r_{\text{c}}(M_{i_2}^j(\theta_{i_2}^j))\right)\right). \quad (6.22)$$

The selection ordering, hence, selects one model over another if the former dominates the latter, or if the former is in the same solution front as the latter but has a greater crowding distance (i.e., occupies a sparser region in the objective space) than the latter. Selection of n_s individuals to be propagated to the next iteration takes place in Step 6 of Algorithm 1.

Additionally, in Step 7 of Algorithm 1, an archive of non-dominated solutions $\mathfrak{A}$ is updated with the newly generated candidate solutions. The archive $\mathfrak{A}$ records the set of all non-dominated solutions explored by GP upto the current iteration, i.e.

Algorithm 6 Crowding-distance-based sorting [7]

Require: Solution front $\mathcal{F} = \{M_i(\theta_i)\}_{i=1}^{n}$, number of performance measures n_{obj}
1: Initialize $l \leftarrow 1$
2: **for each** $M_i(\theta_i) \in \mathcal{F}$ **do**
3: Initialize $d(M_i(\theta_i) \leftarrow 0)$ ▷ Initialize the crowding distance of each model as 0
4: **while** $l \leq n_{\text{obj}}$ **do**
5: $I = (i_1, \ldots, i_n) \leftarrow \text{ORDERING}(\mathcal{F}, l)$ ▷ Compute the ordering of model indices that sorts the models based on performance measure J_l
6: $d(M_{i_1}(\theta_{i_1})) \leftarrow \infty$ ▷ Set crowding distance on models on the ends of the frontier as infinite
7: $d(M_{i_1}(\theta_{i_1})) \leftarrow \infty$
8: **for each** $i \in \{i_2, \ldots, i_{n-1}\}$ **do**
9: $d(M_i(\theta_i)) \leftarrow d(M_i(\theta_i)) + (J_l(M_{i+1}, \theta_{i+1}, \mathcal{D}_N) - J_l(M_{i-1}, \theta[i-1], \mathcal{D}_N))$ ▷ Compute the crowding distance of each individual as the sum of distances between the closest neighbours
10: $l \leftarrow l + 1$
11: $\mathcal{F} \leftarrow \text{SORT}(\mathcal{F}, \{d(M_i(\theta_i))\}_{i=1}^{n})$ ▷ Sort the solution front based on crowding distance of the individuals
12: **return** $\mathcal{F}$

$$\mathfrak{A}^{(j)} = \mathcal{E}(X^{(1)} \cup X^{(2)} \cup \cdots \cup X^{(j)}, \prec_J), \tag{6.23}$$

and can be computed iteratively as

$$\mathfrak{A}^{(j)} = \mathcal{E}(\mathfrak{A}^{(j-1)} \cup X^{(j)}, \prec_J). \tag{6.24}$$

For practical implementation, the archive size is limited to a finite number $m_{\text{a}} \in \mathbb{Z}_{>0}$ of individuals. If the size of the efficient set $\mathcal{E}(\mathfrak{A}^{(j-1)} \cup X^{(j)}, \prec_J)$ is larger than m_{a}, crowding-distance-based ordering is used to select m_{a} individuals from $\mathcal{E}(\mathfrak{A}^{(j-1)} \cup X^{(j)}, \prec_J)$.

6.3 Parameter Estimation for Grammar G_{N}

The description of the algorithm presented in Sect. 6.2 is on a fairly general level, and can be applied for arbitrary user-specified TAG and performance measures. It is worth noting that there is one crucial exception to the general applicability of the algorithm—the parameter estimation step. While the variation operators are defined based on fundamental concepts related to TAG and not on the specific TAG chosen by the user, the parameter estimation step, in general, is dependent on the model set generated by the chosen TAG. While general global optimization techniques such as CMA-ES or Differential Evolution (DE) can be used to solve the parameter estimation problem for a large variety of grammars, the flexibility comes at the cost of added computational complexity. This problem is acerbated by the fact that, in GP, the parameter estimation step must be performed for n_{s} candidate solutions in each of the L iterations.

In this section, we discuss optimization techniques for specific choices proposed in Chaps. 4 and 5. More specifically, we discuss optimization techniques for prediction error minimization (i.e. J_1) and simulation error minimization (i.e. J_2) for the class of P-NARMAX models generated by TAG G_N. Hence, in this section, we address the lower-level optimization problem in (6.5), where we chose the trade-off values $\beta_1 = 1$ and $\beta_2 = 0$.

6.3.1 Prediction Error Minimization

Any model structure M proposed by the GP algorithm can be written as

$$y(k) = \sum_{i=1}^{p} c_i \prod_{j=0}^{n_u} u^{b_{i,j}}(k-j) \prod_{q=1}^{n_\xi} \xi^{d_{i,q}}(k-q) \prod_{r=1}^{n_y} y^{a_{i,r}}(k-r) + \xi(k). \tag{6.25}$$

Based on model (6.25), we can obtain a one-step-ahead predictor model which is a non-linear filter formulated in terms of known input u_m, measured output y_m and the one-step-ahead prediction error $\varepsilon_\mathrm{p}(k)$. The one-step-ahead predictor model for P-NARMAX model (6.25) is given by:

$$\hat{y}(k \mid k-1) = \sum_{i=1}^{p} c_i \prod_{j=0}^{n_u} u_\mathrm{m}^{b_{i,j}}(k-j) \prod_{q=1}^{n_\xi} \varepsilon_\mathrm{p}^{d_{i,q}}(k-q) \prod_{r=1}^{n_y} y_\mathrm{m}^{a_{i,r}}(k-r). \tag{6.26}$$

where $\varepsilon_\mathrm{p}(k) := y_\mathrm{m}(k) - \hat{y}(k|k-1)$ is the one-step-ahead prediction error.

In the literature, several methods have been proposed to estimate model parameters in (6.26) for the squared prediction error cost function J_1. For an overview, see (Billings [4, Chap. 3]). For completeness, in this section, we present an iterative least-squares-based estimator to estimate model parameters in (6.26) for objective function J_1, see Billings and Voon [3] or (Billings [4, Chap. 3]) for more details. This approach is also known as pseudo-linear regression in the SI literature.

Let $\theta_\mathrm{p} = (c_1, \ldots, c_p)^\top$ be the parameter vector for the predictor model (6.26). The predictor (6.26) can be written in matrix form as

$$\hat{Y} = \phi_\mathrm{p}\theta_\mathrm{p} \tag{6.27}$$

where $\hat{Y} = (\hat{y}(n_1|n_1-1), \ldots, \hat{y}(N_\mathrm{est}|N_\mathrm{est}-1))^\top$ is the vector of one-step-ahead predicted outputs, $\phi_\mathrm{p} = (\varphi_{\mathrm{p},1}, \ldots, \varphi_{\mathrm{p},N_\mathrm{est}})^\top$ is the regression matrix with kth row being

$$\varphi_{\mathrm{p},k} = \left(\prod_{j=0}^{n_u} u_\mathrm{m}(k-j)^{b_{1,j}} \prod_{q=1}^{n_\xi} \varepsilon_\mathrm{p}(k-q)^{d_{1,q}} \prod_{r=1}^{n_y} y_\mathrm{m}(k-r)^{a_{1,r}}, \ldots, \right.$$
$$\left. \prod_{j=0}^{n_u} u_\mathrm{m}(k-j)^{b_{p,j}} \prod_{q=1}^{n_\xi} \varepsilon_\mathrm{p}(k-q)^{d_{p,q}} \prod_{r=1}^{n_y} y_\mathrm{m}(k-r)^{a_{p,r}} \right). \quad (6.28)$$

The regression matrix ϕ_p is a function of the unknown prediction error ε_p. Hence, to estimate the model parameters in (6.25), we use an identification procedure that iteratively estimates model parameters θ_p and prediction error ε_p.

Reformulate the predictor model in (6.26) as

$$\begin{aligned} \hat{y}(k \mid k-1) = & f_{uy}(u_\mathrm{m}(k-1), \ldots, u_\mathrm{m}(k-n_u), y_\mathrm{m}(k-1), \ldots, y_\mathrm{m}(k-n_y)) \\ & + f_{uy\xi}(u_\mathrm{m}(k-1), \ldots, u_\mathrm{m}(k-n_u), y_\mathrm{m}(k-1), \ldots, y_\mathrm{m}(k-n_y), \\ & \varepsilon_\mathrm{p}(k-1), \ldots, \varepsilon_\mathrm{p}(k-n_\xi)), \end{aligned} \quad (6.29)$$

where f_{uy} is the part of the model expression without any noise terms and $f_{uy\xi}$ is the part of the model with noise terms. Note that the decomposition in (6.29) is not unique. In order to obtain an initial estimate of the prediction error sequence, we estimate model parameters for the underlying P-NARX predictor model that can be computed by ignoring $f_{uy\xi}$ in (6.29). The P-NARX predictor model can also be written in matrix form

$$\hat{Y} = \phi_{uy}\theta_{uy}, \quad (6.30)$$

where the definitions of ϕ_{uy} and θ_{uy} is similar to the P-NARMAX case. The minimizer of J_1 for the predictor model in (6.30) can be computed as the least squares estimate $\hat{\theta}_{uy}$ of θ_{uy}, and is computed as

$$\hat{\theta}_{uy} = \left(\phi_{uy}^\top \phi_{uy}\right)^{-1} \phi_{uy}^\top \hat{Y}. \quad (6.31)$$

Using the estimate $\hat{\theta}_{uy}$, the initial estimate of the prediction error $\hat{\varepsilon}_\mathrm{p}^{(0)}$ can be computed as

$$\begin{aligned} \hat{\varepsilon}_\mathrm{p}^{(0)}(k) &= y_\mathrm{m}(k) - \hat{y}(k|k-1), \\ &= y_\mathrm{m}(k) - \varphi_{uy,k}\hat{\theta}_{uy}, \end{aligned} \quad (6.32)$$

where $\varphi_{uy,k}$ is the kth row of the regressor matrix.

The initial estimate of prediction error $\hat{\varepsilon}_\mathrm{p}^{(0)}$ can be used to begin the iterative estimation procedure. Let h be the iteration index. In each iteration, the parameter estimate $\hat{\theta}_\mathrm{p}^{(h)}$ can be computed as the linear least squares estimate for the predictor model in (6.27) where the prediction error terms in the regressor matrix ψ_p is replaced by the estimate $\hat{\varepsilon}_p^{(h-1)}$. Based on the new parameter estimate, the prediction error estimate can be updated as

$$\begin{aligned}\hat{\varepsilon}_p^{(h)}(k) &= y_m(k) - \hat{y}(k|k-1),\\ &= y_m(k) - \varphi_{p,k}\hat{\theta}_p^{(h)}.\end{aligned} \tag{6.33}$$

The iterations are stopped when the prediction error (or parameter) estimates converge within some user-specified tolerance level ϵ:

$$\sum_{k=1}^{N_{est}} \left(\hat{\varepsilon}_p^{(h)}(k) - \hat{\varepsilon}_p^{(h-1)}(k)\right)^2 < \epsilon. \tag{6.34}$$

The final estimate $\hat{\theta}_p$ minimizes the cost function J_1 for the grammar G_N. The estimate can be computed efficiently using a sequence of least squares estimators, and typically converges in a small number of iterations (typically less than 10 iterations, see Billings [4]). If the lower-level of the original bi-level problem in (6.2) is simplified as in (6.5), then $\hat{\theta}_p$ is one of the parameter estimates in the feasible set of the parameter space.

6.3.2 Simulation Error Minimization

For the simulation error minimization problem, we again assume that structure of the model proposed by the GP algorithm can be expressed as (6.25). Under the assumption that $\xi(k) \sim \mathcal{N}(0, \sigma_\xi)$, from Theorem 5.10 and Remark 5.6 we get that the $l-$approximate simulation model can be expressed as

$$y_{s,l}(k) = \sum_{i \in P_e} c_i \prod_{j=0}^{n_u+l} u^{b_{i,j}}(k-j) \prod_{r=1}^{n_y} y_{s,1}^{a_{i,r}}(k-l-r) \prod_{q=1}^{n_\xi+l} \sigma_\xi^{d_{i,q}}(d_{i,q}-1)!!. \tag{6.35}$$

In comparison with prediction error minimization, simulation-error-based estimation of P-NARMAX models is a far more challenging non-convex optimization problem. While the predictor model in (6.26) is formulated in terms of known past measured input and output, the simulation model in (6.35) is formulated in terms of the unknown past simulation output. This leads to complex non-linear dependencies between the parameters to be estimated and the simulation output. Nonetheless, the problem of simulation error minimization has been treated in the literature, for example for P-NARX models in Piroddi and Spinelli [25], for decoupled P-NARX models in Karami et al. [18] and for Polynomial Non-linear State-Space (pNLSS) models in Paduart et al. [23]. While the specific details of these approaches vary, the general idea is similar—the model parameters are optimized using a gradient-descent-based optimization technique, where various techniques are used to compute or approximate the Jacobian.

In Farina and Piroddi [10, 11] the authors proposed an approximation to simulation error minimization via an iterative multi-step-ahead prediction error scheme. The proposed Multi-step prediction (MSP)-based algorithm is demonstrated to be robust to initial conditions when compared to direct gradient-descent-based simulation error minimization. The authors also showed that the MSP algorithm achieves faster convergence when compared to other approaches for simulation error minimization (see Farina and Piroddi [11]). However, for the class of P-NARMAX models and P-NARX models that are *not* output-affine, the MSP algorithm proposed in Farina and Piroddi [10] leads to biased estimates. This is due to the fact that the higher-order moments of the noise distribution are not taken into account in the simulation model considered in Farina and Piroddi [10]. In Sect. 5.4, the l−approximate simulation model concept was developed to systematically incorporate the bias introduced due to the higher-order moments of the noise in the computation of the simulation response. This allows us to develop an MSP-based algorithm for the larger class of P-NARMAX models.

In the following, we present (i) a Gauss-Newton-based method, and (ii) a MSP-based method to solve the non-convex simulation error minimization problem for the class of P-NARMAX models.

Direct gradient-based optimization

In this section, we describe the computations involved in computing the simulation-error parameter estimate of (6.35) using the Gauss-Newton method. A similar computational scheme was devised in Paduart et al. [23] for the class of pNLSS models.

Let $P_{\mathrm{e}} = \{i_1, \dots, i_p\}$ be the indices of model terms in (6.35) and let $\theta_{\mathrm{s},l} = (c_{i_1} \cdots c_{i_p})^\top$. Introduce $\beta(\theta_{\mathrm{s},l}) = (\beta_{i_1} \cdots \beta_{i_p})^\top$ where $\beta_i := c_i \prod_{q=1}^{n_\xi+l} \sigma_\xi^{d_{i,q}} (d_{i,q} - 1)!!$, and let $\varphi_i(k, \theta_{\mathrm{s},l}) := \prod_{j=0}^{n_u+l} u^{b_{i,j}}(k-j) \prod_{r=1}^{n_y} y_{\mathrm{s},l}^{a_{i,r}}(k-l-r)$. The simulation output $y_{\mathrm{s},l}(k)$ can be expressed as

$$
\begin{aligned}
y_{\mathrm{s},l}(k) &= \varphi(k, \theta_{\mathrm{s},l}) \beta(\theta_{\mathrm{s},l}), \\
&:= \left(\varphi_{i_1}(k, \theta_{\mathrm{s},l}) \cdots \varphi_{i_p}(k, \theta_{\mathrm{s},l})\right) \beta(\theta_{\mathrm{s},l}).
\end{aligned} \tag{6.36}
$$

The mean squared simulation error of (6.35) is given by

$$
J_2(M, \theta_{\mathrm{s},l}, \mathcal{D}_{\mathrm{est}}) = \frac{1}{N} \sum_{k=n_l+1}^{N_{\mathrm{est}}} \left(y_{\mathrm{m}}(k) - y_{\mathrm{s},l}(k, \theta_{\mathrm{s},l})\right)^2. \tag{6.37}
$$

Let $\hat{Y}_{\mathrm{s},l}(\theta_{\mathrm{s},l}) = \{y_{\mathrm{s},l}(k, \theta_{\mathrm{s},l})\}_{k=1}^{N_{\mathrm{est}}}$ and let ∇J_2 denote the Jacobian of (6.37), i.e., $\nabla J_2 = \frac{\partial}{\partial \theta_{\mathrm{s},l}} J_2(M, \theta_{\mathrm{s},l}, \mathcal{D}_{\mathrm{est}})$. Then, the simulation error parameter estimate $\hat{\theta}_{\mathrm{s},l}$ can be computed using the Gauss-Newton method, starting from an initial estimate $(\hat{\theta}_{\mathrm{s},l})^0$, with the following iterations

$$(\hat{\theta}_{s,l})^j = (\hat{\theta}_{s,l})^{j-1} - \left((\nabla J_2)^\top (\nabla J_2)\right)^{-1} (\nabla J_2)^\top \hat{Y}_{s,l}. \tag{6.38}$$

To perform the iterations in (6.38), we need to compute the Jacobian ∇J_2 for a given parameter estimate $\hat{\theta}_{s,l}$. In the following, we discuss the computation of the Jacobian.

The Jacobian is given by

$$\begin{aligned}
\nabla J_2 &= \frac{\partial}{\partial \theta_{s,l}} J_2(M, \theta_{s,l}, \mathcal{D}_{\text{est}})\bigg|_{\theta_{s,l}=\hat{\theta}_{s,l}}, \\
&= -\frac{2}{N} \sum_{k=n_l+1}^{N_{\text{est}}} \left(y_{\text{m}}(k) - y_{s,l}(k, \theta_{s,l})\right) \frac{\partial}{\partial \theta_{s,l}} y_{s,l}(k, \theta_{s,l})\bigg|_{\theta_{s,l}=\hat{\theta}_{s,l}}
\end{aligned} \tag{6.39}$$

Define $\alpha(k, \theta_{s,l}) := \frac{\partial y_{s,l}}{\partial \theta_{s,l}}$. The Jacobian of the simulated output, given by $\alpha(k, \theta_{s,l})$, can be computed using the chain rule in (6.36), as follows.

$$\begin{aligned}
\alpha(k, \theta_{s,l}) &= \varphi(k, \theta_{s,l}) \frac{\partial \beta(\theta_{s,l})}{\partial \theta_{s,l}} + \frac{\partial \varphi(k, \theta_{s,l})}{\partial \theta_{s,l}} \beta(\theta_{s,l}), \\
&= \varphi(k, \theta_{s,l}) \frac{\partial \beta(\theta_{s,l})}{\partial \theta_{s,l}} + \left(\frac{\partial \varphi_{i_1}(k, \theta_{s,l})}{\partial \theta_{s,l}} \cdots \frac{\partial \varphi_{i_p}(k, \theta_{s,l})}{\partial \theta_{s,l}} \right) \beta(\theta_{s,l}).
\end{aligned} \tag{6.40}$$

Recall that $\beta(\theta_{s,l}) = (\beta_{i_1} \cdots \beta_{i_p})^\top$ where $\beta_i = c_i \prod_{q=1}^{n_\xi+l} \sigma_\xi^{d_{i,q}} (d_{i,q} - 1)!!$. Hence, in (6.40), the term $\frac{\partial \beta(\theta_{s,l})}{\partial \theta_{s,l}}$ can be easily computed for a given estimate $\hat{\theta}_{s,l}$, since

$$\frac{\partial}{\partial \theta_{s,l}} \beta_i = \prod_{q=1}^{n_\xi+l} \sigma_\xi^{d_{i,q}} (d_{i,q} - 1)!! \, \mathbf{1}_i, \tag{6.41}$$

where $\mathbf{1}_i \in [0, 1]^{\dim(\theta_{s,l})}$ is the indicator vector such that the ith component of the vector is 1, and all other components are 0.

Remark 6.2 Note that, in order to compute (6.41), we can use the standard deviation of prediction error ε_{p} as an estimate of σ_ξ. However, this estimate will be accurate only when the predictor model accurately describes the dynamics of the true system. Alternatively, we can incorporate the contributions of σ_ξ in $\theta_{s,l}$, and subsequently estimate $\theta_{s,l} = (c_{i_1} \prod_{q=1}^{n_\xi+l} \sigma_\xi^{d_{i_1,q}} \cdots c_{i_p} \prod_{q=1}^{n_\xi+l} \sigma_\xi^{d_{i_p,q}})^\top$.

Let $R = \{1, 2, \ldots, n_y\}$. Since $\varphi(k, \theta_{s,l})$ is a vector of polynomial regressor terms, the term $\frac{\partial \varphi_i(k, \theta_{s,l})}{\partial \theta_{s,l}}$ in (6.40) can be computed as follows:

$$\frac{\partial \varphi_i(k, \theta_{s,l})}{\partial \theta_{s,l}} = \prod_{j=0}^{n_u+l} u^{b_{i,j}}(k-j) \left(\sum_{\bar{r}=1}^{n_y} \left(\prod_{r \in R \setminus \bar{r}} y_{s,l}^{a_{i,r}}(k-l-r) \right. \right.$$
$$\left. \left. a_{i,\bar{r}} y_{s,l}^{a_{i,\bar{r}}-1}(k-l-\bar{r})\, \alpha(k-l-\bar{r}) \right) \right). \tag{6.42}$$

By combining (6.40), (6.41), and (6.42), we get the equations necessary to compute the Jacobian of the simulated output $y_{s,l}(k)$. Observe that the Jacobian is formulated as a recursive filter for $\{\alpha(k)\}_{k=n_l+1}^{N_{est}}$. Hence, in order to compute the Jacobian, we need an estimate of the initial conditions $\{\alpha(k)\}_{k=1}^{n_l}$. This can be done by including the initial conditions as unknown parameters to be estimated in the non-linear optimization problem. Alternatively, if the Jacobian filter is asymptotically stable, the initial conditions can be assumed to be 0, and the consequent exponentially decaying transient error can be ignored by incorporating a burn-in period in the computation of J_2.

The Multi-step Prediction algorithm for P-NARMAX models

The direct gradient-descent-based approach described in the previous section may converge to local minima, and furthermore, may be slow to converge. To alleviate such problems, we also develop an MSP-based approach to solving the simulation error minimization problem for P-NARMAX models. Unlike Farina and Piroddi [10], the MSP algorithm presented in this section is applicable to the entire class of P-NARMAX models due to the $l-$approximate simulation concept introduced in Sect. 5.4.

In order to describe the concept of the MSP algorithm, we recall some concepts and notations introduced in Chap. 5. Recall that the simulation model, as defined in (5.7), of (6.25) can be seen as the infinite-step-ahead predictor of the model. Also recall the in Sect. 5.4, we proposed the $l-$approximate simulation model concept, that makes it possible to compute an approximate simulation response for P-NARMAX models. The l—approximate simulation model of (6.25) is given by

$$y_{s,l}(k) = f_{s,l}\left(U_0^{n_u+l}(k), {Y_{s,l}}_{l+1}^{n_y+l}(k)\right), \tag{6.43}$$

where $f_{s,l}$ is a polynomial function in terms of past inputs $U_0^{n_u+l}(k) = \{u(k-i)\}_{i=0}^{n_u+l}$ and past simulated outputs ${Y_{s,l}}_{l+1}^{n_y+l}(k) = \{y_{s,l}(k-i)\}_{i=0}^{n_u+l}$. The exact form of $f_{s,l}$ can be found in Theorem 5.10. Let the l—approximate simulation model is parameterized by parameter vector $\theta_{s,l}$.

The basic concept of the MSP algorithm is as follows. Define, for some $\tau \in \mathbb{Z}_{>0}$, such that $\tau \geq l$, the τ-step-ahead predictor of the l—approximate simulation model (6.43), as follows

$$\hat{y}(k|k-\tau) = \hat{f}_\tau\left({U_m}_0^{n_u+\tau}(k), {Y_m}_{\tau+1}^{n_y+\tau}(k)\right), \tag{6.44}$$

where, $\hat{y}(k|k-\tau)$ is the τ-step-ahead prediction, $\hat{f}_\tau$ is the τ-set-ahead predictor (computational aspects will be discussed in the sequel), and ${Y_m}_{\tau+1}^{n_y+\tau}(k) = \{y_m(k-i)\}_{i=\tau+1}^{n_y+\tau}$ is the set of past measured outputs.

Remark 6.3 The predictor (6.44) is the τ-step-ahead predictor of the l-approximate simulation response in (6.43) and should not be confused with the predictor of the P-NARMAX model in (6.25).

The predictor (6.44) allows us to approximate $l-$approximate simulation (6.43) using measured data. As $\tau \to \infty$, the τ-step-ahead prediction output approximates arbitrarily well the $l-$approximate simulation response $y_s(l)$, i.e.,

$$\lim_{\tau \to \infty} \hat{y}(k|k-\tau) = y_{s,l}(k). \tag{6.45}$$

Hence, in the MSP algorithm, the simulation model parameters $\theta_{s,l}$ is approximated by a sequence of parameter estimates $(\hat{\theta}_{s,l})_\tau$ for $\tau = l, l+1, \ldots \infty$, that optimize the cost function

$$(\hat{\theta}_{s,l})^*_\tau = \arg\min J_\tau(\theta_{s,l}), \tag{6.46}$$

where $J_\tau(\theta_{s,l})$ is the 2-norm of the squared τ-step-ahead prediction error cost function. In practice, a stopping criteria based on convergence of parameter estimates or maximum number of iterations is used.

Remark 6.4 In principle, the prediction model 6.44 can be computed by recursively replacing the output terms model expression in (6.43), and re-formulating the subsequent model as a non-linear filter in terms of measured inputs and outputs. However, since the prediction model in (6.44) is polynomial, performing multiple recursive substitutions of the output terms may result in polynomial growth of the order and the length of the model, making it infeasible to compute for large values of τ. A feasible approach for computing the τ-step-ahead predictor is discussed in the following subsection.

Remark 6.5 The role of parameter τ in the MSP algorithm is similar to the role of parameter l in the $l-$approximation method for computing the simulation model. The crucial difference between the two is that the $l-$approximation scheme determines the accuracy of a *simulation model*, while parameter τ ultimately determines the accuracy of a *prediction model*.

In the following, we discuss the computational structure of the MSP algorithm. For $\tau = l$, the predictor model of (6.43) can be written in matrix form as

$$\hat{Y}_l(\theta_{s,l}) = \phi_l \theta_{s,l}, \tag{6.47}$$

where $\hat{Y}_l = (\hat{y}(l+1|1), \ldots, \hat{y}(N_{\text{est}}|N_{\text{est}} - l))^\top$ is the vector of l-step-ahead predicted outputs, $\phi_l = (\varphi_l(l+1), \ldots, \varphi_l(N_{\text{est}}))^\top$ is the regression matrix. Rows $\varphi_l(k)$ consist of monomials in u_m and y_m and can be derived from (6.35). Computation of ϕ_τ is discussed in detail in the following section. Similarly, the matrix-form of (6.44) for $\tau > l$ can be computed by recursively substituting output terms in (6.44), and can be written as

$$\hat{Y}_\tau(\theta_{s,l}) = \phi_\tau \vartheta_\tau(\theta_{s,l}), \tag{6.48}$$

where $\vartheta_\tau(\theta_{s,l})$ is the non-linear parameterization (due to the recursive substitutions of the output terms) of the τ-step-ahead predictor model w.r.t $l-$approximate simulation model parameters $\theta_{s,l}$, and ϕ_τ is the corresponding regressor matrix.

For a given τ, the optimization problem (6.46) can written as

$$(\hat{\theta}_{s,l})^{*}_{\tau} = \arg\min \frac{1}{N_{\text{est}} - \tau}(Y_{\text{m}}(\tau) - \hat{Y}_{\tau}(\theta_{s,l}))^{\top}(Y_{\text{m}}(\tau) - \hat{Y}_{\tau}(\theta_{s,l})), \tag{6.49}$$

where $Y_{\text{m}}(\tau) = (y_{\text{m}}(\tau+1), \ldots, y_{\text{m}}(N_{\text{est}}))^{\top}$ is a vector of measured outputs of the estimation dataset $\mathcal{D}_{\text{est}}$. Gauss-Newton method is used to solve the non-linear least squares problem in (6.49).

The MSP algorithm consists of two loops—the outer loop iterates over prediction time-horizon τ, while the inner loop corresponds to Gauss-Newton iterations required for solving the optimization problem in (6.49) for a fixed time horizon τ. Denote the parameter estimate for the ith Gauss-Newton iteration corresponding to time horizon τ as $(\hat{\theta}_{s,l})^{i}_{\tau}$. The algorithm is initialized with the linear least squares estimate $(\hat{\theta}_{s,l})^{*}_{l}$ (see (6.47)). Set $\tau = l + 1$ and $(\hat{\theta}_{s,l})^{0}_{\tau} = (\hat{\theta}_{s,l})^{*}_{\tau-1}$. For each time horizon τ, the Jacobian is given by

$$\begin{aligned}
\nabla_{\theta_{s,l}} J_{\tau}(\theta_{s,l}) &= \frac{2}{N_{\text{est}} - \tau}\left(\hat{Y}_{\tau}(\theta_{s,l})^{\top} - Y_{\text{m}}(\tau)^{\top}\right) \overbrace{\nabla_{\theta_{s,l}} \hat{Y}_{\tau}(\theta_{s,l})}^{\Psi_{\tau}(\theta_{s,l}):=} \\
&= \frac{2}{N_{\text{est}} - \tau}\left(\hat{Y}_{\tau}(\theta_{s,l})^{\top} - Y_{\text{m}}(\tau)^{\top}\right) \Psi_{\tau}(\theta_{s,l}).
\end{aligned} \tag{6.50}$$

The computation of Ψ_{τ} is discussed in the following subsection. The Gauss-Newton iterations for a given time-horizon τ is given by

$$\begin{aligned}
(\hat{\theta}_{s,l})^{i+1}_{\tau} = (\hat{\theta}_{s,l})^{i}_{\tau} + \left(\Psi_{\tau}((\hat{\theta}_{s,l})^{i}_{\tau})^{\top}\Psi_{\tau}((\hat{\theta}_{s,l})^{i}_{\tau})\right)^{-1} \Psi_{\tau}((\hat{\theta}_{s,l})^{i}_{\tau})^{\top} \\
\left(Y_{\text{m}}(\tau) - \hat{Y}_{\tau}((\hat{\theta}_{s,l})^{i}_{\tau})\right)
\end{aligned} \tag{6.51}$$

The Gauss-Newton method is said to have converged when

$$|J_{\tau}((\hat{\theta}_{s,l})^{i}_{\tau}) - J_{\tau}((\hat{\theta}_{s,l})^{i-1}_{\tau})| \leq \epsilon_{s}, \tag{6.52}$$

where ϵ_{s} is a user-specified tolerance threshold. Alternatively, the Gauss-Newton algorithm terminates after reaching a maximum number of iterations (to be specified by the user). The converged parameter estimate for time horizon τ is denoted by $(\hat{\theta}_{s,l})^{*}_{\tau}$. After convergence of the inner loop, set $\tau = \tau + 1$ and initialize $(\hat{\theta}_{s,l})^{0}_{\tau} = (\hat{\theta}_{s,l})^{*}_{\tau-1}$, and repeat the Gauss-Newton method. The outer loop is terminated when the following criterion is satisfied

$$|J_{\tau}((\hat{\theta}_{s,l})^{8}_{\tau}) - J_{\tau}((\hat{\theta}_{s,l})^{*}_{\tau-1})| \leq \epsilon_{s}, \tag{6.53}$$

or when the outer loop exceeds the maximum number of iterations. The final parameter estimate is denoted as $\theta_{s,l}[][*]$.

Computation of the multi-step-ahead prediction output and its Jacobian

Computation of the multi-step-ahead prediction model in (6.44) for large time-horizons may be infeasible since the model expression may grow polynomially in length and degree due to the numerous recursive substitutions. In order to circumvent this problem, the authors in Farina and Piroddi [11] proposed an iterative scheme to compute the multi-step-ahead prediction output $\hat{Y}_\tau(\theta_{s,l})$ and its Jacobian $\Psi_\tau(\theta_{s,l})$. In this approach, for a given parameter vector $\hat{\theta}_{s,l}$, the τ-step-ahead prediction output $\hat{Y}_\tau(\hat{\theta}_{s,l})$ and Jacobian $\Psi_\tau(\hat{\theta}_{s,l})$ are computed recursively based on $\hat{Y}_l(\hat{\theta}_{s,l})$, $\hat{Y}_{l+1}(\hat{\theta}_{s,l}), \ldots, \hat{Y}_{\tau-1}(\hat{\theta}_{s,l})$.

For a given parameter vector $\hat{\theta}_{s,l}$, the $l-$step-ahead predicted output is given by the linear-in-the-parameters formulation

$$\hat{y}(k|k-l) = \varphi_l(k)^\top \hat{\theta}_{s,l}. \tag{6.54}$$

where $\varphi_l(k)$, $(k > l)$ are the rows of matrix ϕ_l and contain monomials in u_m and y_m. Hence, $\hat{Y}_l(\hat{\theta}_{s,l})$ can be computed as a matrix inner-product. Based on the $\hat{Y}_l(\hat{\theta}_{s,l})$, the $(l+1)$-step-ahead prediction output can be expressed as

$$\hat{y}(k|k-l-1) = \varphi_{l+1}(k, \hat{\theta}_{s,l})^\top \hat{\theta}_{s,l}, \tag{6.55}$$

where $\varphi_{l+1}(k, \hat{\theta}_{s,l})$, $(k > l+1)$ contains monomials in $u_m(k), \ldots, u_m(k - n_u - l - 1)$, $y_m(k-l-1), \ldots, y_m(k-l-n_y)$, and $\hat{y}(k-l|k-l-1)$, and can be computed from $\varphi_l(k)$ using the following substitution

$$\varphi_{l+1}(k, \hat{\theta}_{s,l}) = \varphi_l(k)|_{\forall j<(l+1): y_m(k-j)=\hat{y}(k-j|k-l-1)}. \tag{6.56}$$

Observe that, although the $(l+1)$-step-ahead prediction $\hat{y}(k|k-l-1)$ depends non-linearly on $\hat{\theta}_{s,l}$, the non-linear dependency is hidden in the regressor $\varphi_{l+1}(k)$, which can be computed if $\varphi_l(k)$ is known. Hence, using $\hat{Y}_l(\hat{\theta}_{s,l})$, predicted outputs $\hat{Y}_{l+1}(\hat{\theta}_{s,l})$ can be computed as a matrix inner-product.

This concept can be generalized to compute the τ-step-ahead prediction output $\hat{y}(k|k-\tau)$ iteratively, through the following sequence of matrix inner-products

$$\begin{aligned}
\hat{y}(k|k-l) &= \varphi_l(k)^\top \hat{\theta}_{s,l}, \\
\hat{y}(k|k-l-1) &= \varphi_{l+1}(k, \hat{\theta}_{s,l})^\top \hat{\theta}_{s,l}, \\
&\vdots \\
\hat{y}(k|k-\tau) &= \varphi_\tau(k, \hat{\theta}_{s,l})^\top \hat{\theta}_{s,l},
\end{aligned} \tag{6.57}$$

where $\varphi_\tau(k, \hat{\theta}_{s,l})$ is derived from $\varphi_l(k), \ldots, \varphi_l(k, \hat{\theta}_{s,l})$ by performing the following substitution

$$\varphi_\tau(k) = \varphi_l(k)|_{\forall j<\tau: y_m(k-j)=\hat{y}(k-j|k-\tau)}. \tag{6.58}$$

The variable substitution in (6.58) replaces measured output terms $y_m(k-j)$ in $\varphi_l(k)$ that lie within the prediction horizon $k-(\tau+1), \ldots, k$ with predicted output terms $\hat{y}(k-j|k-\tau)$. This implies that the vector $\varphi_\tau(k)$ consists of monomials in $u_m(k), \ldots, u_m(k-n_u-l), y_m(k-\tau), \ldots, y_m(k-l-n_y)$, and $\hat{y}(k-l-1|k-\tau), \ldots, \hat{y}(k-\tau+1|k-\tau)$. In this manner, performing numerous recursive symbolic substitutions in the non-linear model to compute (6.44) can be avoided.

For a given $\hat{\theta}_{s,l}$, the Jacobian of the predicted outputs can also be computed iteratively, as follows. Let $\Psi_{\tau,k}$ denote the Jacobian of the predicted output $\hat{y}(k|k-\tau)$, i.e.,

$$\Psi_{\tau,k} = \nabla_{\theta_{s,l}} \hat{y}(k|k-\tau). \tag{6.59}$$

Since the derivatives of measured inputs u_m and outputs y_m with respect to $\theta_{s,l}$ is 0, we can compute the Jacobian of the l-step-ahead predicted as

$$\Psi_{l,k} = \varphi_l(k)^\top. \tag{6.60}$$

For $\tau > l$, we get

$$\Psi_{l+1,k} = \varphi_{l+1}(k, \hat{\theta}_{s,l})^\top + \hat{\theta}_{s,l}^\top \nabla_{\theta_{s,l}} \varphi_{l+1}(k, \theta_{s,l})\big|_{\theta_{s,l}=\hat{\theta}_{s,l}},$$

$$\vdots$$

$$\Psi_{\tau,k} = \varphi_\tau(k, \hat{\theta}_{s,l})^\top + \hat{\theta}_{s,l}^\top \nabla_{\theta_{s,l}} \varphi_\tau(k, \theta_{s,l})\big|_{\theta_{s,l}=\hat{\theta}_{s,l}}. \tag{6.61}$$

Recall that $\varphi_{\tau,\hat{\theta}_{s,l}}(k)$ consists of monomials in $u_m(k), \ldots, u_m(k-n_u-l), y_m(k-\tau), \ldots, y_m(k-l-n_y)$, and $\hat{y}(k-l-1|k-\tau), \ldots, \hat{y}(k-\tau+1|k-\tau)$. Hence, $\Psi_{\tau,k}$ can be computed iteratively since for all $j \in [l+1, \tau-1]$, we have $\nabla_{\theta_{s,l}} \hat{y}(k-j|k-\tau) = \Psi_{\tau-j,k-j}$.

6.4 Analysis of the Algorithm

In this section, we analyse the proposed identification approach along two lines—the asymptotic behaviour of the proposed algorithm and the computational complexity of the same.

The identification methodology presented in Sect. 6.2 was formulated in a general setting for an arbitrary TAG G and arbitrary performance measures. To make the analysis concrete and tractable, we make the following choices:

- The model set of the identification problem is specified by TAG G_N, which generates the class of P-NARMAX models,

- The general multi-criteria identification problem in (6.2) is simplified to the form of (6.5), i.e., the lower-level multi-objective optimization problem is simplified to multiple single-objective optimization problem. The necessity of such a simplification is discussed in Sect. 6.1.2.
- The multi-objective optimization problem (6.5) is formulated in terms of the four performance measures proposed in Chap. 5, i.e., $J(M, \theta, \mathcal{D}_N)=\{J_i(M, \theta, \mathcal{D}_N)\}_{i=1}^4$, and $J_{\text{cont}}(M, \theta, \mathcal{D}_N) = \{J_i(M, \theta, \mathcal{D}_N)\}_{i=1}^2$.

6.4.1 Asymptotic Analysis

In this section, we perform an asymptotic analysis of the proposed algorithm. The focus of this analysis is not on the internal details of the population being evolved, but on the evolution of the population as a whole. The analysis presented here is based on work presented in Rudolph and Agapie [28], and is extended to incorporate parameter estimation in Step 4 of Algorithm 1.

Recall that a maximum limit of m_d is imposed on the depth of the trees explored by GP. Let the set of models with corresponding derived tree-depth less than m_d be $\mathcal{M}_\text{d}(G_\text{N}) \subset \mathcal{M}(G_\text{N})$. Let $\mathcal{F}^*$ be the efficient set of solutions in the model set $\mathcal{M}_\text{d}(G_\text{N})$ with respect to the multi-criteria identification problem in (6.5).

In order to understand the asymptotic behaviour of the proposed algorithm, we ask ourselves the following sequence of questions:

A. How likely is it for GP to propose candidate model structures that correspond to solutions that belong to the Pareto-efficient set $\mathcal{F}^*$?
B. Do the parameter estimation algorithms discussed in Sect. 6.3 converge to the global minima of their corresponding objective functions?
C. If a model $M(\theta) \in \mathcal{F}^*$ is successfully identified in generation j of the algorithm, can it successfully be retained in archive $\mathfrak{A}^{(j)}$ in the limit $j \to \infty$?

The three questions listed above will demarcate the *three stages of the analysis* presented in the sequel. However, before we begin, it is useful to define the notions of (i) convergence for the multi-criteria identification problem and (ii) persistence of excitation for the generalized model set specified by a TAG.

We study the evolution of the archive of estimated Pareto-efficient solutions $\mathfrak{A}^{(j)}$ as the number of iterations j increases. The distance between the estimated Pareto-efficient set $\mathfrak{A}^{(l)}$ to the true Pareto-efficient set $\mathcal{F}^*$ is measured using the following metric

$$\delta_{\mathcal{F}^*}(\mathfrak{A}^{(j)}) = \left|\mathfrak{A}^{(j)}\right| - \left|\mathfrak{A}^{(j)} \cap \mathcal{F}^*\right|, \tag{6.62}$$

where $|A|$ is the cardinality of set A. The metric corresponds to the number of models in the estimated Pareto-front $\mathfrak{A}^{(j)}$ that does not belong to the true Pareto-front $\mathcal{F}^*$. Note that while the set $\mathcal{F}^*$ may be uncountably infinite, the set $\mathfrak{A}^{(j)} \cap \mathcal{F}^*$ is, at most, finite because the size of the archive is limited to $\left|\mathfrak{A}^{(j)}\right| < m_\text{a}$.

Based on the distance metric in (6.62), we can define the notion of convergence for the multi-criterion identification algorithm as follows.

Definition 6.1 The estimated Pareto-front $\mathfrak{A}^{(j)}$ of the multi-criteria identification algorithm in Algorithm 1 is said to converge to $\mathcal{F}^*$ if,

$$\lim_{j\to\infty} \delta_{\mathcal{F}^*}(\mathfrak{A}^{(j)}) \to 0, \qquad \text{w.p. } 1. \tag{6.63}$$

Remark 6.6 Observe that the notion of convergence proposed in Definition 6.1 avoids the issue arising due to the fact that the cardinality of set $\mathcal{F}^*$ is typically larger than that of $\mathfrak{A}^{(j)}$. The notion of convergence proposed here only requires the estimated Pareto-efficient set $\mathfrak{A}^{(j)}$ to be a subset of $\mathcal{F}^*$.

To develop an asymptotic analysis of the parameters estimated in the proposed identification method, we also need to introduce a notion of *persistence of excitation* w.r.t to the model set $\mathcal{M}_\text{d}(G)$. Due to the fact that the TAG-based model set may span models of varying complexity, it is possible to have multiple models in the model set that are indistinguishable under the given set of performance measures for any input $u \in \mathcal{U}$, i.e., the model set $\mathcal{M}_\text{d}(G)$ is not *identifiable* under the specified performance measures. However, observe that, while the model set of the overall identification method is $\mathcal{M}_\text{d}(G)$, the parameters estimation step only functions at the lower-level of the bi-level optimization problem in (6.5), where the model structure M is fixed. This observation leads us to the following weaker notion of persistence of excitation.

Definition 6.2 Let u be a quasi-stationary signal, $M \in \mathcal{M}_\text{d}(G)$ be a model structure parameterized by $\theta \in \Theta(M)$, where domain Θ is determined by the structure of M, and let $J_{\text{cont},\theta}(u)$ be the mapping from input u to the image of J_cont for a given θ. Input u is *weakly persistently exciting* for the model set $\mathcal{M}_\text{d}(G)$ if, for any $M \in \mathcal{M}_\text{d}(G)$, and for all θ_1, θ_2 in the corresponding parameter space $\Theta(M)$ for which $J_{\text{cont},\theta_1}(\cdot) \neq J_{\text{cont},\theta_2}(\cdot)$, input u satisfies

$$J_{\text{cont},\theta_1}(u) \neq J_{\text{cont},\theta_2}(u) \tag{6.64}$$

The notion of persistence of excitation introduced here ensures that for any two models in the model set that *can* be differentiated (i.e., do not lead to the identical mappings from $\mathcal{U}$ to the image of J_cont), there exists at least one input sequence $u \in \mathcal{U}$ that differentiates the two models.

In order to develop an asymptotic analysis of the model estimates obtained from Algorithm 1, we make the following assumption.

Assumption 6.1 The conceptual infinitely long dataset $\mathcal{D}_\infty$ is obtained by exciting the true system S by a weakly persistently exciting input $\{u(k)\}$ for the model set $\mathcal{M}_\text{d}(G_\text{N})$.

Equipped with the notions of convergence and persistence of excitation, we now present the three stages of the asymptotic analysis of the proposed identification framework (based on the three questions listed in the prequel).

Stage A

In this part of the analysis, we study the behaviour of GP population $\{X^{(j)}\}$ as a finite Markov process. Due to the imposed limit m_{d} on the depth of trees generated from G_{N}, the maximum number of (possibly non-unique) model structures that can be generated, in the worst case,[7] is $n_{\mathrm{T}} = |\mathcal{I}| \times \frac{1-|\mathcal{A}|^{m_{\mathrm{d}}}}{1-|\mathcal{A}|}$, where $|\mathcal{I}|$ and $|\mathcal{A}|$ are the number of initial and auxiliary trees in G_{N}, respectively. In each iteration, the proposed identification algorithm proposes a population $X^{(j)}$ from the set of all possible population of model structures $S_{\mathrm{M}} := \mathcal{M}_{\mathrm{d}}(G_{\mathrm{N}})^{n_{\mathrm{s}}}$. The population $X^{(j)}$ is obtained entirely from the population $X^{(j-1)}$ via the crossover, mutation and selection operators. Moreover, the stochastic choices made in the crossover and mutation steps remain independent over the number of iterations. Hence, the evolution of the population X can be described as a time-homogeneous Markov process, since the probability distribution of $X^{(j)}$ is completely determined by $X^{(j-1)}$, i.e.,

$$p\left(X^{(j)} \mid X^{(j-1)}, \ldots, X^{(0)}\right) = p\left(X^{(j)} \mid X^{(j-1)}\right). \tag{6.65}$$

The state-space of the Markov process, i.e., the set of all possible values that X can take, is S_{M}. The transition matrix P of the finite time-homogeneous Markov process is the collection of the transition probabilities $(p_{X,X'})_{X,X' \in S_{\mathrm{M}}}$. Let $\pi(j)$ be the probability distribution over the state-space at iteration j. The probability distribution at the next iteration can then be computed as

$$\pi(j+1) = P\pi(j). \tag{6.66}$$

In each iteration, the new population of proposed model structures is obtained by applying the crossover and mutation operators. Hence, the probability distribution $\pi(j+1)$ can also be computed as

$$\pi(j+1) = P_{\mathrm{m}} P_{\mathrm{c}} \pi(j), \tag{6.67}$$

where P_{c} and P_{m} are the transition matrices of the crossover and mutation operators, respectively.

The properties of transition matrices P_{c} and P_{m} allow us to study the limit behaviour of the Markov chain model of the proposed algorithm. To do so, we introduce the following notions.

Definition 6.3 *(Stochastic matrix)* A *stochastic matrix* is a square matrix in which all entries are non-negative and the sum of entries in each column is 1.

Definition 6.4 *(Column-allowable transition matrix,* Rudolph and Agapie [28]*)* A stochastic matrix is said to be *column-allowable* if each column of the matrix has at least one positive entry.

---

[7] This is based on the assumption that each auxiliary tree in $\mathcal{A}$ can adjoin to each syntactic tree generated from the grammar. This is usually not the case.

Definition 6.5 *(Irreducible transition matrix,* Rudolph [26]*)* The transition matrix P of a finite time-homogeneous Markov chain is said to be *irreducible* if there exist *no* equivalence transformations that reduce P to the form

$$\begin{pmatrix} C & 0 \\ R & T \end{pmatrix}, \tag{6.68}$$

where C, T are square matrices.

By definition, transition matrices are stochastic matrices. A Markov chain $\{X^{(j)}\}$ with an irreducible transition matrix implies that it is possible to reach any state $X' \in S_{\mathrm{M}}$ from any initial state $X \in S_{\mathrm{M}}$ in a finite number of transitions. More formally, for any pair $X, X' \in S_{\mathrm{M}}$, there exists an $m \in \mathbb{Z}_{>0}$ such that

$$p\left(X^{(j+m)} = X' \mid X^{(j)} = X\right) > 0. \tag{6.69}$$

Based on these notions, we obtain the following result.

Lemma 6.6 *The Markov chain corresponding to the sequence of population $\{X^{(j)}\}$ in Algorithm 1 with state space S_{M} is irreducible for $p_c \in [0, 1)$ and $p_m \in (0, 1]$.*

Proof See Appendix. □

It should be noted that crossover and mutation rates in Lemma 6.6 are sufficient conditions, but not necessary, to establish irreducibility of the Markov Chain $\{X^{(j)}\}$. Since Markov chain $\{X^{(j)}\}$ can be shown to be homogeneous and irreducible, we can use Lemma 1 of Rudolph and Agapie [28], given below, to describe its limit behaviour.

Lemma 6.7 *([28]) A homogeneous Markov chain with finite state space and irreducible transition matrix visits every state infinitely often with probability one regardless of the initial distribution.*

From Lemmas 6.6 and 6.7, we infer that the proposed Algorithm 1 generates all model structures in $\mathcal{M}_{\mathrm{d}}(G_{\mathrm{N}})$, and consequently, in $\mathcal{F}^*$, with probability 1 as $j \to \infty$. This inference relates to merely the search for appropriate model structures, i.e., model structures in $\mathcal{F}^*$, in a finite, albeit large, discrete search space. In other words, the inference relates to the asymptotic behaviour of the search algorithm corresponding to the upper-level optimization problem of the overall multi-criteria identification problem in (6.5).

Stage B

We now turn our attention towards parameter estimation algorithms that solve the lower-level optimization problems. In the given setting, the lower-level optimization problem is formulated in terms of mean-squared prediction error J_1, and mean-squared simulation error J_2. For any candidate model structure $M \in \mathcal{M}_{\mathrm{d}}(G)$ proposed by GP, we can examine the mean squared error (MSE) (prediction or simulation) achieved by the estimated model $M(\hat{\theta})$. For this, let the true, unknown DGS be

S parameterized by θ_0. Let the candidate model structure be $M(\theta)$ and the parameter estimated for a given dataset $\mathcal{D}_N$ be $\hat{\theta}$. The model M may be a prediction or simulation model, and correspondingly, the parameter estimate $\hat{\theta}$ may be prediction error or simulation-error-based estimate. For a given input u, let y_0 be the the response of the DGS and $y(\hat{\theta})$ be the model response (prediction or simulation). The MSE can be expressed as

$$\mathrm{MSE}(y_0 - y(\hat{\theta})) = E_\xi[(y(\hat{\theta}) - y_0)^2]. \tag{6.70}$$

Under the assumption that $\{y(\hat{\theta})\}$ is a quasi-stationary process, the MSE can be decomposed as follows [13]:

$$\begin{aligned}
\mathrm{MSE}(y_0 - y(\hat{\theta})) &= E_\xi\left[\left(y_0 - E_\xi[y(\hat{\theta})] + E_\xi[y(\hat{\theta})] - y(\hat{\theta})\right)^2\right], \\
&= E_\xi\left[\left(\left(y_0 - E_\xi[y(\hat{\theta})]\right) - \left(y(\hat{\theta}) - E_\xi[y(\hat{\theta})]\right)\right)^2\right], \\
&= E_\xi\left[\left(y_0 - E_\xi[y(\hat{\theta})]\right)^2\right] + E_\xi\left[\left(y(\hat{\theta}) - E_\xi[y(\hat{\theta})]\right)^2\right] - \\
&\qquad 2E_\xi\left[\left(y_0 - E_\xi[y(\hat{\theta})]\right)(y(\hat{\theta}) - E_\xi[y(\hat{\theta})])\right], \\
&= \underbrace{\lim_{N\to\infty} \frac{1}{N} \sum_{k=1}^{\infty} \left(y_0(k) - E_\xi[y(k,\hat{\theta})]\right)^2}_{\text{bias}^2} + \underbrace{E_\xi\left[\left(y(\hat{\theta}) - E_\xi[y(\hat{\theta})]\right)^2\right]}_{\text{var}}.
\end{aligned} \tag{6.71}$$

Hence the MSE of the model estimate can be decomposed into two components—bias and variance. Bias refers to a systematic error in the estimated model and variance refers to variability of the model (or parameter) estimate. Under Assumption 6.1, the variance of the parameter estimate reduces to 0. The bias component of the error may be caused due to a structural error in the model structure M or a systematic error in the estimator, and are typically dependent on the nature of the excitation signal. Let $\mathcal{U}$ be the desired class of inputs and $u \in \mathcal{U}$. Then the bias error, averaged over all possible inputs $u \in \mathcal{U}$ can be decomposed into the following components:

$$\begin{aligned}
E_u[(y_0 - E_\xi[y(\hat{\theta})])^2] &= E_u[(y_0 - y^* - E_\xi(y(\hat{\theta})) + y^*)], \\
&= \underbrace{E_u[(y_0 - y^*)^2]}_{\text{structural bias}} + \underbrace{E_u[(E_\xi[y(\hat{\theta})] - y^*)^2]}_{\text{estimation bias}},
\end{aligned} \tag{6.72}$$

where y^* corresponds to the response of the hypothetical average model estimate $E_u[M(\hat{\theta})]$ over the input class $\mathcal{U}$ for the fixed model structure M. The first term in (6.72) can be attributed to a structural error in model structure M. The second term can be attributed to a systematic estimation error, e.g., when the parameter estimator is biased or is susceptible to be converge to a local minima.

Equation (6.72) reveals the essential challenge for parameter estimation within the proposed identification framework. For a given identification problem, the efficient set $\mathcal{F}^*$ consists of a large number of models with varying model structures. While some of these model structures may be complex enough to describe the DGS S, other model structures in $\mathcal{F}^*$ will be too simple to model the dynamics present in the dataset. For such model structures in $\mathcal{F}^*$, the structural bias component in (6.72) cannot be reduced to 0, and is reflected in the MSE (prediction or simulation) achieved by the corresponding models. Hence, in order to ensure that the model estimate $M(\hat{\theta})$ converges to $\mathcal{F}^*$, in terms of the distance metric in (6.62), the parameter estimator, for both simulation error and prediction error, is required to converge to the global minima. In other words, for any model structure $M \in \mathcal{M}_{\mathrm{d}}(G_{\mathrm{N}})$ (or at least $\mathcal{F}^*$), the second term in (6.72) should be 0 in expectation.

Based on this discussion, we make the following assumption to simplify the overall analysis of the proposed identification method.

Assumption 6.2 Under Assumption 6.1, for any model structure $M \in \mathcal{F}^*$, parameter estimates $\hat{\theta}_{\mathrm{p}}$ (for prediction error) and $\hat{\theta}_{\mathrm{s}}$ (for simulation error) converge to the corresponding global minima with probability (w.p.) 1.

Considering that the problems of prediction error minimization and simulation error minimization are non-linear for the class of P-NARMAX model, Assumption 6.2 is a strong assumption. Global optimization techniques such as GA or CMA-ES can be used to obtain unbiased estimates asymptotically. Unfortunately, this comes at a great computational cost.

In Sect. 6.3, we presented algorithms for optimizing model parameters w.r.t objective functions J_1 and J_2 for the class of P-NARMAX models. The proposed estimation algorithms are far more computationally efficient than global optimization techniques. Furthermore, the convergence properties of the estimators are also well studied in the literature. Unfortunately, the proposed estimators are not guaranteed to converge for all P-NARMAX models, as discussed below.

An iterative least-squares method was presented in Sect. 6.3.1 to estimate model parameters P-NARMAX models based on 1-step-ahead prediction error. In Ljung [21], the author described the recursive algorithm using an equivalent ODE system, and derived the conditions required for convergence of the algorithm to a local minima for linear ARMAX models. Essentially, the algorithm was found to converge asymptotically if the noise model of the DGS is invertible. In Chen and Billings [5], the analysis was extended to the case of P-NARMAX models. The authors showed that the analysis can be directly extended to the set of P-NARMAX models that can be decomposed as

$$y(k) = f\left(y(k-1), \ldots, y(k-n_y), u(k), \ldots, u(k-n_u)\right) + H(q^{-1})\xi(k), \quad (6.73)$$

where $H(q^{-1})$ is a linear filter and f is a polynomial non-linear function. Extending the result to the general class of P-NARMAX models, however, is far more involved since a general P-NARMAX model may contain multiplicative terms of input, output and noise, which complicates the notion of invertibility of the noise model. Hence, Chen and Billings [5] defined a notion of *m invertibility*, which makes it possible to extend the conditions for convergence to the case of P-NARMAX models.

The multi-step prediction algorithm described in Sect. 6.3.2 was proposed and studied in Farina and Piroddi [10, 11]. Convergence properties of the MSP algorithm can be found in Theorem 3.1 in [9]. A key assumption required for the validity of the results is that the autonomous response of model M to non-zero initial conditions decays exponentially to 0, i.e., the model M does not have infinite memory. This excludes, for e.g., chaotic systems.

Stage C

In Part A, we established the sufficient conditions required for the GP algorithm operating in search space $\mathcal{M}_d(G_N)$ to visit each model structure in $\mathcal{F}^*$ infinitely often w.p. 1. In Part B, we assumed that for each model structure M in $\mathcal{F}^*$ the parameter estimates for the mean-squared simulation and prediction error loss functions converge to the respective global minima. By combining these two elements, we get that as $j \to \infty$, the proposed algorithm identifies each model structure in $\mathcal{F}^*$ and estimates the corresponding model parameters w.p. 1. It remains to be shown that once a model $M(\theta)$ in $\mathcal{F}^*$ is identified, it enters and subsequently remains in the archive $\mathfrak{A}^{(j)}$ as $j \to \infty$.

The need to preserve non-dominated solutions in the population is well recognized in MOEA literature. An EA that preserves non-dominated solutions in the population deterministically (or w.p. 1) is said to have an *elite preserving strategy*, or in short, an *elitist strategy*. Rudolph [27] demonstrated that elitism is a crucial property required to proof convergence of an MOEA to a subset of Pareto-efficient solutions in an finite search space. As a consequence, many MOEAs with elitist strategies have been since developed, e.g., Strength Pareto evolutionary algorithm (SPEA) [34] and Pareto archived evolution strategy (PAES) [19].

In Step 7 of Algorithm 1, the archive $\mathfrak{A}^{(j)}$ stores the set of efficient solutions in the set $\mathfrak{A}^{(j-1)} \cup X^{(j)}$. The set of efficient solutions is computed using the non-dominated sorting algorithm in Algorithm 5 (see Deb et al. [7]), which is an elitist algorithm.

We can now assemble the three stages of reasoning discussed above, in order to obtain a general result for the asymptotic properties of the proposed identification method. Recall the following conclusions from the three stages of the analysis:

(i) In Stage A of the reasoning, we showed that asymptotically, for certain ranges of crossover and mutation rates, the proposed algorithm is likely to explore all model structures in $\mathcal{F}^*$ with probability 1.
(ii) In Stage B, we determined the parameter estimation properties required in the general setting of TAG-based model set specification.
(iii) In Stage C, we determined the properties required from the selection operator in order to retain model estimates that belong to the Pareto-efficient set $\mathcal{F}^*$.

Based on these conclusions, we now obtain the following result.

Theorem 6.8 *For the model set $\mathcal{M}_{\mathrm{d}}(G_{\mathrm{N}})$, under Assumptions 6.1 and 6.2, the archive $\mathfrak{A}^{(j)}$ of the identification method in Algorithm 1 satisfies*

$$\lim_{j\to\infty} \delta_{\mathcal{F}^*}(\mathfrak{A}^{(j)}) \to 0, \qquad \text{w.p. } 1, \tag{6.74}$$

if $p_c \in [0, 1)$ and $p_m \in (0, 1]$.

Proof See Appendix. □

Theorem 6.8 establishes the sufficient conditions required in order to guarantee asymptotic convergence (in the sense of (6.63)) of the proposed identification algorithm. Unlike classical identification techniques such as PPEM, the proposed identification method converges *not* to a single optimal solution, but to an entire set of optimal (i.e., non-dominated) solutions. In doing so, we do not require the true DGS to be part of the model set in order for our analysis to hold.

It should be noted that while we make a strong claim with regards to solving the structure-determination problem (i.e., the upper-level optimization problem in (6.5)), the analysis rests on the assumption that the parameter estimation techniques used to solve the lower-level optimization problem in (6.5) converge to the global minima. While it is reasonable to require that the parameter estimates converge to the global minima in order to obtain convergence for the broader problem of determining non-dominated models in the model set, it is nonetheless a strong assumption in the context of non-linear system identification.

Remark 6.7 It should be pointed out that Theorem 6.8 only provides sufficient conditions for convergence as defined in Definition 6.1. Furthermore, the notion of convergence used here only ensures that the models in archive $\mathfrak{A}^{(j)}$ are non-dominated, and do not consider the diversity of models in the archive. In practice, the diversity is ensured by rank-ordering models that have the same Pareto-dominance rank using crowding-distance-based ordering (see details in Sect. 6.2.2). Furthermore, in practice, duplicate models are not included in the archive.

6.4.2 Computational Complexity

The objective of Algorithm 1 is to perform automated model structure selection and subsequent parameter estimation, such that minimal user intervention is required during model estimation, while affording the user the ability to choose a suitable model *a posteriori* based on the estimated Pareto front. While the proposed identification approach makes it easier for the user to obtain data-driven model estimates, it comes with the cost of a higher demand for computational resources. In this section, we look at the computational complexity of Algorithm 1. Empirical run-time figures for various datasets will be provided in Chap. 7.

The most computational-intensive step in the proposed algorithm is the parameter estimation step, i.e., Step 4 in Algorithm 1. The complexity of the parameter estimation step depends on the lower-level problem of the bi-level identification problem. In this section, we study the run-time complexity of the algorithm for the class of P-NARMAX models and performance measures $\{J_i(M, \theta, \mathcal{D}_N)\}_{i=1}^2$.

For grammar G_N, prediction error minimization involves a sequence of linear least squares operations, see Sect. 6.3.1. The run-time complexity of each least squares operation is given by $O(N \dim(\theta))$, where N is the length of the dataset, and $\dim(\theta)$ is the length of parameter vector θ. Let m_p be the maximum number of least squares iterations required for the prediction error minimization algorithm to converge or time-out. The computational complexity to estimate parameter θ for an individual $M(\theta) \in \mathcal{M}(G_\text{N})$ is $O(m_\text{p} N_\text{est} \dim(\theta))$.

The complexity of each step in simulation error minimization for a given time-horizon τ is given below

- Computation of τ-step-ahead predicted outputs $\hat{y}(k|k-\tau)$ has time-complexity $O(N_\text{est}\tau \dim(\theta_{\text{s},l}))$.
- Computation of Jacobian Ψ_τ requires $O(N_\text{est}\tau(\dim(\theta_{\text{s},l}))^2)$ time.
- The parameter update step in (6.51) requires $O((\dim(\theta_{\text{s},l}))^3)$ computations for matrix inversion and $O(N_\text{est}(\dim(\theta_{\text{s},l}))^2)$ for matrix multiplication.

Since $N_\text{est} \gg \dim(\theta_{\text{s},l})$, the dominant computation is $O(N_\text{est}\tau(\dim(\theta_{\text{s},l}))^2)$. This step is repeated for $\tau = 1, \ldots, m_\tau$, where m_τ is the maximum prediction horizon for which the model estimates converge or the algorithm times-out. Furthermore, let m_s be the maximum number of Gauss-Newton iterations required for each prediction horizon τ. The overall complexity for simulation-based parameter estimation for an individual $M(\theta) \in \mathcal{M}(G_\text{N})$ is $O(m_\text{s} N_\text{est} m_\tau^2 (\dim(\theta_{\text{s},l}))^2)$.

As one would expect, parameter estimation based on simulation error is the most time-consuming step for each individual in an iteration. This step is repeated for each individual in the population, and for each iteration of the GP run. Hence, the overall computational complexity of the algorithm is $O(n_\text{s} L m_\text{s} N_\text{est} m_\tau^2 (\dim(\theta_{\text{s},l}))^2)$, where n_s is the size of the population and L is the maximum number of iterations in the GP algorithm.

6.5 Discussion

In this chapter, the concepts and methodologies introduced in Chap. 3, 4 and 5 were integrated to formulate an identification approach that is aligned with the main research goal of this dissertation, see Chap. 1. The main features of the proposed identification framework are listed below.

- Based on the discussion in Sect. 2.5, we formalized the identification problem as a multi-criteria bi-level optimization problem in (6.2). The formulation of the optimization problem was made possible by the TAG-based model set concept introduced in Sect. 4.3.1 and Pareto-dominance-based ordering of candidate solutions discussed in Sect. 5.2.

- The general identification problem essentially consists of two inter-connected problems—model structure determination and model parameter estimation. The proposed multi-criteria optimization problem in (6.2) encapsulates these two problems into a single hierarchical structure. The upper-level optimization problem relates to the model structure determination, and the lower level corresponds to the model parameter estimation problem.
- The general identification problem, as formulated in (6.2), consists of multi-criteria optimization problems on both levels, making it a challenging problem to solve. Hence, the problem is simplified in Eqs. (6.3) and (6.5) by replacing the lower-level multi-objective problem by a number of single-objective optimization problems. Essentially, we discretize the number of possible trade-offs between objectives in J_{cont}.
- Genetic programming forms the foundation of the proposed identification algorithm to solve the multi-criteria problem. The evolutionary search explores the discrete space of model structures generated by TAG, and hence is associated with the upper-level optimization problem. Variation operators are adapted to ensure that the rules of grammar encoded in the given TAG are preserved during the search. Furthermore, the selection operator is modified to select models based on Pareto-dominance and crowding-distance-based ordering.
- It should be pointed out that the search for model structures is formulated on the basic building blocks of any TAG, and is independent of the choice of TAG. Hence, the proposed algorithm can address the model structure determination problem for any model class, and even representation, that can be formulated as a TAG.
- Within each iteration of GP, a parameter estimation step is incorporated to optimize model parameters w.r.t the performance measures in the lower-level optimization problem. Unlike the evolutionary search for model structure, the parameter estimation algorithms used will typically depend on the performance measures and the specific choice of TAG.
- The result of the identification method is an estimate of the efficient set of models that lie on the estimated Pareto-front. The Pareto-front estimate informs the user of the possible trade-offs that can be made between the various modelling objectives. Based on this information, the user may select *a posteriori* a model that achieves the trade-off that is suitable for use-case.

The purpose of the identification method developed in this chapter is to automate the system identification process. Nonetheless, the user may choose to customize the algorithm, for example, to introduce prior understanding of the dynamics of the DGS or to express preferential bias for model structures. From the user's perspective, the identification method can be adapted in the following ways.

- The user may introduce prior knowledge of the dynamics of the DGS in the TAG that specifies the model set. The prior information may be formulated at a high level, for example, restricting the search to linear models of arbitrary complexity, or may be more detailed, for example, by seeding the initial population with an preliminary model estimate.

- The user may incorporate additional modelling objectives that are relevant for the ultimate use-case of the model.

In Sect. 6.4.1, we analyzed the the asymptotic behaviour of the algorithm and showed that, under some assumptions on the parameter estimation algorithms, the estimated Pareto-efficient set of models converges to a subset of the true Pareto-efficient set. While the analysis is primarily academic in nature, it does reveal several important aspects of the algorithm. Part A of the analysis reveals that the types and rates crossover and mutation operator may play a crucial role in determining the convergence properties of the algorithm. While the analysis only address asymptotic convergence, it may be plausible that the crossover and mutation rates also influence the rate of convergence. This will be examined via simulation studies in Chap. 7. Part B of the analysis reveals the parameter estimation algorithms are required to converge to the global optimum of the optimization problem to assure convergence. And finally, Part C of the analysis highlights the importance of the archive updating rule used in the algorithm.

Finally, in Sect. 6.4.2, we examined the run-time complexity of the proposed algorithm. As one would expect, the proposed algorithm is far more computationally demanding than most tradition identification methodologies. However, in general, the additional computational time can be easily offset considering that the proposed identification method automates the identification task for the user. This is especially beneficial for more involved non-linear ideis ntification problems. Furthermore, the computations involved can easily be parallelized over multiple CPU cores, thereby making it possible to scale the method for larger population sizes or datasets.

Appendix 1: Mathematical Proofs

Proof of Lemma 6.6

To prove that the Markov chain $\{X^{(j)}\}$ is irreducible, we must prove that the transition matrix P is irreducible. Since the transition from population $X^{(j)}$ to $X^{(j+1)}$ takes place via the crossover and mutation operators, as shown in (6.67), we can equivalently prove that the matrix product $P_{\mathrm{m}}P_{\mathrm{c}}$ is irreducible.

We first analyse the mutation operator. The mutation operation is applied to a derivation tree ρi with probability p_m. Consider any two individuals $\rho, \rho' \in L_{\mathrm{D}}(G_{\mathrm{N}})$. Based on the facts that (i) there exist two types of mutation operation—deletion and insertion (see Sect. 6.2.2), and (ii) grammar G_{N} has only one initial tree $\eta = \langle V, E, r\rangle$ with $l(r) = S$ (see Sect. 4.3.2), we can conclude that there exists at least one sequence of mutation operations for which the derivation tree ρ' is reachable starting from derivation tree ρ. This sequence of mutations can be realized by first deleting all nodes, except the root node, in ρ using deletion-type mutation, and subsequently introducing the nodes that result in ρ' using insertion-type mutation. Hence, if the probability of mutation is strictly greater than 0, there is some non-zero probability that derivation tree ρ' can be reached from ρ after a finite number of mutations.

The arguments presented here for individuals ρ and ρ' can be trivially extended to two populations X and X' in the state-space S_{M}. Hence the mutation operation is irreducible for $p_m \in (0, 1]$.

Subsequently, we analyse the crossover operator. The crossover operator is applied to pairs of individuals in the population with probability p_c. This implies that the probability of two individuals remaining unchanged after crossover operation is $1 - p_c$. Hence the probability of a population X of size n_{s} remaining unchanged after crossover is $(1 - p_c)^{\frac{n_{\mathrm{s}}}{2}}$ (assuming n_{s} is even). If p_c is strictly less than 1, we can conclude that the diagonal entries of P_{c} are strictly positive, and hence P_{c} is column-allowable. Applying Lemma 2 from [28], we can conclude that the product $P = P_{\mathrm{m}}P_{\mathrm{c}}$ is irreducible for $p_c \in [0, 1)$ and $p_m \in (0, 1]$.

Proof of Theorem 6.8

Lemmas 6.6 and 6.7 establish that as $j \to \infty$, all model structures in $\mathcal{F}^*$ are generated by the GP algorithm w.p. 1 for $p_c \in [0, 1)$ and $p_m \in (0, 1]$. Assumptions 6.1 and 6.2 ensure that for all model structures in $\mathcal{F}^*$, the parameter estimates converge to the global minima. Finally, in Step 7 of Algorithm 1, archive $\mathfrak{A}^{(j)}$ is updated using the following elitist strategy

$$\mathfrak{A}^{(j)} = \mathcal{E}(\mathfrak{A}^{(j-1)} \cup X^{(j)}, \prec_J). \tag{6.75}$$

The archive $\mathfrak{A}^{(j-1)}$ consists of all non-dominated solutions estimated upto iteration $j - 1$ (upto a maximum of m_{a} models). Assume, without loss of generality, that in iteration j a model $M(\theta)$ that belongs to the efficient set $\mathcal{F}^*$ is estimated in population $\tilde{X}^{(j)}$. Using the elitist update rule in (6.75), the model $M(\theta)$ is included in archive $\mathfrak{A}^{(j)}$, and simultaneously, all models in archive $\mathfrak{A}^{(j-1)}$ that are dominated by $M(\theta)$ are deleted from the archive. As $j \to \infty$, each model in $\mathcal{F}^*$ is estimated infinitely often w.p. 1, and subsequently incorporated in archive $\mathfrak{A}^{(j)}$, while dominated solutions are deleted from archive $\mathfrak{A}^{(j)}$. Hence, we get

$$\begin{aligned} \lim_{j \to \infty} \delta_{\mathcal{F}^*}(\mathfrak{A}^{(j)}) &= \left|\mathfrak{A}^{(j)}\right| - \left|\mathfrak{A}^{(j)} \cap \mathcal{F}^*\right|, \\ &\to 0, \qquad \text{w.p. } 1. \end{aligned} \tag{6.76}$$

References

1. Ashlock D, McGuinness C, Ashlock W (2012) Representation in evolutionary computation. In: IEEE World Congress on Computational Intelligence. Springer, pp. 77–97
2. Basar T, Selbuz H (1979) Closed-loop Stackelberg strategies with applications in the optimal control of multilevel systems. IEEE Trans Autom Control 24(2):166–179
3. Billings S, Voon W (1984) Least squares parameter estimation algorithms for non-linear systems. Int J Syst Sci 15(6):601–615
4. Billings SA (2013) Nonlinear system identification: NARMAX methods in the time, frequency, and spatio-temporal domains. Wiley, New York

5. Chen S, Billings S (1989) Recursive prediction error parameter estimator for non-linear models. Int J Control 49(2):569–594
6. Colson B, Marcotte P, Savard G (2007) An overview of bilevel optimization. Ann Oper Res 153(1):235–256
7. Deb K, Pratap A, Agarwal S, Meyarivan T (2002) A fast and elitist multiobjective genetic algorithm: Nsga-ii. IEEE Trans Evolution Comput 6(2):182–197
8. Emmerich MT, Deutz AH (2018) A tutorial on multiobjective optimization: fundamentals and evolutionary methods. Na Comput 17(3):585–609
9. Farina M, Piroddi L (2010) An iterative algorithm for simulation error based identification of polynomial input-output models using multi-step prediction. Int J Control 83(7):1442–1456
10. Farina M, Piroddi L (2011) Simulation error minimization identification based on multi-stage prediction. Int J Adapt Control Signal Process 25(5):389–406
11. Farina M, Piroddi L (2012) Identification of polynomial input/output recursive models with simulation error minimisation methods. Int J Syst Sci 43(2):319–333
12. Fisk C (1984) Game theory and transportation systems modelling. Transp Res Part B Methodol 18(4–5):301–313
13. Friedman J, Hastie T, Tibshirani R (2001) The elements of statistical learning, vol 1. Springer Series in Statistics New York
14. Hansen P, Jaumard B, Savard G (1992) New branch-and-bound rules for linear bilevel programming. SIAM J Sci Statistical Comput 13(5):1194–1217
15. Hoai NX, McKay R (2001) A framework for tree adjunct grammar guided genetic programming. In: Proceedings of the Post-graduate ADFA Conference on Computer Science (PACCS'01), pp 93–99
16. Hoai NX, McKay RI, Essam D, Chau R (2002) Solving the symbolic regression problem with tree-adjunct grammar guided genetic programming: the comparative results. In: Proceedings of the 2002 Congress on Evolutionary Computation (CEC'02) (Cat. No. 02TH8600), vol 2. IEEE, pp 1326–1331
17. Ishibuchi H, Tsukamoto N, Nojima Y (2008) Evolutionary many-objective optimization: a short review. In: IEEE Congress on Evolutionary Computation (IEEE World Congress on Computational Intelligence). IEEE, pp 2419–2426
18. Karami K, Westwick D, Schoukens J (2019) Identification of decoupled polynomial narx model using simulation error minimization. In: 2019 American Control Conference (ACC). IEEE, pp 4362–4367
19. Knowles JD, Corne DW (2000) Approximating the nondominated front using the pareto archived evolution strategy. Evolution Comput 8(2):149–172
20. Koza JR (1992) Genetic programming: on the programming of computers by means of natural selection, vol 1. MIT Press, Cambridge
21. Ljung L (1979) Convergence of recursive estimators. IFAC Proc Vol 12(8):131–144
22. Ljung L (1998) System identification. In: Signal analysis and prediction. Springer, pp 163–173
23. Paduart J, Lauwers L, Swevers J, Smolders K, Schoukens J, Pintelon R (2010) Identification of nonlinear systems using polynomial nonlinear state space models. Automatica 46(4):647–656
24. Pintelon R, Schoukens J (2012) System identification: a frequency domain approach. Wiley, New York
25. Piroddi L, Spinelli W (2003) An identification algorithm for polynomial narx models based on simulation error minimization. Int J Control 76(17):1767–1781
26. Rudolph G (1994) Convergence analysis of canonical genetic algorithms. IEEE Trans Neural Networks 5(1):96–101
27. Rudolph G (1998) Evolutionary search for minimal elements in partially ordered finite sets. In: International conference on evolutionary programming. Springer, pp 345–353
28. Rudolph G, Agapie A (2000) Convergence properties of some multi-objective evolutionary algorithms. In: Proceedings of the 2000 Congress on Evolutionary Computation (CEC00) (Cat. No. 00TH8512), vol 2. IEEE, pp 1010–1016
29. Sadigh AN, Mozafari M, Karimi B (2012) Manufacturer-retailer supply chain coordination: a bi-level programming approach. Adv Eng Software 45(1):144–152

30. Safdarian A, Fotuhi-Firuzabad M, Lehtonen M (2014) A distributed algorithm for managing residential demand response in smart grids. IEEE Trans Ind Inform 10(4):2385–2393
31. Simaan M, Cruz JB (1973) On the Stackelberg strategy in nonzero-sum games. J Optim Theory Appl 11(5):533–555
32. Sinha A, Malo P, Deb K (2017) A review on bilevel optimization: from classical to evolutionary approaches and applications. IEEE Trans Evolution Comput 22(2):276–295
33. Zhu M, Martinez S (2011) Stackelberg-game analysis of correlated attacks in cyber-physical systems. In: Proceedings of the 2011 American Control Conference. IEEE, pp 4063–4068
34. Zitzler E, Thiele L (1999) Multiobjective evolutionary algorithms: a comparative case study and the strength pareto approach. IEEE Trans Evolution Comput 3(4):257–271

Chapter 7
Experimental Results

The casino is the only human venture I know where the probabilities are known, Gaussian, and almost computable.

Nassim Nicholas Taleb

In Chap. 6, we presented the proposed identification framework and a theoretical analysis of its asymptotic behaviour and computational complexity. In this chapter, we evaluate the performance of the proposed identification approach on a number of identification problems, categorized into three types: an academic example, experimental case studies and benchmark examples.

The academic example is used to empirically evaluate the performance and convergence properties of the proposed algorithm, and to evaluate the influence of some key algorithm hyper-parameters on the estimated result.

Next, we consider the problem of identifying three physical systems: a pendulum system, a thermo-mechanical system and a weaving machine. The three physical case studies pose varying challenges in terms of dynamics to be identified, including complex non-linearities and large, unknown delays in the system.

Finally, the performance of the proposed identification method is compared to that of other non-linear identification techniques proposed in the literature based on three benchmark examples: the silverbox system, the coupled drives system and a Wiener-Hammerstein benchmark example.

Some key observations derived from the obtained results are discussed.

7.1 Introduction

As stated in Chap. 1, the over-arching objective of the present dissertation is to automate the process of system identification and enable the user to estimate dynamical models that are better aligned to the ultimate use-case of the model. To achieve this, over the course of Chaps. 4, 5 and 6, we developed a multi-objective identification framework based on the concepts of TAG, GP and multi-objective optimization. The

D. Khandelwal, *Automating Data-Driven Modelling of Dynamical Systems*,
Springer Theses, https://doi.org/10.1007/978-3-030-90343-5_7

proposed identification algorithm minimizes the need for user-intervention during the identification process by incorporating the model-structure-determination problem within the optimization problem being solved. On the other hand, the algorithm still allows the user to introduce preferences, in terms of desired model structures and identification criteria. In Sect. 6.4, the proposed algorithm was studied theoretically to verify the asymptotic behaviour and the computational cost. The stage is now set to study the performance of the proposed algorithm empirically, and verify whether the proposed identification framework achieves the overall research objective, and more specifically, the research goals laid down in Chap. 1.

In this chapter, we apply the multi-criteria identification approach proposed in Chap. 6 to a collection of identification problems. The identification problems are categorized into three types—academic example, experimental case studies and benchmark problems. We use an academic example to study the performance of the proposed algorithm in a setting where the DGS is known. Subsequently, the algorithm is used to model data obtained from real physical systems. The chosen case studies emulate a variety of challenges encountered in modelling physical systems. Finally, the performance of the algorithm is examined and compared with other specialized non-linear system identification methods with the help of benchmark datasets available online. All computations described in this chapter were performed on a PC equipped with an Intel® Core™ i7 CPU with 4 cores, 3.40 GHz frequency and 16 GB RAM.

In the sequel, we discuss some of the recurrent aspects of the identification experiments that will be discussed in this chapter.

Tree Adjoining Grammar

The initial and auxiliary trees of the most general TAG used in this chapter are illustrated in Fig. 7.1. The proposed TAG extends beyond the class of P-NARMAX by including sin, cos and abs functions of the input, output and noise terms. Note that the trigonometric and absolute-value non-linearities are introduced *without* scaling of their arguments, thereby retaining the linear-in-the-parameters structure of the models. The following grammars will be used in the case studies:

- The full grammar G_{AT} with initial trees $\{\eta_i\}_{i=1}^{4}$ and auxiliary trees $\{\alpha_i\}_{i=1}^{8}$ in Fig. 7.1.
- The grammar G_{trig} with initial trees $\{\eta_i\}_{i=1}^{3}$ and auxiliary trees $\{\alpha_i\}_{i=1}^{8}$ in Fig. 7.1. Grammar G_{N} extends beyond the class of P-NARMAX by including trigonometric (sin and cos) transformations of input, output and noise terms.
- The P-NARMAX grammar G_{N} with initial trees $\{\eta_1\}$ and auxiliary trees $\{\alpha_i\}_{i=1}^{7}$ in Fig. 7.1.

Of course,[1] as discussed in Sect. 4.3.3, it is possible to extend the grammar, and hence the resulting model set, by introducing suitable elementary trees in the TAG used for identification. Similarly, it is also possible to restrict the TAG to specialized model structures, whose choice is guided by prior knowledge or user preference.

[1] The superscript *NA* in auxiliary tree α_8 refers to a *null adjunction* constraint on the root node, which prohibits the adjunction of any auxiliary tree at that location. See [6] for details.

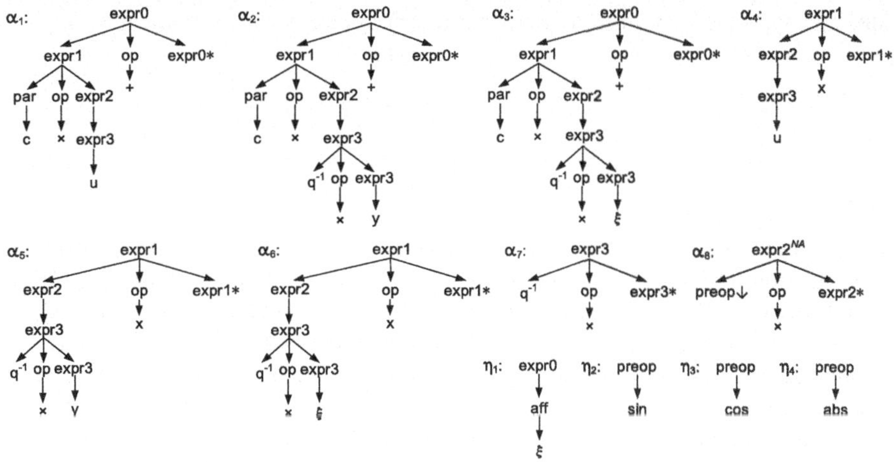

Fig. 7.1 Initial trees $\{\eta_i\}_{i=1}^{4}$ and auxiliary trees $\{\alpha_j\}_{j=1}^{8}$ of the of the over-arching TAG G_{AT}

Table 7.1 General hyper-parameters of the algorithm and their default values

Hyper-parameter	Notation	Default value
Population size	n_{s}	100
Maximum GP iterations	L	150
Maximum adjunctions	m_{d}	150
Maximum initial adjunctions	m_{id}	50
Tree adjoining grammar	G	G_{AT}
l-approximation simulation recursions	l	1

Hyper-Parameters

While the objective of the proposed identification approach is to minimize user-intervention during the identification procedure, there are nonetheless some hyper-parameters of the algorithm that must be selected upfront. While some of these hyper-parameters such as grammar G, population size n_{s} and maximum tree depth m_{d} can be chosen fairly conservatively such that they do not strongly influence the results of the algorithm, other hyper-parameters such as probability of crossover p_c and mutation p_m may play a critical role in determining the quality of results obtained.

In Table 7.1, we list the hyper-parameters that are not critical to performance of the algorithm, when selected conservatively, along with their default values. These hyper-parameter values will be used for the identification experiments presented in the subsequent sections, unless otherwise specified.

The hyper-parameters n_{s}, L, m_{d} and m_{id} can be adapted to a particular identification experiment based on prior information about the system. For example, if the

Table 7.2 Hyper-parameters that determine the dynamics of the evolutionary search

Hyper-parameter	Notation
Crossover rate	p_c
Mutation rate	p_m

true system is expected to be complex in terms of the tree-depth of the corresponding derivation tree, the derivation tree depth m_{id} can be suitably adjusted to reflect this information. Similarly, self-adaptive mechanisms can also be implemented for these hyper-parameters, e.g., see [18]. The parameter l determines the accuracy of the simulation response, and can be increased for higher simulation accuracy, when required.

As suggested in Lemma 6.6, the hyper-parameters p_c and p_m, listed in Table 7.2, play a vital role in determining the dynamics of the evolutionary search. To study the influence of p_c and p_m on the performance of the proposed algorithm, we perform Monte-Carlo (MC) simulations for the academic example in Sect. 7.2. Based on the simulation study, we select values for p_c and p_m that will be used for all other identification studies considered in this chapter.

Identification Criteria

In Chap. 6, we formulated the identification problem as a bi-level optimization problem, see (6.2). The upper-level optimization problem was formulated in terms of the set-valued identification criteria $J(M, \theta, \mathcal{D}_N)$, and the lower-level problem was formulated in terms of the subset of the performance measures $J_{\text{cont}}(M, \theta, \mathcal{D}_N)$ that depend on the value of parameter θ. In Chap. 5, we proposed four performance measures for the multi-criteria identification problem: (i) $J1$ based on prediction error, (ii) $J2$ based on simulation error, (iii) $J3$ based on parametric complexity and (iv) $J4$ based on dynamic complexity. Based on these performance measures, the bi-level optimization problem was simplified to (6.5), where the upper-level optimization problem is formulated in terms of $J(M, \theta, \mathcal{D}_N) = \{Ji(M, \theta, \mathcal{D}_N)\}_{i=1}^{4}$ and the lower-level optimization problem is formulated in terms of $J1(M, \theta, \mathcal{D}_N)$ and $J4(M, \theta, \mathcal{D}_N)$.

For all identification results discussed in this chapter, we use the set-valued performance measure $J(M, \theta, \mathcal{D}_N) = \{Ji(M, \theta, \mathcal{D}_N)\}_{i=1}^{4}$ for the upper-level optimization problem (similar to (6.5)). However, we simplify the lower-level optimization problem by optimizing model parameters only for prediction error, i.e., $J1(M, \theta, \mathcal{D}_N)$. This is done to reduce the overall algorithm complexity from $O(n_s L m_s N_{\text{est}} m_\tau^2 (\dim(\theta_{s,l}))^2)$ to $O(n_s L m_p N_{\text{est}} \dim(\theta))$ (see Sect. 6.4.2), since, in the parameter estimation step, we only make use of the estimation technique presented in Sect. 6.3.1. This reduction in algorithm complexity comes at a price; for model structures that are too simple to describe the DGS, the estimated models will be tuned for prediction error alone, and will have non-optimal simulation performance. For model structures that can describe the DGS without overfitting, we expect estimated models to also have good simulation performance, as will be demonstrated in Sect. 7.2. Note that simulation error minimization *is* included in the set-valued performance measure J in the upper-level optimization problem. This ensures that the selection of proposed candidate models for subsequent iterations takes into account the simulation performance of the estimated model.

Data Collection or Generation

As discussed in Sect. 6.2.2, the proposed algorithm ideally makes use of three independent datasets:

- *Estimation dataset* $\mathcal{D}_{\text{est}}$ for estimating model parameters that minimize the lower-level objectives;
- *Validation dataset* $\mathcal{D}_{\text{val}}$ for computing the upper-level objectives and additionally to estimate the generalization error of the estimated models;
- *Test dataset* $\mathcal{D}_{\text{test}}$ to estimate the generalization error of the final Pareto-front estimate.

In the experimental results presented in the sequel, whenever possible, we make use of three independent datasets as described above. However, for certain experiments (or benchmark datasets), it was not possible to obtain or derive the requisite number of datasets. In such cases, we make ad-hoc decisions to design a work-around. For example, when we have access to only two independent datasets, we may use $\mathcal{D}_{\text{est}}$ to estimate model parameters that optimize the lower-level criterion, i.e. $J1$, and also compute the upper-level criteria $\{Ji\}_{i=1}^{4}$ for the estimated model parameters. In other words, we set $\mathcal{D}_{\text{est}} = \mathcal{D}_{\text{val}}$. Recall that the purpose of dataset $\mathcal{D}_{\text{val}}$ in the proposed identification method was to prevent over-fitting in models that were over-parameterized, by performing cross-validation on the parameter estimates using a dataset independent of $\mathcal{D}_{\text{est}}$. This function is no longer performed when we set $\mathcal{D}_{\text{est}} = \mathcal{D}_{\text{val}}$. Nonetheless, this is not an issue, since

- In the lower-level problem, dataset $\mathcal{D}_{\text{est}}$ is used to estimate model parameters that optimize prediction error alone. Hence, in the upper-level optimization problem, simulation error can be used as a proxy to test the generalization capability of the estimated model structure and parameters.
- The presence of complexity measures $J3$ and $J4$ in the multi-objective criteria keeps the population bloating in check (see discussion in Sect. 3.3).

For each identification experiment in this chapter, we will describe the measured (or simulated) data available, and how it is utilized.

7.2 Academic Example

In this section, we perform an empirical study of the proposed multi-criteria identification approach using a simulation example. In particular, we empirically evaluate the estimated Pareto-front obtained from the algorithm and the ability to recover the structure of the true DGS from finite measurement data and finite number of GP iterations. Furthermore, the effect of varying crossover and mutation probabilities is studied using MC simulations.

System, Data and Hyper-Parameters

The DGS S is a polynomial IO model given by

$$y(k) = 0.5y(k-1) - 0.3u^3(k-5) + 0.3u(k-1), \tag{7.1}$$

with additive measurement noise ξ acting on the measured output y_{m}, i.e.,

$$y_{\mathrm{m}}(k) = y(k) + \xi(k), \tag{7.2}$$

The three simulated datasets $\mathcal{D}_{\mathrm{est}}$, $\mathcal{D}_{\mathrm{val}}$ and $\mathcal{D}_{\mathrm{test}}$, each consisting of 2501 samples, are generated using a normally distributed excitation signal $u_{\mathrm{m}}k \sim \mathcal{N}(0, 1)$ and i.i.d Gaussian noise $\xi(k) \sim \mathcal{N}(0, \sigma_\xi)$ with $\sigma_\xi = 0.05$. The resulting signal-to-noise ratio (SNR) is approximately 56 dB, where the SNR is computed as

$$\mathrm{SNR} = 20 \log_{10} \frac{\sigma_y}{\sigma_\xi}, \tag{7.3}$$

where σ_y is the standard deviation of output y. Hence the dataset contains low measurement noise.

Remark 7.1 Note that system (7.1) has an output-error noise structure. However, models generated by TAG are only optimized for one-step-ahead prediction error and not for simulation error (see Sect. 7.1). The mismatch in noise structure is likely to cause bias in the parameter estimates. However, we expect the bias to be small due to the low noise in the dataset.

System S has a fairly simple structure, with only three model terms. The structure is nonetheless complex enough for a simple random search to be ineffective. This will be verified in the sequel via simulation results.

The hyper-parameter values presented in Table 7.1 are used for all simulations presented in this section. In addition, a range of values are used for the crossover and mutation probabilities, see Table 7.3. As discussed in Sect. 7.1, the values for population size n_{s}, maximum GP iterations L and maximum adjunctions m_{d} are chosen fairly conservatively. We make this choice deliberately to test the capabilities of the proposed methodology. Furthermore, we also make a conservative choice for the grammar by choosing TAG G_{AT}. This would typically be a choice made when there exists no prior preference for the desired model set. For the purpose of this analysis, TAG G_{AT} ensures that the generated model structures not only include the model structure that corresponds to that of DGS S, but also model structures that are too simple or too complex with respect to S. For the probabilities of crossover and mutation, we choose a range of values in $[0, 1]$. In the sequel, we present results from several runs of the proposed algorithm for each of the proposed values of p_c and p_m.

Table 7.3 Algorithm hyper-parameters for the academic example

Hyper-parameter	Value
Probability of crossover p_c	$\{0, 0.2, 0.4, 0.6, 0.8, 1\}$
Probability of mutation p_m	$\{0, 0.2, 0.4, 0.6, 0.8, 1\}$
Grammar	G_{AT}

Simulation Results—Random Search

The proposed identification approach leverages the abundance of computational resources to minimize user-intervention in the identification cycle. A common criticism of such approaches is that a random search of model structures may achieve a performance similar to that of EC-based method. The criticism is not unfounded since it can be reasoned that arguments used in the asymptotic analysis of the proposed algorithm in Sect. 6.4 can also be used for a random search, where model structures are randomly sampled, with uniform distribution, from the model set $\mathcal{M}_{\mathrm{d}}(G_{\mathrm{AT}})$. However, for finite number of iterations, the performance of the GP-based algorithm is far superior to that of a simple random-search approach. To verify this, in this section, we present the results obtained by a random search of model structures generated by TAG. The overall approach is exactly the same as in Algorithm 1 in Chap. 6 with one key difference: in each iteration, new models are generated randomly from the TAG instead of by applying variation operators on the existing population. The initialization scheme proposed in Sect. 6.2.2 is used to generate the new model structures. In the next section, the results obtained by the GP-based approach for a variety of crossover and mutation probabilities will be presented.

The results obtained by random search are illustrated in Fig. 7.2. Figure 7.2a depicts the evolution of minimum and average simulation and prediction performance of the population as a function of number of iterations. We can observe that both prediction and simulation performance remain fairly stagnant throughout the 150 iterations, making only sporadic and intermittent improvements. Effectively, the algorithm does not "learn" from model structures that perform better in a relative sense, and hence the fitness of the overall population does not improve as the algorithm progresses. The best results achieved for mean-squared simulation and prediction error are 0.44 and 0.52, respectively. In contrast, the theoretical minimum is given by $\sigma_\xi^2 = 0.0025$. Hence, the performance of the random search is two order of magnitude away from the theoretical solution.

Since we make use of 4 performance measures, the estimated Pareto-front lies in a four-dimensional space. The projections of the estimated Pareto-front are depicted in Fig. 7.2b–e. The estimated Pareto-fronts are fairly sparse, since there exists no mechanism in the random search approach to "learn" from the dynamics of the estimated non-dominated solutions.

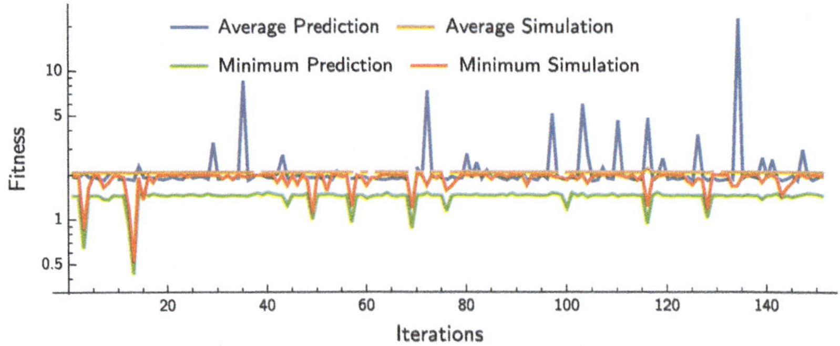

(a) Evolution of fitness with respect to number of iterations.

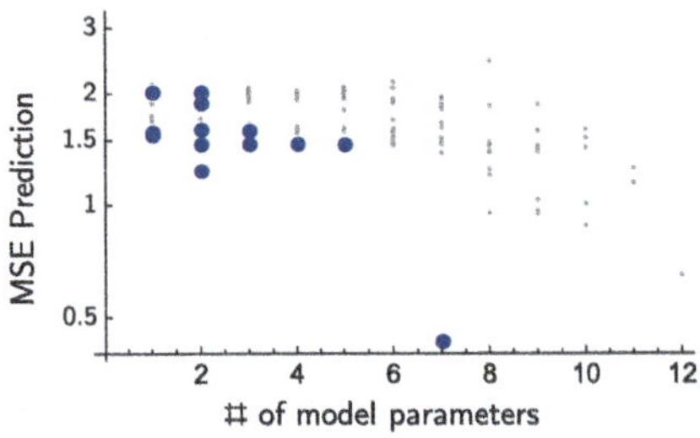

(b) Pareto front evaluated on data $\mathcal{D}_{\text{test}}$ - prediction error vs parametric complexity. Blue dots represent non-dominated solutions, while the smaller grey dots represent dominated solutions.

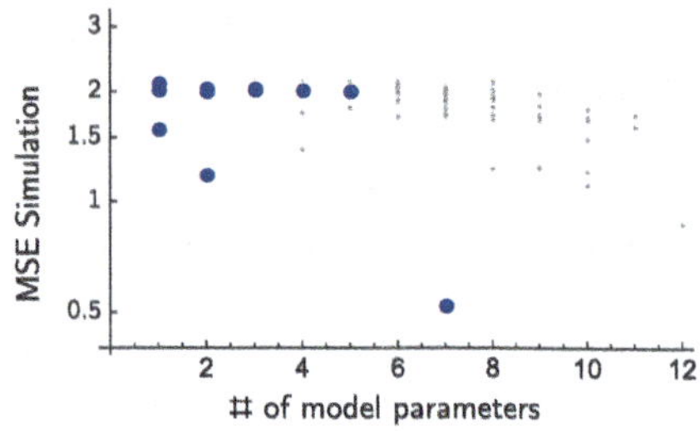

(c) Pareto front evaluated on data $\mathcal{D}_{\text{test}}$ - simulation error vs parametric complexity.

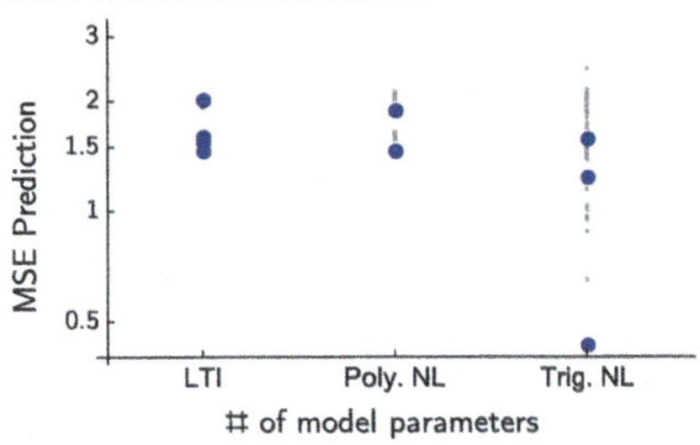

(d) Pareto front evaluated on data $\mathcal{D}_{\text{test}}$ - prediction error vs dynamic complexity.

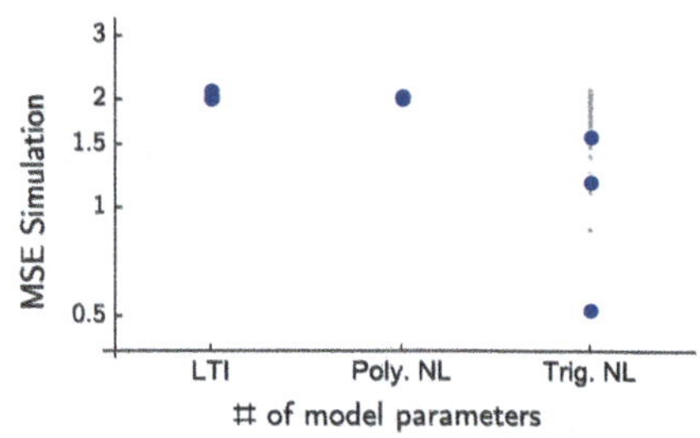

(e) Pareto front evaluated on data $\mathcal{D}_{\text{test}}$ - simulation error vs dynamic complexity.

Fig. 7.2 Numerical results for the academic example—random search using grammar G_{trig}

Simulation Results—Influence of p_c and p_m

With the performance obtained by simple random search serving as a baseline, we now return to the empirical study of the proposed identification method. In this section, we study the influence of hyper-parameters p_c and p_m on the performance of the proposed algorithm. The experimental approach adopted is as follows. For fixed crossover rate $p_c = 1$, we perform MC simulations with 10 runs of the proposed identification approach for each of the mutation rates listed in Table 7.3. Subsequently, based on the results obtained, we select a single mutation rate and perform MC simulations with 10 runs for each crossover rate listed in Table 7.3. Ideally, to get complete insight and statistically sound results, we should perform MC simulations for each unique pair of p_c and p_m in Table 7.3 with many more runs in each simulation. However, since each simulation takes approximately 1–3 h, we conform to the more restricted strategy. In order to compare the results across the various simulations, we use the following metrics:

- Convergence rate for prediction (or simulation) error—this is computed as the percentage of runs in a MC simulation in which the minimum prediction (or simulation) error is equal to the theoretical minimum (i.e., the noise floor) of σ_ξ^2. To take into account the effects of finite data-length, we include a tolerance of 5%. The convergence rate serves as a measure for how frequently the proposed algorithm could estimate model structures and parameters that describe the true DGS for the given mutation and crossover probabilities.
- Average iterations to convergence—this is computed as the average of the number of iterations taken, among the runs that converged to the true minimum in each simulation, to converge (in the sense of item 1). Average iterations to converge describe how quickly the proposed algorithm recovered model structures that could describe the true DGS in the cases where the algorithm did converge. Naturally, this metric is useful only for simulations that achieved a high rate of convergence.
- Average run-time—this is the time taken to run the proposed identification method averaged over the 10 runs of the MC simulation.

The results of MC simulations for various (p_c, p_m) pairs are collected in Table 7.4. As described earlier, for the first set of simulations, we fix $p_c = 1$ and vary mutation probability p_m. In Table 7.4, we can observe that the proposed algorithm converges 9 out of 10 times, for both mean squared prediction and simulation error, for $p_m = 1, 0.8, 0.4, 0.2$. On the other hand, the worst convergence rate is obtained for $p_m = 0$, indicating that low mutation rates adversely affect the convergence of the proposed algorithm. This is also in line with Theorem 6.8. On the other hand, we observe that it is possible for the proposed algorithm to achieve good convergence rates for $p_c = 1$, suggesting that the corresponding sufficient condition of $p_c \in [0, 1)$ in Theorem 6.8 may be conservative.

In Figs. 7.3 and 7.4 we plot the simulation results obtained for (p_c, p_m) values of $(1, 0.8)$ and $(1, 0)$, respectively. To prevent a clutter of figures, the simulation results for other (p_c, p_m) pairs are presented in Appendix. In each figure (including the ones in Appendix):

- Sub-plot (a) depicts the average time taken per iteration of the algorithm (in black) along with time consumed by each of the 10 runs in the simulation (in light blue). In general, the computation time in each iteration depends on the complexity (both parametric and dynamic) of the candidate model structures proposed in that iteration. Since the maximum tree depth m_d is greater than the maximum tree depth at initialization m_id, we expect the computation time per iteration to increase (not

Table 7.4 Summary of results for the academic example

(p_c, p_m)	Covergence Pred. error (%)	Convergence Sim. err. (%)	Average iterations to convergence (Pred. err.)	Average iterations to convergence (Sim. error)	Avg. run-time (for 150 iters) (h)
(1, 1)	90	90	46.1	53.7	2.81
(1, 0.8)	90	90	73.7	79.1	2.07
(1, 0.6)	70	70	53.3	61.6	2.81
(1, 0.4)	90	90	60.1	66.7	3.16
(1, 0.2)	90	90	65.7	81.2	2.04
(1, 0)	30	30	13.7	32.3	2.69
(0.8, 0.8)	80	80	46.5	54.4	2.3
(0.6, 0.8)	80	80	59.4	63.5	2.78
(0.4, 0.8)	90	90	63.8	72.1	1.62
(0.2, 0.8)	70	70	96	100.6	1.26
(0, 0.8)	40	40	113.5	119	2.34

necessarily monotonically) as the proposed identification method explores more complex models over the course of the proposed algorithm.

- Sub-plot (b) illustrates the distribution of best mean-squared simulation and prediction error achieved among the models in the archive (i.e., $\mathfrak{A}^{(L)}$) in each of the 10 runs. If, for the given values of p_c and p_m, the proposed identification algorithm converges to *a* solution, within $L = 150$ iterations, we expect the distributions in sub-plot (b) to have small variances. Moreover, if the proposed algorithm can successfully estimate the true model structure regularly over the MC runs, the distributions in sub-plot (b) will be centered around MSE $= 0.0025$ approximately. Distributions with long tails indicate outliers, e.g., when all but 1 of the MC runs converges to a solution.
- Sub-plots (c) and (d) depict the evolution of minimum mean-squared prediction and simulation error, respectively, in each run of the simulation (distinguished by different colours). Over the course of the proposed algorithm, we expect the estimated mean squared error to improve whenever GP proposes a candidate model structure that can describe (a part of) the dynamics of the true system. Hence, we expect the improvements in MSE to occur in discrete jumps.
- Sub-plots (e)–(h) depict 2-d projections of the 4-dimensional Pareto front estimated in each run of the simulation (again, distinguished by different colours). Each of these sub-plots depict an accuracy versus complexity curve, where the accuracy may be measured in terms of MSE prediction or simulation error, and the complexity can be measured in terms of dynamic or parametric complexity.

Remark 7.2 In sub-plot (b) in Figs. 7.4 and 7.5, the distribution plots seem to indicate that negative mean squared error is achieved. This is merely due to the high variance in performance of the estimated models, and can be confirmed by sub-plots (c) and (d) of the corresponding Figures.

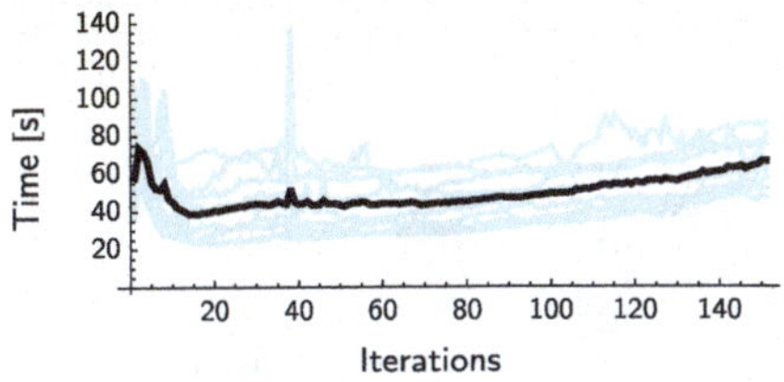

(a) Average time taken per generation for each GP run (light blue) and mean over the 10 runs (black).

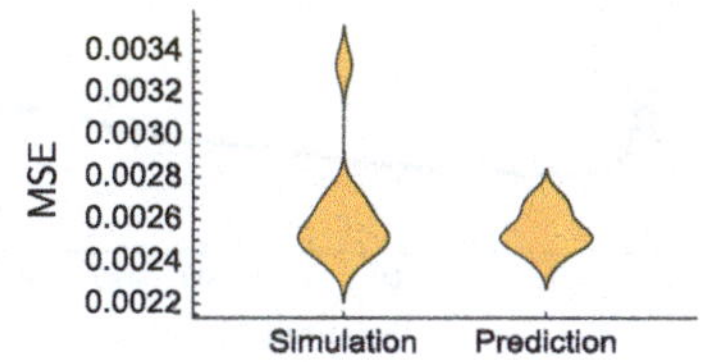

(b) Distribution of the minimum MSE prediction and simulation error evaluated on $\mathcal{D}_{\text{test}}$ over the 10 runs.

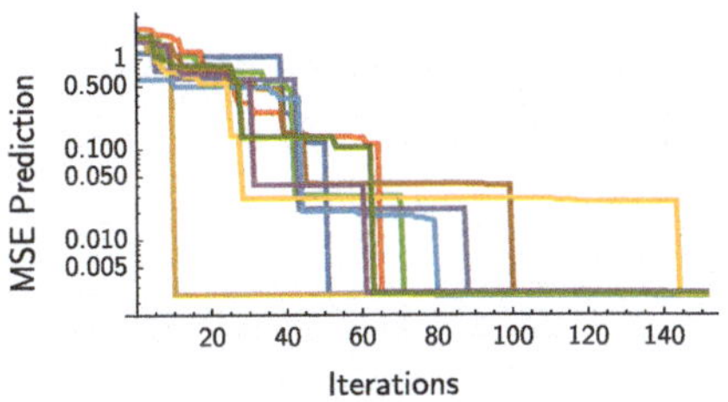

(c) Evolution of minimum prediction error w.r.t GP iterations.

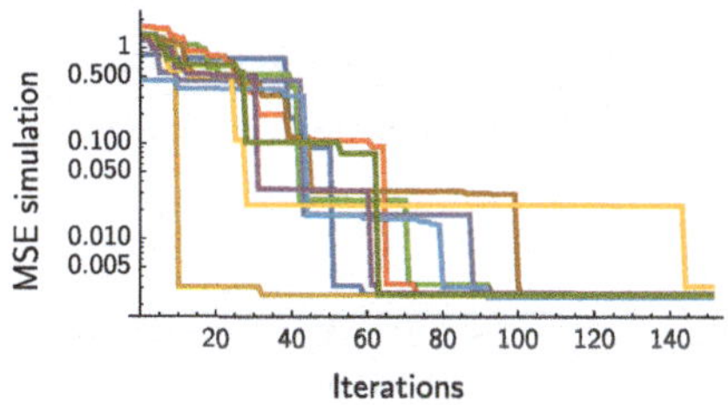

(d) Evolution of minimum simulation error w.r.t GP iterations.

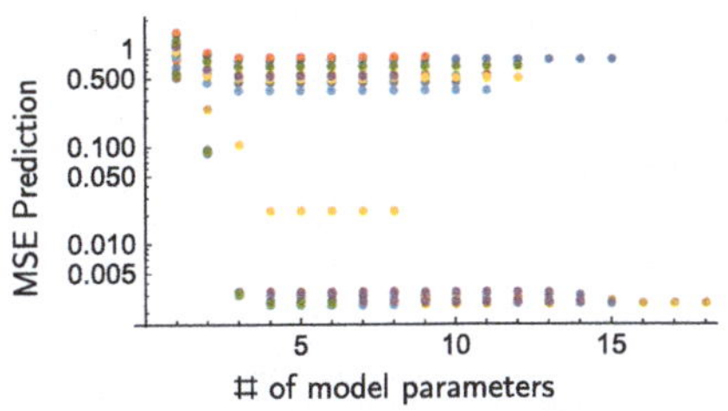

(e) Pareto front evaluated on data $\mathcal{D}_{\text{test}}$ - prediction error vs parametric complexity.

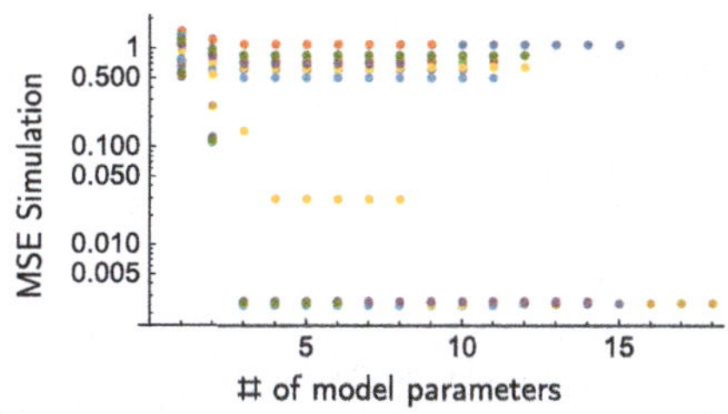

(f) Pareto front evaluated on data $\mathcal{D}_{\text{test}}$ - simulation error vs parametric complexity.

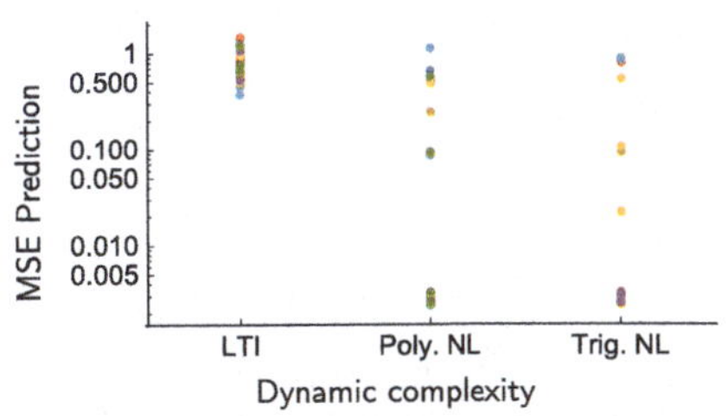

(g) Pareto front evaluated on data $\mathcal{D}_{\text{test}}$ - prediction error vs dynamic complexity.

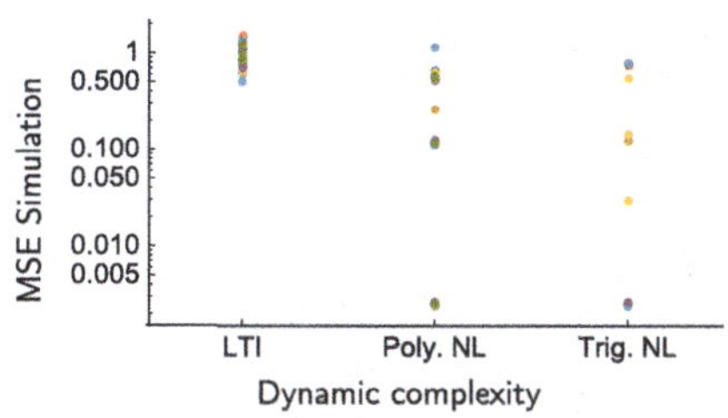

(h) Pareto front evaluated on data $\mathcal{D}_{\text{test}}$ - simulation error vs dynamic complexity.

Fig. 7.3 MC simulation results for the academic example with $p_m = 0.8$ and $p_c = 1$

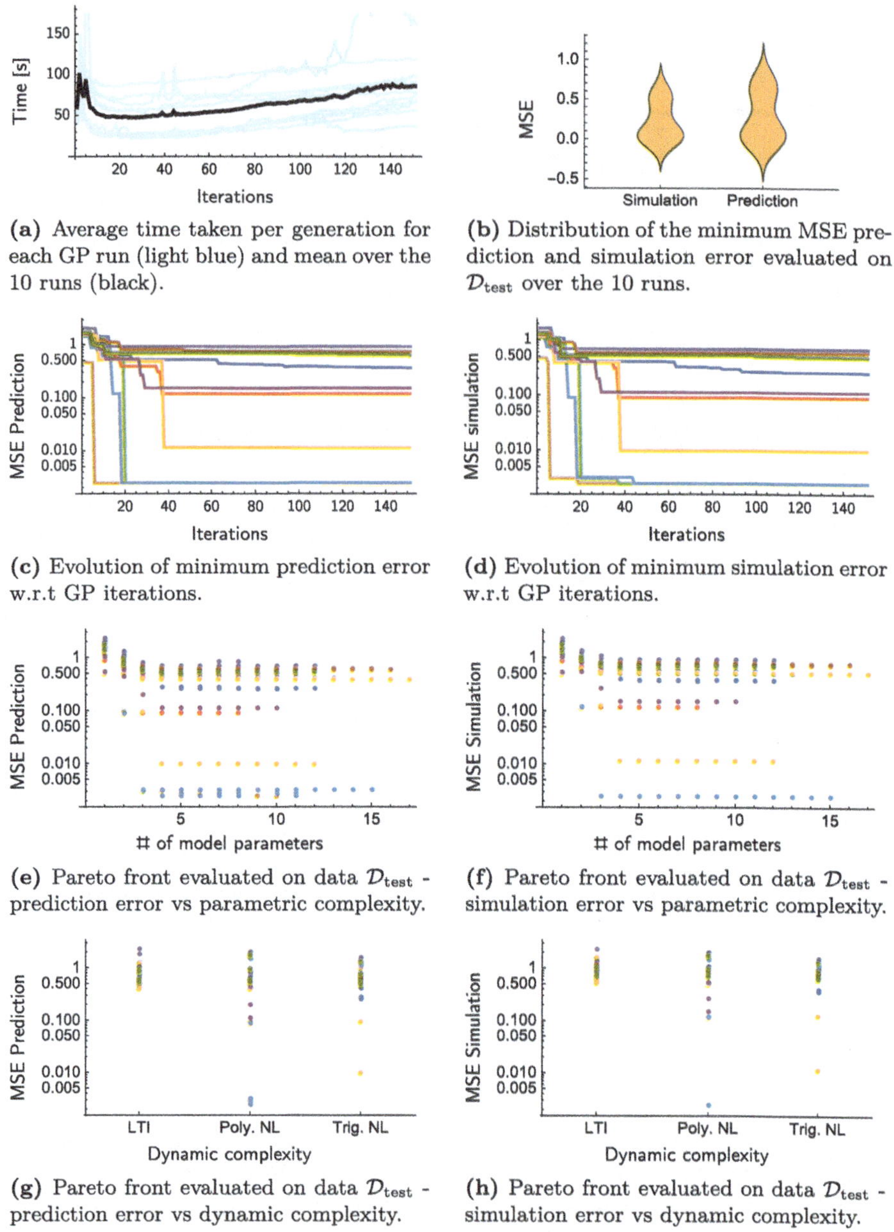

(a) Average time taken per generation for each GP run (light blue) and mean over the 10 runs (black).

(b) Distribution of the minimum MSE prediction and simulation error evaluated on $\mathcal{D}_{\text{test}}$ over the 10 runs.

(c) Evolution of minimum prediction error w.r.t GP iterations.

(d) Evolution of minimum simulation error w.r.t GP iterations.

(e) Pareto front evaluated on data $\mathcal{D}_{\text{test}}$ - prediction error vs parametric complexity.

(f) Pareto front evaluated on data $\mathcal{D}_{\text{test}}$ - simulation error vs parametric complexity.

(g) Pareto front evaluated on data $\mathcal{D}_{\text{test}}$ - prediction error vs dynamic complexity.

(h) Pareto front evaluated on data $\mathcal{D}_{\text{test}}$ - simulation error vs dynamic complexity.

Fig. 7.4 MC simulation results for the academic example with $p_m = 0$ and $p_c = 1$

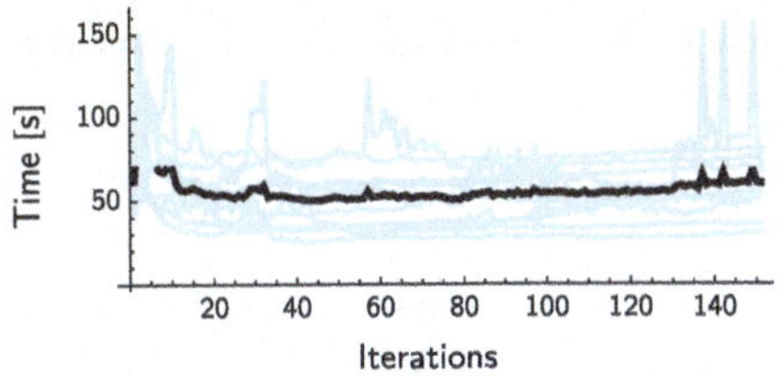

(a) Average time taken per generation for each GP run (light blue) and mean over the 10 runs (black).

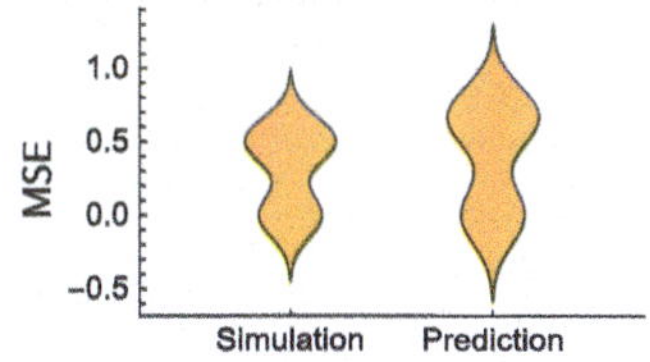

(b) Distribution of the minimum MSE prediction and simulation error evaluated on $\mathcal{D}_{\text{test}}$ over the 10 runs.

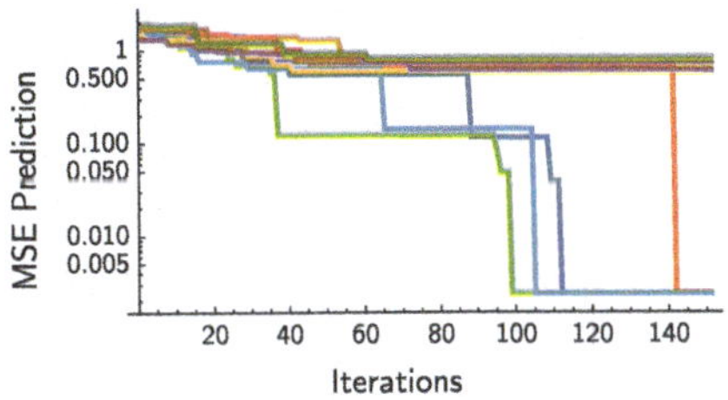

(c) Evolution of minimum prediction error w.r.t GP iterations.

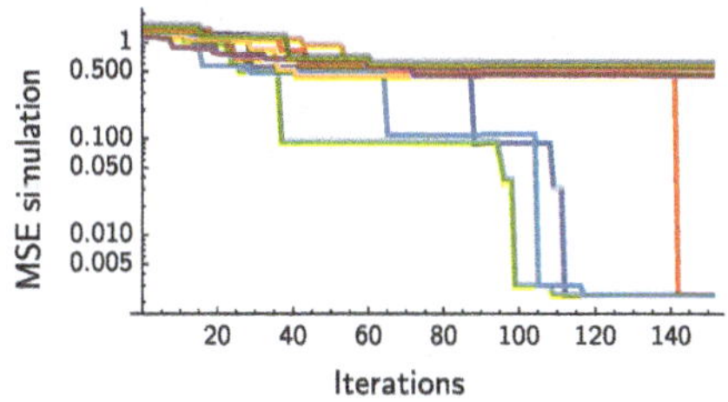

(d) Evolution of minimum simulation error w.r.t GP iterations.

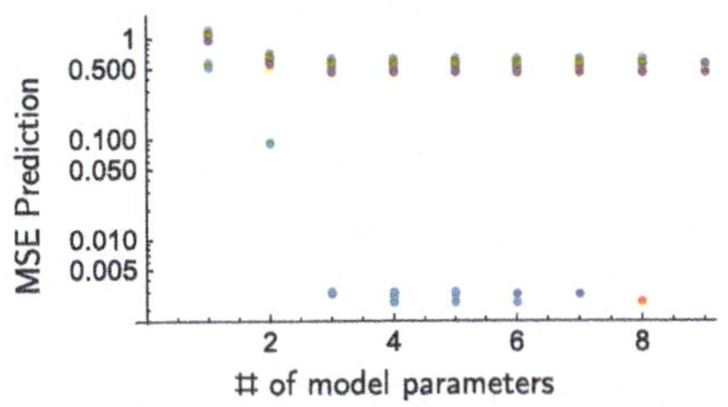

(e) Pareto front evaluated on data $\mathcal{D}_{\text{test}}$ - prediction error vs parametric complexity.

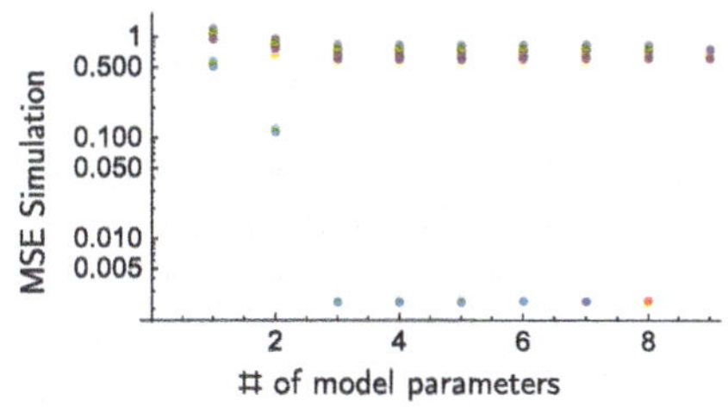

(f) Pareto front evaluated on data $\mathcal{D}_{\text{test}}$ - simulation error vs parametric complexity.

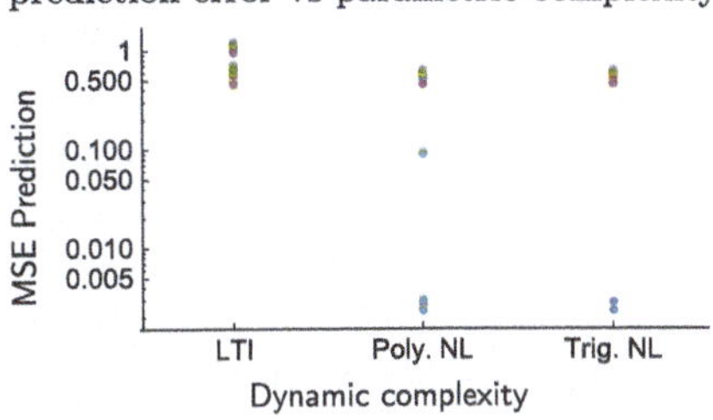

(g) Pareto front evaluated on data $\mathcal{D}_{\text{test}}$ - prediction error vs dynamic complexity.

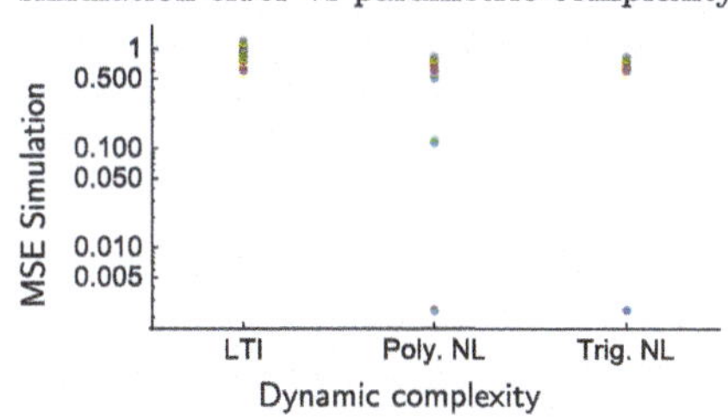

(h) Pareto front evaluated on data $\mathcal{D}_{\text{test}}$ - simulation error vs dynamic complexity.

Fig. 7.5 MC simulation results for the academic example with $p_m = 0.8$ and $p_c = 0$

Table 7.5 Number of times the correct model structure is identified for each (p_c, p_m) value pair

(p_c, p_m)	No. of times correct model structure is identified
(1, 1)	9
(1, 0.8)	9
(1, 0.6)	8
(1, 0.4)	8
(1, 0.2)	8
(1, 0)	3
(0.8, 0.8)	8
(0.6, 0.8)	7
(0.4, 0.8)	9
(0.2, 0.8)	7
(0, 0.8)	3

By comparing the distributions plots in sub-plot (b) of Figs. 7.3, 7.4, 7.20, 7.21, 7.22 and 7.23, in Appendix, we observe that the best minimum mean-squared error, for both prediction and simulation, we obtained for $p_m = 0.8$. This can also be confirmed by comparing the evolution of the minimum prediction and simulation error estimates as a function of GP iterations, in sub-plots (c), (d) of the corresponding Figures. Sub-plots (e), (f) of the corresponding Figures demonstrate that estimated Pareto-front typically consist of two levels—the upper level corresponding to the linear models in the estimated Pareto-efficient set and the lower level corresponding to the non-linear models in the efficient set. This is also confirmed in sub-plots (g), (h), which illustrate that the polynomial non-linear model class is rich enough to describe the dynamics of the DGS.

Based on these results, we select $p_m = 0.8$ for the next set of simulations for varying crossover rates. The results are summarized in Table 7.4. Again, the algorithm achieves poor convergence for low values of crossover probability. The simulation results obtained for (p_c, p_m) values of $(1, 0.8)$ and $(0, 0.8)$, respectively, are plotted in Figs. 7.3 and 7.5, while the results of the other (p_c, p_m) pairs are collected in Figs. 7.24, 7.25, 7.26 and 7.27 in Appendix. We observe similar features in these Figures in comparison with the Figures for varying probability of mutation.

Discussion

In contrast to the result obtained by random search, depicted in Fig. 7.2, the results obtained from the proposed algorithm demonstrate that the identification method developed in Chap. 6 can effectively "learn" from the models that perform comparatively better in the population, and thereby obtain a more uniform estimate of the Pareto-front. The results obtained for varying values of p_c and p_m suggest that low values of these hyper-parameters adversely affect the convergence rate of the algorithm, confirmed by the high variance of distributions of the MSE error depicted in sub-plot (b) of Figs. 7.4 and 7.5 (also see Table 7.4). On the other hand, for high values of p_c and p_m, the proposed identification is likelier to converge to the true solution, in terms of mean squared prediction and simulation error, as can be seen in sub-plot (b) of Figs. 7.3, 7.20, 7.21, 7.22, 7.24, 7.25 and 7.26 in Appendix. Besides p_c and p_m, all other algorithm hyper-parameters, listed in Table 7.1, are fairly generic and easy to interpret, and can be easily adapted by the user based on prior knowledge. Hence, besides the task of data collection and pre-processing, minimal user-interaction was required to generate the results discussed in this section. While the proposed methodology requires high computational time, as seen in the last column of Table 7.4, for many challenging real-world applications, this is a small price to pay and can be easily offset by the reduction in interaction with an expert user. Moreover, the visualizations of estimated Pareto-front in sub-plots (e)–(h) in these Figures clearly indicate the possible trade-offs to the user, thereby simplifying the task of model selection considerably. For example, based on the Pareto-front estimated for $p_c = 1$ and $p_m = 0.8$, we can conclude that the most accurate (for both prediction and simulation) *and* parsimonious model in the model set $\mathcal{M}_{\mathrm{d}}(G_{\mathrm{AT}})$ is the following polynomial model with 3 parameters:

$$y(k) = 0.5y(k-1) - 0.3u(k-5)^3 + 0.3u(k-1) + \xi(k). \tag{7.4}$$

Notice that the structure of the identified model (in the simulation setting) is identical to that of the true system (7.1). The number of times the correct model structure was successfully recovered by the proposed identification method, for different (p_c, p_m) value pairs, is listed in Table 7.5.

Based on the discussion presented in this section, we select crossover rate $p_c = 1$ and mutation rate $p_m = 0.8$ for *all subsequent identification problems* discussed in this chapter.

Remark 7.3 Recall that, in Theorem 6.8, we established that the sufficient conditions required for asymptotic convergence of the proposed identification methodology are that crossover and mutation rates satisfy $p_c \in [0, 1)$ and $p_m \in (0, 1]$, respectively. While the results tabulated in Table 7.4 confirm that the convergence rate of the proposed method decreases for mutation rates close 0 (at least for a finite number of iterations), no such effect is observed for crossover rates close to 1. Hence, the results in Table 7.4 suggest that the condition $p_c \in [0, 1)$ in Theorem 6.8 may not be necessary in order to obtain asymptotic convergence of the proposed method.

7.3 Experimental Case Studies

In this section, we look at three physical systems that pose a range of challenges in terms of data-driven modelling.

The first system is a pendulum system realized as mass attached to a disc mounted on the axle of a DC motor. It is well known that for small perturbations around the equilibrium, the dynamics of a pendulum can be well approximated as a linear system. However, for excitation signals with higher amplitudes, the dynamics become severely non-linear. Such dynamics are often observed in motion systems, making this an interesting application for data-driven modelling. For this relatively simple case-study, we investigate whether the proposed identification method can recover the model structure obtained from first principles approach, without any prior knowledge.

The second system is a thermo-mechanical modelling problem. The DGS is a mechanical positioning system designed to achieve high positioning accuracy at a Point-Of-Interest (POI). In order to achieve high positioning accuracy, one must take into account deformations induced in the system due to heating of the motor coils during operation. Hence, the modelling objective is to model the disturbances introduced at a POI in the system due to the thermal load caused by heating in the motor coils. Unlike mechanical systems, thermal dynamics are relatively slow, and hence there exists a large delay in the system between the input applied at the motor coil location, and the output located at the POI. Hence, in this case study, we investigate whether the proposed identification method can automatically detect the large delay in the system.

The third system is a drive-train of a weaving machine. The system consists of several mechanical components in translational and rotational motion, leading to complex non-linear dynamics. In such as scenario, it is challenging to hypothesize a suitable parametric non-linear model structure for the system, and expert understanding of the mechanics of the system is required to derive a first-principles model.

The results obtained for each of the case studies are tabulated in Table 7.6. The table includes

- the best mean-squared prediction error (MSE_p) achieved, on the test data-set, among all models in the estimated Pareto-efficient set,
- the best mean-squared simulation error (MSE_s) achieved, on the test data-set, among all models in the estimated Pareto-efficient set,
- the best fit ratio of prediction error BFR_p, computed as

$$\text{BFR}_\text{p} = \left(1 - \frac{\sum_{k=n_1+1}^{N_\text{test}} \varepsilon_\text{p}(k|k-1)^2}{\sum_{k=n_1+1}^{N_\text{test}} \left(y_\text{m}(k) - \frac{1}{N_\text{test}-n_1}\sum_{k=n_1+1}^{N_\text{test}} y_\text{m}(k)\right)^2}\right) \times 100\,(\%) \quad (7.5)$$

- the best fit ratio of simulation error BFR_s, computed as

$$\text{BFR}_\text{s} = \left(1 - \frac{\sum_{k=n_1+1}^{N_\text{test}} \varepsilon_\text{s}(k)^2}{\sum_{k=n_1+1}^{N_\text{test}} \left(y_\text{m}(k) - \frac{1}{N_\text{test}-n_1}\sum_{k=n_1+1}^{N_\text{test}} y_\text{m}(k)\right)^2}\right) \times 100\,(\%) \quad (7.6)$$

- the run-time of the algorithm.

Table 7.6 Overview of results for the experimental studies

Case study	MSE_p (test)	MSE_s (test)	BFR_p (%)	BFR_s (%)	Estimation time (≈ hrs.)
Pendulum	9.24×10^{-6}	8.88×10^{-3}	99.87	96.15	2
Thermal setup	1.4×10^{-5}	5.9×10^{-5}	99.63	99.28	6.3
Weaving machine (experiment 1)	5.27×10^{-6}	1.14×10^{-3}	99.73	41.17	1.5
Weaving machine (experiment 2)	9.83×10^{-4}	5.73×10^{2}	100	99.0	1.6

7.3.1 *Pendulum*

First Principles Modelling

The pendulum system is implemented as a voltage-controlled DC motor attached to a circular disc with an unbalanced mass. The system is shown in Fig. 7.6. Due to the relatively simple nature of the pendulum system, a first-principles model of the system can be easily derived. Let m be the effective mass of unbalanced disc acting at the center of gravity, l be the distance from the center of the disc to the center of gravity and J be the moment-of-inertia of the center of mass w.r.t to the axis passing perpendicular through the disc. The input applied to the system is the voltage u applied across the motor terminals and the output y of the system is the angular displacement of the center of gravity. Let R, L be the resistance and inductance of the DC motor, respectively, and i be the current flowing through the circuit. The dynamics of the electric circuit, in continuous time, is given by

Fig. 7.6 The pendulum setup

$$\frac{\mathrm{d}i(t)}{\mathrm{d}t} = -\frac{R}{L}i(t) - \frac{c_1}{L}\frac{\mathrm{d}y(t)}{\mathrm{d}t} + \frac{y(t)}{L}, \tag{7.7}$$

where the last term in the differential equation is the back e.m.f. arising due to the loading of the motor and c_1 is a suitable co-efficient. Using the backward-Euler approximation, for sampling-time h, the dynamics of the circuit can be expressed in approximately in discrete-time as

$$\left(1 + \frac{hR}{L}\right) i(k) - i(k-1) = -\frac{c_1}{L}y(k) + \frac{c_1}{L}y(k-1) + \frac{h}{L}u(k). \tag{7.8}$$

The dynamics of the pendulum in continuous-time is given by

$$J\frac{\mathrm{d}^2 y(t)}{\mathrm{d}t^2} = c_2 i(t) - c_3\frac{\mathrm{d}y(t)}{\mathrm{d}t} + mgl\sin(y(t)), \tag{7.9}$$

where the first term in the differential equation relates to the torque produced by the motor, the second term related to the friction in the system and the last term relates to the dynamics of the pendulum. Again, using the backward-Euler method and re-arranging the terms, we get

$$i(k) = \left(\frac{J}{h^2} + \frac{c_2}{h}\right) y(k) - \frac{mgl}{h^2}\sin(y(k)) - \left(\frac{2J}{h^2} + \frac{c_2}{h}\right) y(k-1) + \frac{J}{h^2}y(k-2). \tag{7.10}$$

The overall system can be obtained by eliminating the current terms $i(k)$ and $i(k-1)$ from the two equations (using substitution). For simplicity, we give here the parametric form of the overall system:

$$\begin{aligned} &y(k) + \theta_1\sin(y(k)) + \theta_2 y(k-1) + \theta_3\sin(y(k-1)) + \theta_4 y(k-2) \\ &+\theta_5 y(k-3) + \theta_6 u(k) = 0. \end{aligned} \tag{7.11}$$

System Identification Using the Proposed Method

To collect experimental data to identify the system, a torque can be applied to the mass by applying an excitation signal to the motor and the angular displacement can be measured (in radians) as the output of the system. The sampling frequency of the system 50 Hz. Three datasets are collected and used as follows:

- Estimation dataset $\mathcal{D}_{\text{est}}$ is used to estimate the model parameters that minimize the lower-level objective, i.e., prediction error.
- Validation dataset $\mathcal{D}_{\text{val}}$ is used to compute the multi-objective fitness of estimated models.
- Test dataset $\mathcal{D}_{\text{test}}$ is used to estimate the generalization error of the final Pareto-set of models returned by Algorithm 1 in Chap. 6.

The inputs used to generate the three datasets are uniformly distributed random signals in the range $[-A_{\max}, A_{\max}]$, with the maximum amplitude $A_{\max}$ being 15, 10 and 12 V for the estimation, validation and test datasets, respectively. All excitation signals were subsequently low-pass filtered using a discrete-time filter with transfer function

$$H(z^{-1}) = \frac{0.1z^{-1}}{1 - 0.9z^{-1}}, \tag{7.12}$$

and bandwidth 0.85 Hz. Each input sequence has length $N = 1000$ samples, where the first $N_t = 50$ samples are set to 0 to remove any transients. The estimation and test dataset, due to their higher amplitudes, include the flipping of the pendulum by more than π radians, while the validation dataset does not include the flipping behaviour. This is done to ensure that the system is sufficiently excited to exhibit the non-linearities present in the system.

Since the system exhibits rotational dynamics, it is reasonable to include trigonometric functions such as sine and cosine in the grammar. Hence, we use G_{trig} to define the model set. Furthermore, the crossover and mutation rates are set to 1 and 0.8, respectively. This choice is based on the results discussed in Sect. 7.2.

The results obtained from the proposed identification method are presented in Fig. 7.7. In Fig. 7.7a, simulation and one-step-ahead prediction responses corresponding to the best models estimated for simulation and prediction error, respectively, are plotted along with the measured output of the test dataset. It can be observed that both models capture the flipping behaviour (seen at the end of the dataset) in both prediction and simulation. The 2-D projections of the final Pareto-front are depicted in Fig. 7.7b, c and the final Pareto-front evaluated on independent test dataset $\mathcal{D}_{\text{test}}$ is depicted in Fig. 7.7d, e. The following observations can be made based on the estimated efficient set.

- The Pareto-fronts in Fig. 7.7b, d have the general trend that is expected—the mean-squared prediction error initially decreases as the model complexity (number of parameters) increases, and subsequently stabilizes beyond a certain level of complexity. The information present in these plots enable a user to select the level of parametric complexity that is appropriate for the ultimate use-case of the estimated model.
- Unlike the prediction error based Pareto-fronts, the simulation error based Pareto-fronts in Fig. 7.7c, e have an unusual shape. This is most likely due to two reasons—(i) model parameters are optimized only for prediction error, and (ii) simulation error performance of a model is more sensitive to errors in the model structure.
- In Figs. 7.7c, e, we can clearly observe that models containing trigonometric non-linearities clearly out-perform those that do not. While the difference in performance between the various model structures may decrease if the parameters are also optimized for simulation error, it will certainly not disappear. This is due to the fact that an LTI model cannot possibly describe the non-linear dynamics of a pendulum. Similarly, while a polynomial model may achieve arbitrarily good approximation of the dynamics of the system for a specific class of excitation signals, such approximate models are significantly more complex (in terms of number of model parameters).

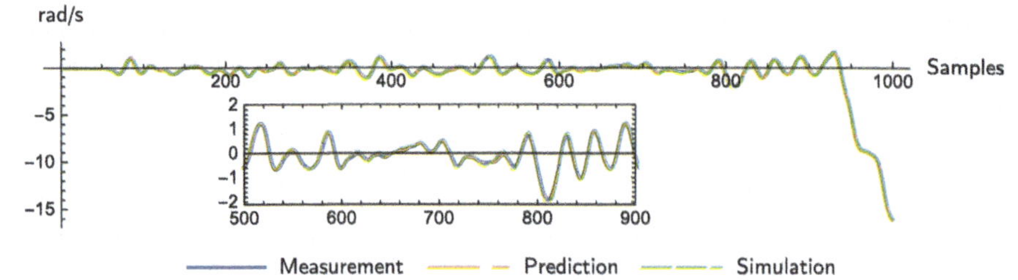

(a) Measured output compared with the prediction and simulation responses computed from models achieving the best prediction and simulation performance on test data $\mathcal{D}_{\text{test}}$.

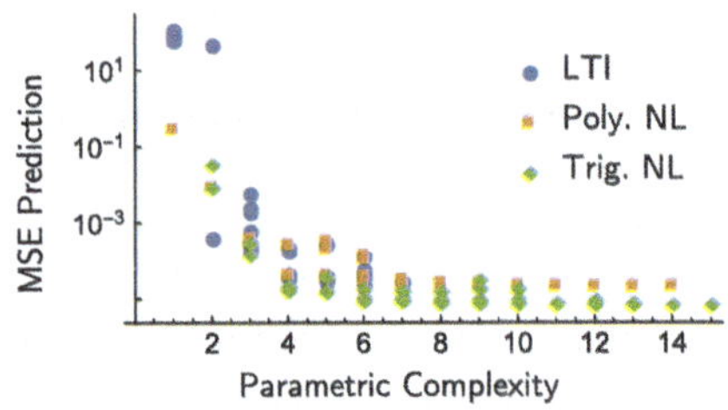

(b) Pareto front evaluated on data $\mathcal{D}_{\text{val}}$ - prediction vs complexity.

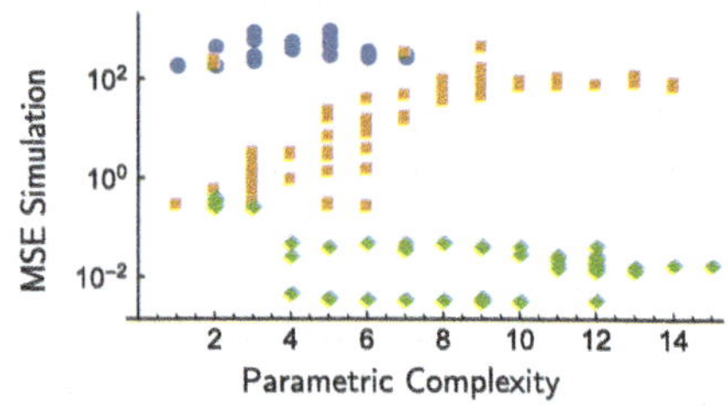

(c) Pareto front evaluated on data $\mathcal{D}_{\text{val}}$ - simulation vs complexity.

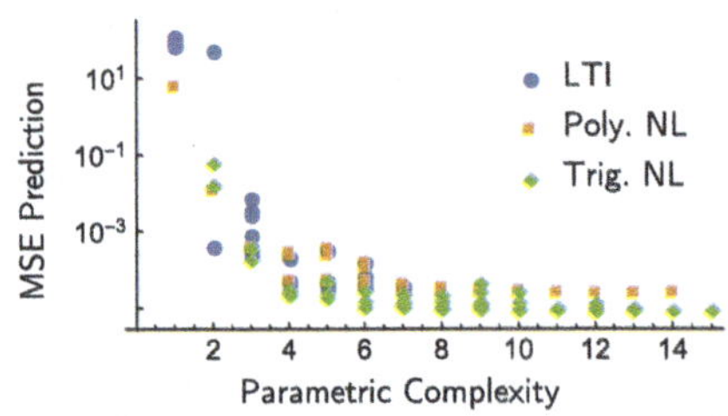

(d) Pareto front evaluated on data $\mathcal{D}_{\text{test}}$ - prediction vs complexity.

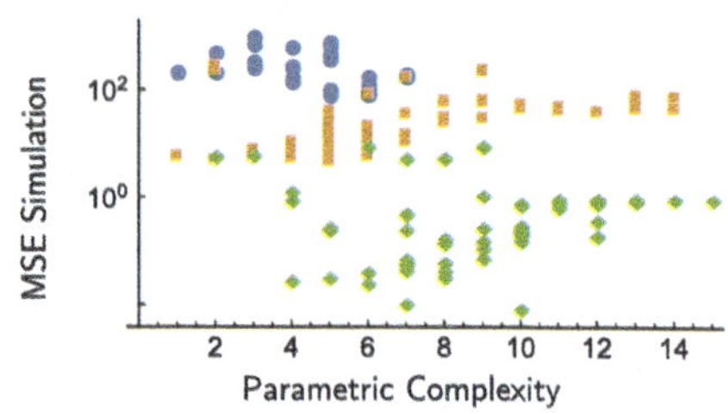

(e) Pareto front evaluated on data $\mathcal{D}_{\text{test}}$ - simulation vs complexity.

Fig. 7.7 Numerical results for the pendulum case-study

The best quality measures achieved by models that belong to the Pareto-front on the test dataset are reported in Table 7.6. Based on the Pareto-front projections depicted in Figs. 7.7d, e, we observe that no significant improvement in prediction, or simulation, is observed beyond a parametric complexity of 8. The Pareto-optimal model with 8 parameters that achieves the best simulation performance is

$$\begin{aligned} y(k) = &-0.073\sin(y(k-1)) - 0.478y(k-3) + 1.478y(k-1) + 0.004u(k-3) \\ &+ 0.008u(k-2) + 0.003u(k) + 0.371\xi(k-1) + \xi(k). \end{aligned} \tag{7.13}$$

The estimated model (7.13) has a structure that is similar to that of the first-principles model in (7.11). However, while a first-order approximation was used to compute the discrete-time model (7.11), it is likely that higher-order terms are also included in the model estimate obtained from the proposed algorithm. The model estimate (7.13) can easily be used in a variety of applications, e.g., it is possible to design a feedback-linearizing controller for this system, since the model reveals the exact nature of the non-linearity in the system.

The pendulum case study illustrates that the modelling approach could automatically build model structures to describe the non-linear dynamics of the system without prior knowledge. This would not be possible within the paradigm of linear time-invariant systems. Furthermore, the case-study also illustrates the utility of the estimated Pareto-front in enabling the user to make informed choices.

7.3.2 Thermal Setup

The thermal setup is a positioning system in which a mass is actuated by two motors so as to align a POI to a set reference. During actuation, the electric coils of the motors heat up, leading to thermal deformations in the system. For highly accurate positioning, the deformations caused due to the thermal disturbance must be taken into account. Hence, in order to develop a model-based controller to compensate for the thermal-load induced disturbances, we need to estimate a model that describes the dynamics of these deformations. The dominant dynamics of the system is due to thermal conduction and convection of heat from the motor coils to the POI and due to thermo-elastic deformations, the former being a slow phenomena. Hence, there exists a large delay in the system. From the perspective of automatic data-driven modelling, we investigate if the proposed identification method can automatically estimate the delay in the system without prior specification.

A schematic of the system is depicted in Fig. 7.8. The motor coils cannot be independently actuated, i.e., the same voltage is applied across moth motors coils. The displacement of the POI is measured by a 1-D displacement sensor. For the identification task, the input excitation is the voltage applied to the motor coils, and the output of the system is the deformation measured at the POI.

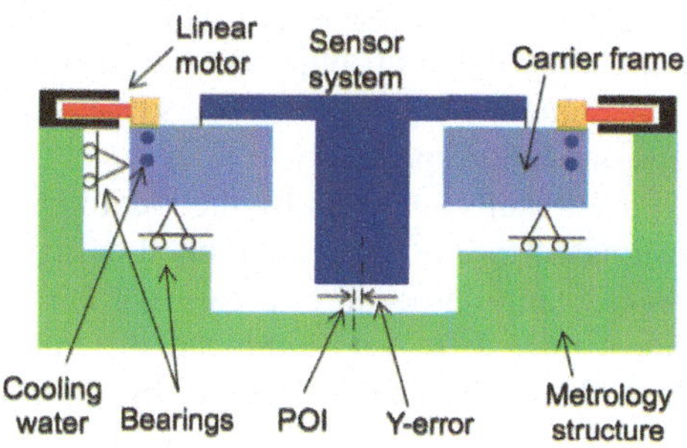

Fig. 7.8 Illustration of the thermal setup

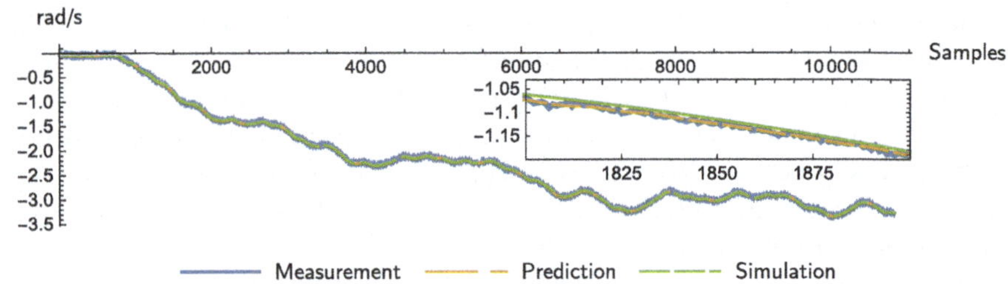

(a) Measured output compared with the prediction and simulation responses computed from models achieving the best prediction and simulation performance on estimation data $\mathcal{D}_{\text{est}}$.

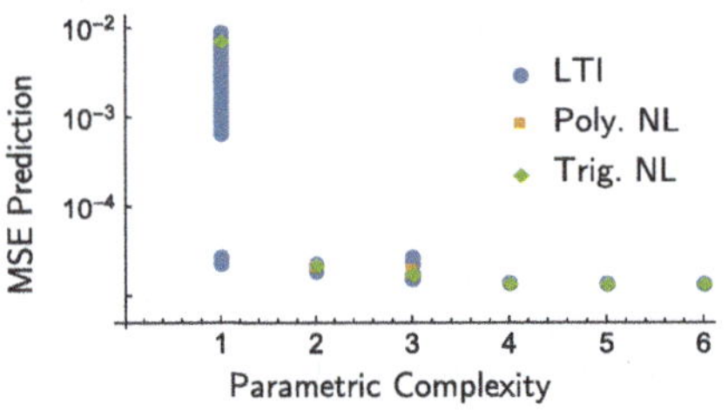

(b) Pareto front evaluated on data $\mathcal{D}_{\text{est}}$ - prediction vs parametric complexity.

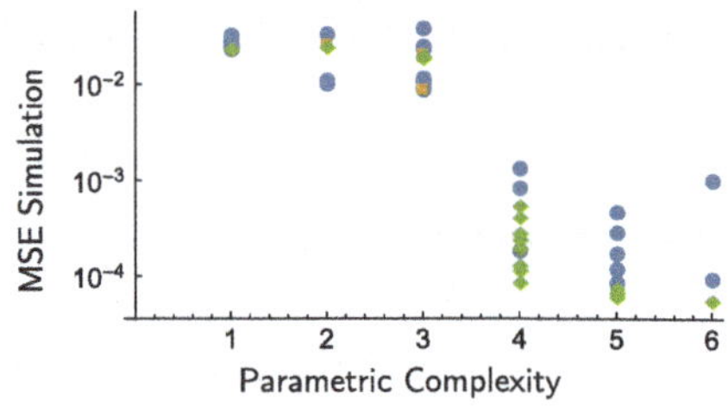

(c) Pareto front evaluated on data $\mathcal{D}_{\text{est}}$ - simulation vs parametric complexity.

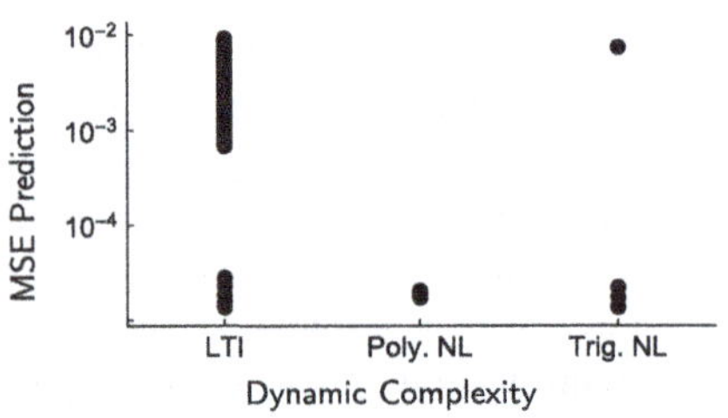

(d) Pareto front evaluated on data $\mathcal{D}_{\text{est}}$ - prediction vs dynamic complexity.

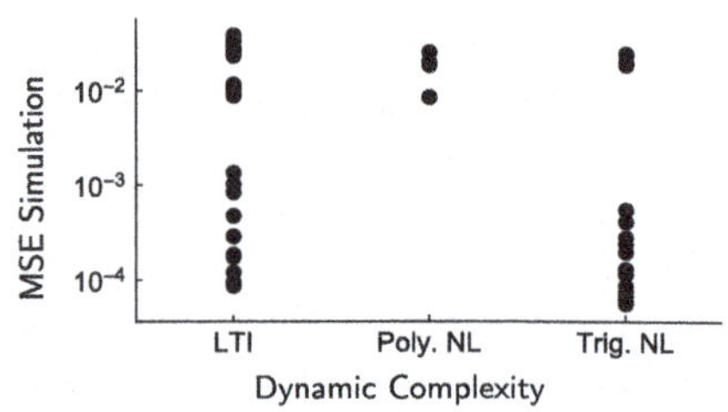

(e) Pareto front evaluated on data $\mathcal{D}_{\text{est}}$ - simulation vs dynamic complexity.

Fig. 7.9 Numerical results for the thermal case-study

The system is excited with a PRBS signal ranging between 0 and 2 V at a sampling rate 1 Hz. The deformations are measured in μm. Only one dataset $\mathcal{D}_{\text{est}}$ of $N = 10{,}810$ samples is measured. The same dataset is used for parameter estimation and computing the multi-objective fitness of the models.

Grammar G_{trig} is used for this case-study. The crossover and mutation rate is set at 1 and 0.8, respectively.

The results achieved by the identification algorithm are plotted in Fig. 7.9. The estimated pareto-fronts are plotted in Fig. 7.9b–e. The Pareto-front estimated by the identification procedure reveals that the dynamics of the system are predominantly linear since there is no significant difference between the best performance achieved by LTI models and that achieved by non-linear models (see Figs. 7.9d–e). Moreover,

Table 7.7 Comparison of performance measures for the thermal case study

Method	MSE_p	MSE_s	BFR_p (%)	BFR_s (%)
PEM	1.8×10^{-5}	1.8×10^{-5}	99.57	99.57
Proposed	1.4×10^{-5}	5.9×10^{-5}	99.63	99.28

the identified models were also able to estimate the delay in the system automatically. For e.g., a Pareto-optimal model with 6 parameters estimated by the algorithm is

$$\begin{aligned} y(k) &= -0.00008u(k-54) - 0.00014u(k-8) + 0.0001u(k) \\ &\quad - 0.0377y(k-25) + 1.0377y(k-1) - 0.9628\xi(k-1) + \xi(k), \end{aligned} \tag{7.14}$$

which models a delay of 54 samples in the first input term. Note that estimating large delays corresponds to large number of adjunctions of auxiliary tree β_7 (see Fig. 7.1) in the model. As a consequence, the identified models also reached the maximum limit of 150 adjunctions during the evolutionary search. Hence, the results obtained could be further improved by increasing the limit of maximum adjunctions for this specific case study. Nonetheless, the results achieved by the proposed identification method are comparable to state-of-the-art linear identification methods [8]. For comparison, the same dataset was used to identify linear models of OE structure[2] using PEM [8] method. As an initial step, the delay in the system was estimated (using the `delayest.m` command in MATLAB). Subsequently, linear models were estimated for a number of model structures, varying model orders ranging from 1 to 10. No significant improvement in performance was observed for models with more than 2 poles, corresponding to the model:

$$\begin{aligned} y(k) &= 1.18 \times 10^{-5}u(k-40) - 2.18 \times 10^{-5}u(k-41) + 1.992y(k-1) \\ &\quad - 0.9918y(k-2) + \xi(k). \end{aligned} \tag{7.15}$$

A comparison of the best performance measures achieved by the pareto-optimal models identified using the proposed method and the best LTI model (corresponding to 10 poles) achieved using PEM method is presented in Table 7.7. The one-step-ahead prediction and simulation response of the corresponding models is plotted in Fig. 7.9a. The best performance measures achieved are also summarized in Table 7.6.

[2] An Auto-Regressive models with eXogenous inputs (ARX) model structure was also used with varying model order upto 10. However, no significant difference in performance was obtained w.r.t OE model structures.

7.3.3 Weaving Machine

The weaving machine is a mechanical system that converts rotational motion to translational motion. The system is pictured in Fig. 7.10a, and schematic of the system is illustrated in Fig. 7.10b. The motor applies a torque T_m, that is converted to torque T_d on the crank via a system of mechanical gears. The crank-shaft system is connected to a body of mass m_s, and the rotational motion of the crank is converted to translational motion of the mass. Finally, the mass connected to the end of the crank shaft is also attached to a non-linear spring. Let θ_l be the angular displacement of the crank and x_s be the displacement of the mass. More details of the system can be found in [20]. The modelling task is to identify a model for the translational displacement x_s of the mass based on the torque T_m applied by the motor.

For this case study, measurements for the identification task are obtained from a high-fidelity first principles model simulator of the system provided by Flanders Make (see [20]) rather than the system itself. Three experiments are performed on the model, at a sampling rate of 250 Hz, resulting in the three datasets $\mathcal{D}_{\text{est}}$, $\mathcal{D}_{\text{val}}$ and $\mathcal{D}_{\text{test}}$ used for the identification task. The excitation signal used in each of the experiments is a band-limited random-phase multi-sine signal in the frequency range of $[0, 20]$ Hz with a uniform distribution on the phase for each point in the discrete frequency grid.

TAG G_{AT} is used to characterize the model set of the algorithm. The crossover and mutation rate of the algorithm are set to 1 and 0.8, respectively.

The results obtained from the proposed identification methodology are summarized in Fig. 7.11. In Fig. 7.11a, the model responses of the best model estimates in the archive (for prediction and simulation) are compared with the independent test dataset $\mathcal{D}_{\text{test}}$. We can observe that, while the prediction output of the model estimate that achieves the best prediction performance closely follows the measured output, the model estimate with the best simulation performance completely misses the dynamics of the system, and only captures the average trend. Figures 7.11b, c depict the 2-D projections of the estimated Pareto-front, and Fig. 7.11e and f depict the performance of the same models evaluated on the test dataset. Figure 7.11b, d illustrate that the prediction performance of the estimated models improve upto models with

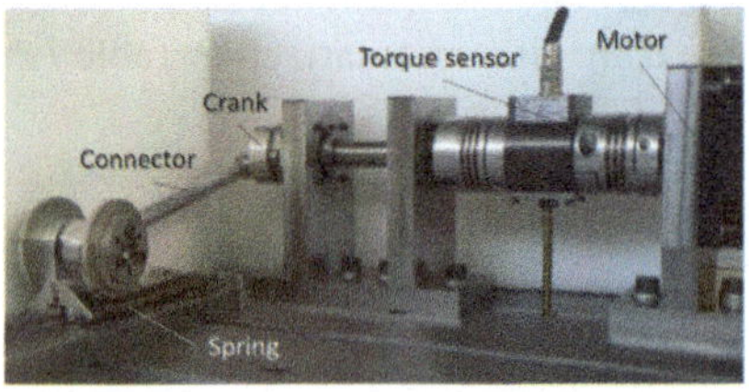

(a) The weaving machine system.

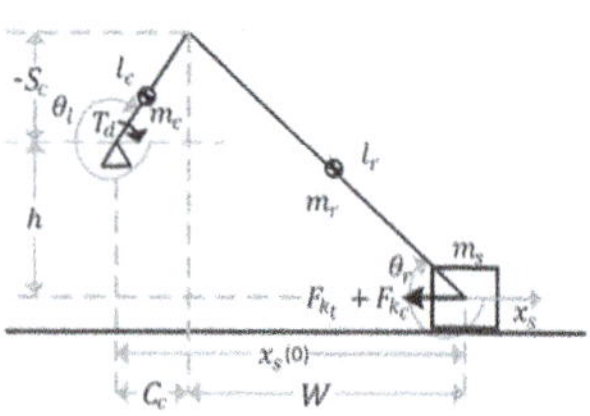

(b) A schematic of the weaving machine system.

Fig. 7.10 Illustrations of the weaving machine case study

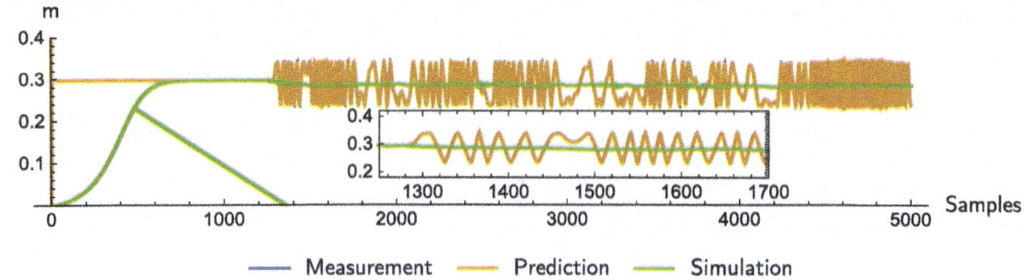

(a) Measured output compared with the prediction and simulation responses computed from models achieving the best prediction and simulation performance on test data $\mathcal{D}_{\text{test}}$.

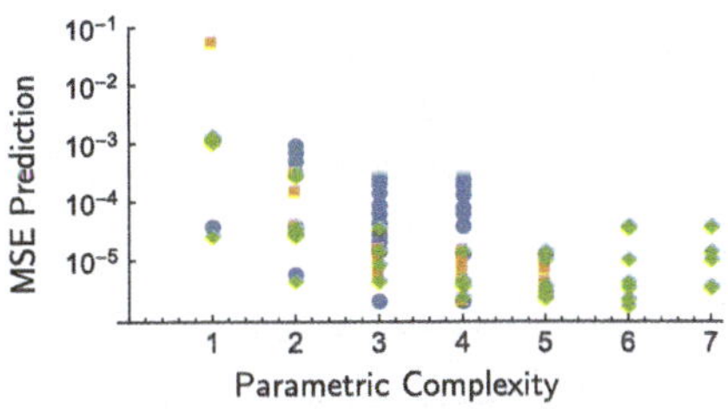

(b) Pareto front evaluated on data $\mathcal{D}_{\text{val}}$ - prediction vs parametric complexity.

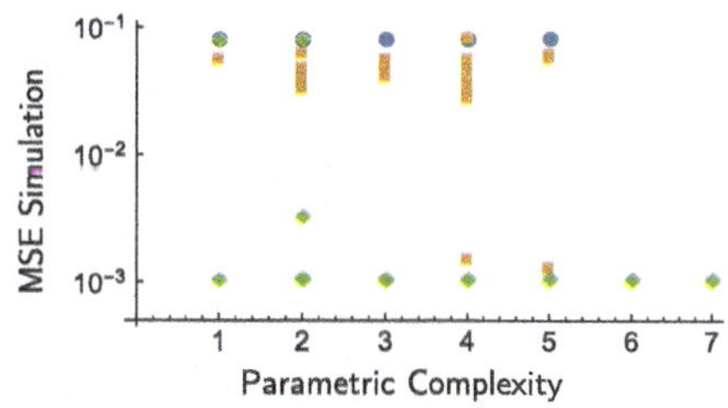

(c) Pareto front evaluated on data $\mathcal{D}_{\text{val}}$ - simulation vs parametric complexity.

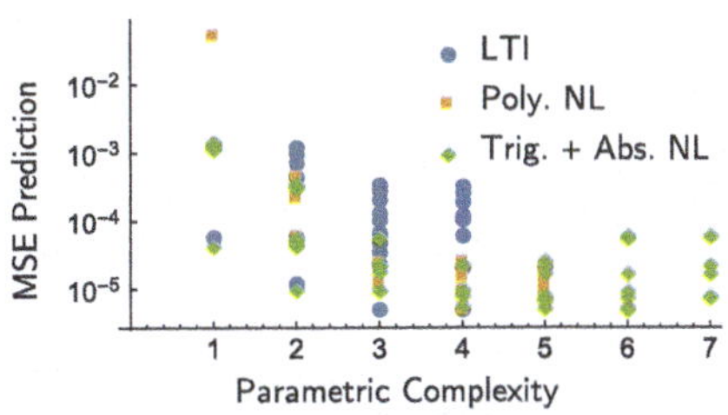

(d) Pareto front evaluated on data $\mathcal{D}_{\text{test}}$ - prediction vs parametric complexity.

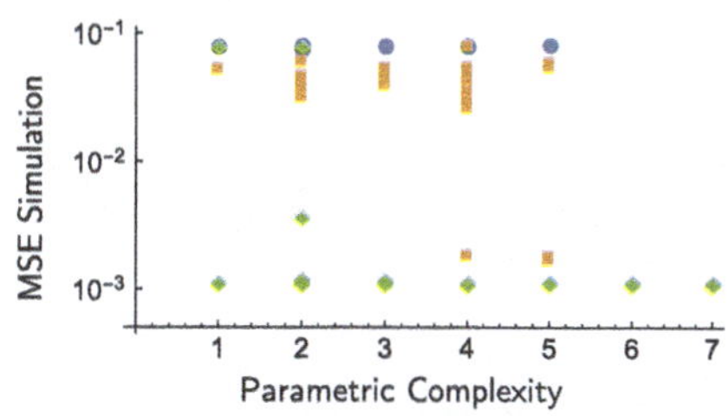

(e) Pareto front evaluated on data $\mathcal{D}_{\text{test}}$ - simulation vs parametric complexity.

Fig. 7.11 Numerical results for the weaving machine case study

4 parameters, and subsequently stabilizes. However, the simulation performance of the estimated models in Fig. 7.11c, e does not follow any discernible trend. This suggests that there is a structural error in the estimated models, and since the model parameters are only estimated for prediction error, we obtain good estimates only for the prediction case.

In order to investigate the cause for poor simulation performance of the model, we perform a second identification experiment. In the schematic in Fig. 7.10b, we can observe that the dynamics of the system can be completely described by the dynamics between torque T_m and the angular displacement θ_l of the crank, since the position of mass m_s is merely a static non-linear function of θ_l. Hence, for the second identification experiment, torque T_m is chosen to be the input signal and angular displacement θ_l is the output measured from the system.

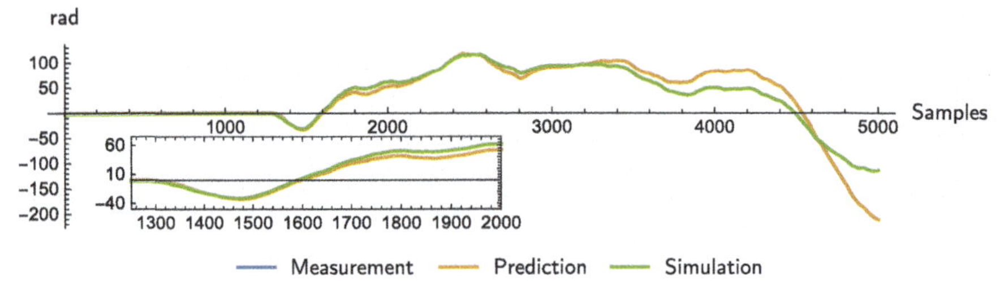

(a) Measured output compared with the prediction and simulation responses computed from models achieving the best prediction and simulation performance on test data $\mathcal{D}_{\text{test}}$.

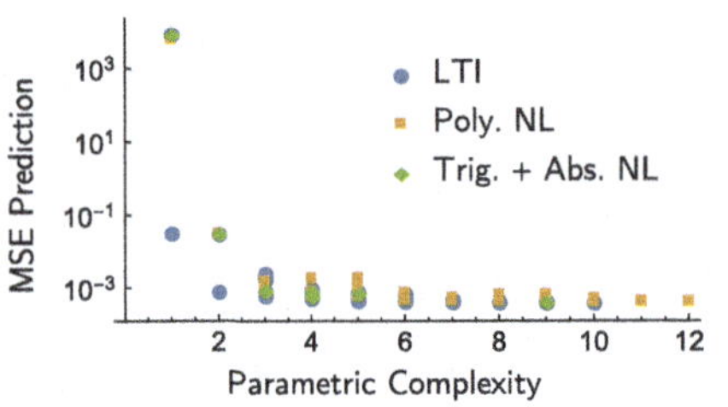

(b) Pareto front evaluated on data $\mathcal{D}_{\text{val}}$ - prediction vs parametric complexity.

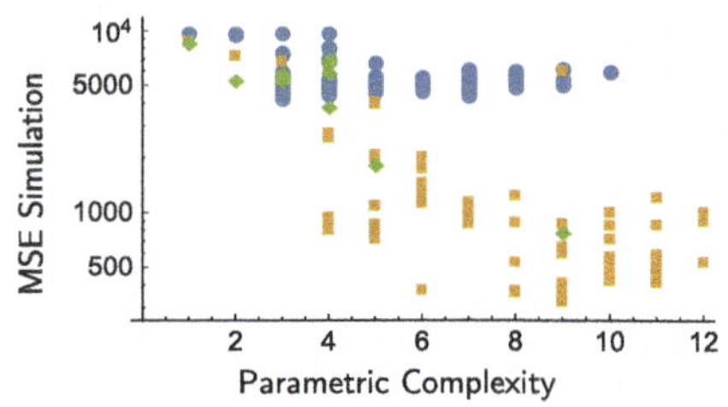

(c) Pareto front evaluated on data $\mathcal{D}_{\text{val}}$ - simulation vs parametric complexity.

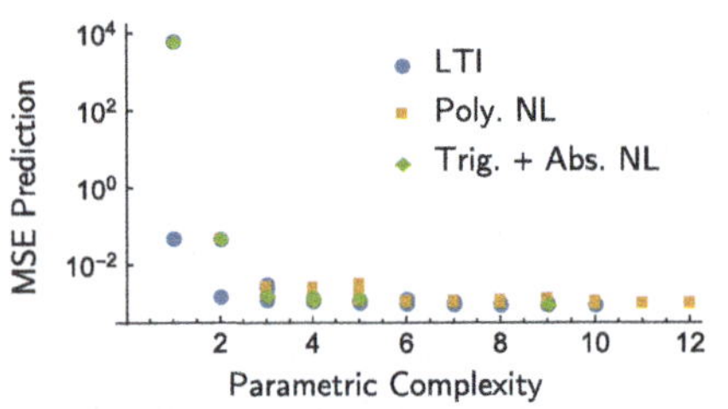

(d) Pareto front evaluated on data $\mathcal{D}_{\text{test}}$ - prediction vs parametric complexity.

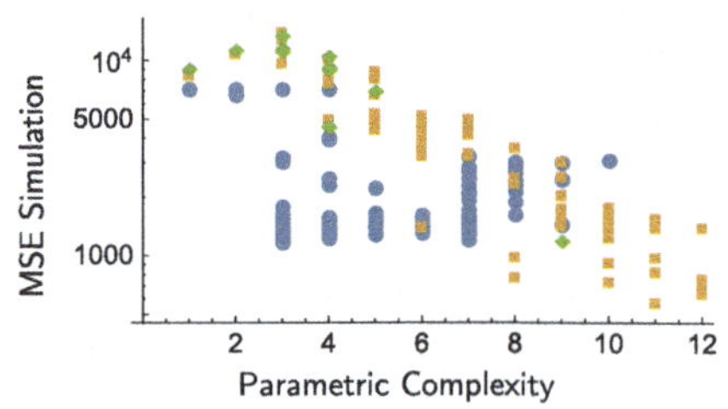

(e) Pareto front evaluated on data $\mathcal{D}_{\text{test}}$ - simulation vs parametric complexity.

Fig. 7.12 Numerical results for the second identification experiment for the weaving machine case study

The results of the second identification experiment are plotted in Fig. 7.12. The model responses obtained from the best prediction and simulation model estimated by the proposed algorithm are compared with the test dataset. Again, the prediction response closely matches the measured output of the system. Unlike the previous experiment, the simulation response does capture the dynamics of the system, however, the model response begins to drift away from the measurement as time progresses. This indicates that there is still some structural error in the estimated model. Nonetheless, the simulation performance of the estimated models for the present experiment configuration is significantly better than that for the previous configuration. In other words, the quality of the model estimate is improved by simply altering the system variable being measured. The 2-D projections of the Pareto-front and the performance of the Pareto-efficient models evaluated on the test dataset are depicted in Fig. 7.12b–e, respectively. Comparing Fig. 7.12c, e, we can observe that the simulation models do not generalize well to the independent test dataset, suggesting that there is plenty of room to improve the quality of model estimates for simulation error.

7.4 Benchmark Examples

In this section, we use the proposed identification method to estimate models of non-linear system identification benchmark datasets. Developing models for benchmark datasets allow us to compare the results obtained from the proposed identification method with those obtained from other non-linear system identification techniques.

7.4.1 Silverbox

The first non-linear benchmark dataset we look at is the so-called *silverbox* data-set proposed in [22]. The silverbox system in an electronic implementation of a mass-spring-damper system with a non-linear spring. The data-set, measured at a sampling rate of 610.35 Hz, consists of 2 parts. The excitation signal used in the first 40,000 samples is a low-pass filtered Gaussian noise signal with a bandwidth 200 Hz and a linearly increasing amplitude ranging from 0 to a maximum value of about 0.3 V. The data-set is plotted in Fig. 7.13. The excitation signal in the second part of the data-set consists of 10 realizations of a random odd multi-sine signal (see [22] for details). The dataset is utilized as follows:

- The first 9 realizations of the multi-sine data as estimation data-set $\mathcal{D}_{\text{est}}$ to estimate model parameters,
- The last realization of the multi-sine data is used as validation data-set $\mathcal{D}_{\text{val}}$ to evaluate the multi-objective fitness of the proposed models, and
- The Gaussian excitation part of the data-set is used as a test data-set $\mathcal{D}_{\text{test}}$ that is independent of the remaining data used during the identification procedure, and is used to evaluate the performance of the Pareto-front obtained at the end of the identification algorithm.

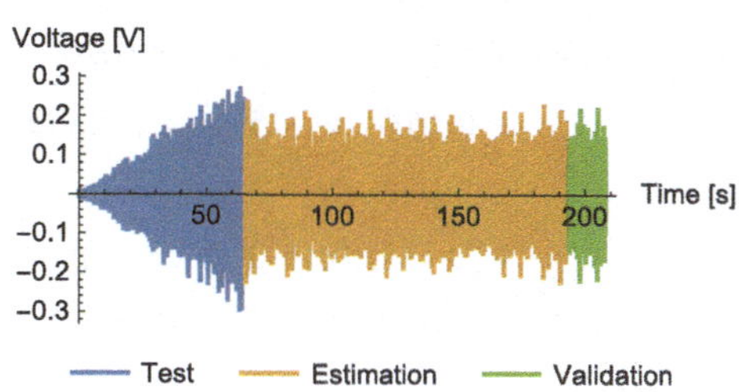

Fig. 7.13 Silverbox measured data

Based on the general information about the system provided in [22] we choose the P-NARMAX grammar G_N for the identification task. Again, the probabilities of crossover and mutation is chosen to be 1 and 0.8, respectively. It should be noted that no other information specific to the benchmark example was incorporated into the identification procedure.

The results obtained are plotted in Fig. 7.14. Figure 7.14c, d depict 2-D projections of the final pareto-front obtained from the identification procedure. Figure 7.14e, f depict the projections of the same Pareto-front evaluated on the independent test dataset $\mathcal{D}_{\text{test}}$. The Pareto-front informs the user that in order to obtain a very accurate model of the silverbox system, the LTI framework is insufficient, and one must model it as a non-linear system. Furthermore, the Pareto-front also informs the user of the trade-off that can be made between the model parametric complexity and model accuracy.

From Fig. 7.14c–f we observe that beyond the complexity level of 6 (i.e. 6 model parameters), the improvement in both performance measures is only gradual. Figure 7.14a depicts the prediction and simulation errors (scaled up by a factor of 10) of best models estimated for prediction and simulation, respectively. We observe in Fig. 7.14a that the latter half of the test data-set requires the model to extrapolate since the magnitude of the input is greater than that in the estimation and validation set. This is the part of the dataset that accrue the most errors for both prediction and simulation error.

The best RMS simulation and prediction error achieved by the model estimates in archive $\mathfrak{A}^{(L)}$, on the test dataset, is presented in Table 7.8. The table also lists results obtained from a number of other non-linear system identification methods, as found in the literature. The performance of the proposed approach is comparable to that of state-of-the-art non-linear system identification methods. The best result till date was achieved by the pNLSS identification method, see [13]. While the pNLSS method produces a single model, in this case with 37 parameters, the proposed approach provides a Pareto-front of models, enabling the user to choose the performance-complexity trade-off *a-posteriori*. Furthermore, in the proposed method, the user is not required to make critical choices, while in the case of pNLSS identification, the user must make choices that may strongly influence the result, such as the order of non-linearity considered and the initialization method. For example, from Fig. 7.14f we can conclude that, for simulation performance, no significant improvement is obtained beyond a parametric complexity of 15. The Pareto-efficient model with 15 parameters that achieves the best simulation performance, is

$$\begin{aligned} y(k) = {} & 0.043u(k-3) + 0.101u(k-2) + 0.380u(k-1) + 0.047u(k) \\ & + 0.016\xi(k-5) - 4.281y(k-1)^3 - 2.239y(k-3)y(k-1)^2 \\ & + 7.973y(k-2)y(k-1)^2 - 5.695y(k-2)^2y(k-1) \\ & + 2.111y(k-3)y(k-2)y(k-1) + 0.009y(k-2)y(k-1) \\ & + 1.329y(k-1) + 0.00134492y(k-6) - 0.0872614y(k-4) \\ & - 0.80891y(k-2) + \xi(k). \end{aligned} \tag{7.16}$$

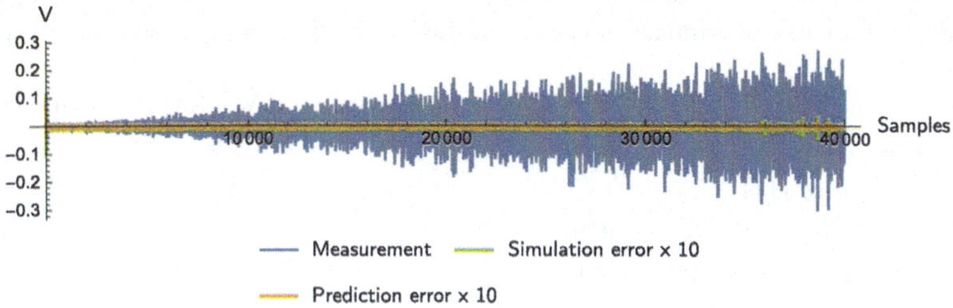

(a) Measured output compared with the prediction and simulation errors computed from models achieving the best prediction and simulation performance on test data $\mathcal{D}_{\text{test}}$. The errors are scaled up by a factor of 10 to make them visible.

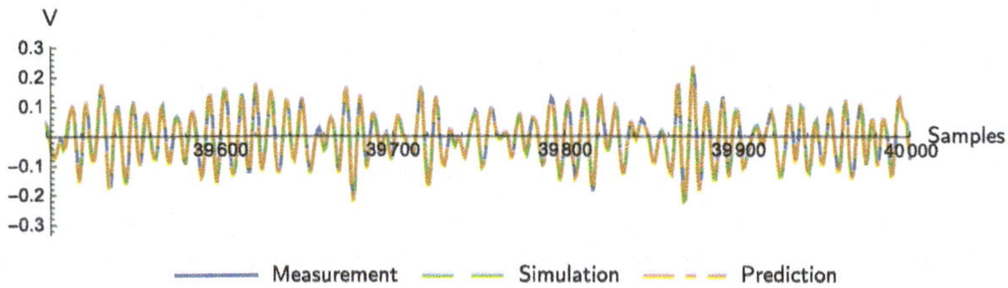

(b) Measured output compared with the prediction and simulation responses computed from models achieving the best prediction and simulation performance on test data $\mathcal{D}_{\text{test}}$, and zoomed in to the end of the dataset.

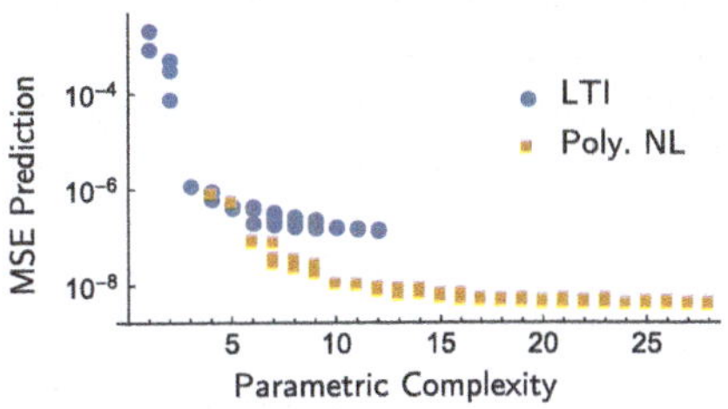

(c) Pareto front evaluated on data $\mathcal{D}_{\text{val}}$ - prediction vs parametric complexity.

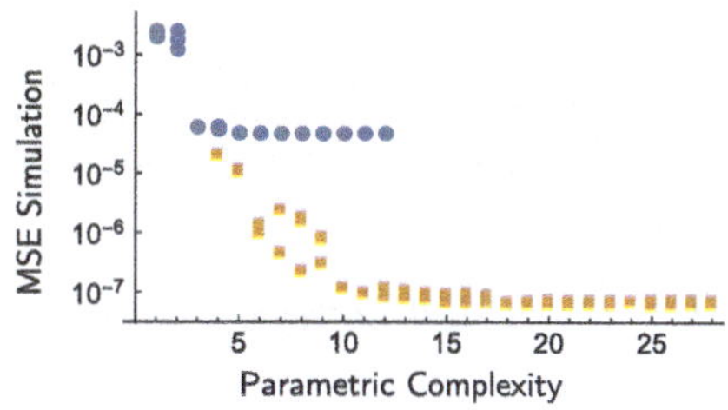

(d) Pareto front evaluated on data $\mathcal{D}_{\text{val}}$ - simulation vs parametric complexity.

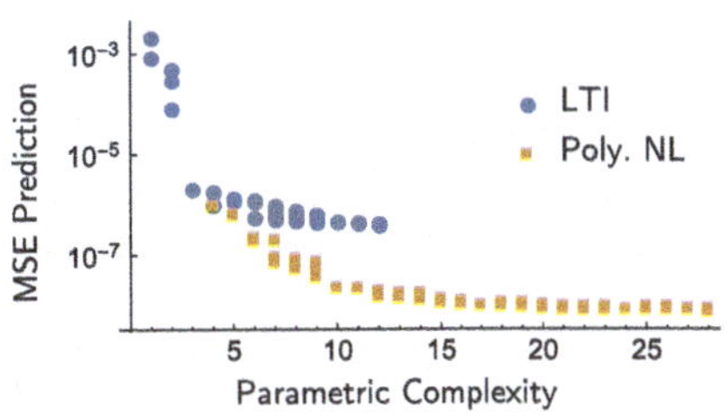

(e) Pareto front evaluated on data $\mathcal{D}_{\text{test}}$ - prediction vs parametric complexity.

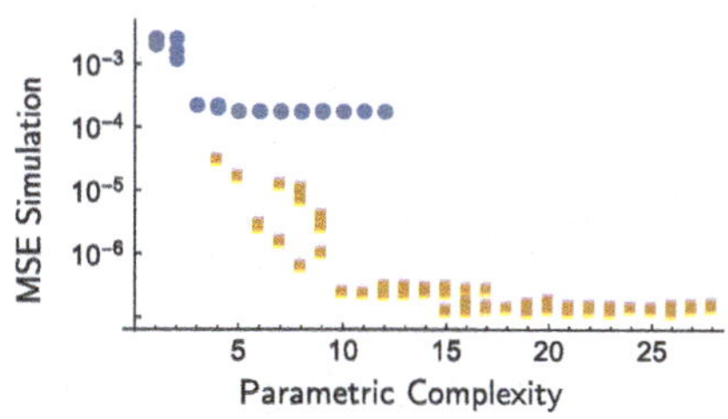

(f) Pareto front evaluated on data $\mathcal{D}_{\text{test}}$ - simulation vs parametric complexity.

Fig. 7.14 Numerical results for the Silverbox benchmark

Table 7.8 Performance of estimated models computed on the silverbox test data-set and compared with literature

Identification method	Test RMS simulation (mV)	Val. RMS simulation (mV)
Proposed method	**0.36**	**0.09**
Best Linear Approx. [11]	13.5	6.9
pNLSS [13]	**0.26**	–
Hjalmarsson and Schoukens [5]	0.96	1.4
Sigmoidal network models [9]	0.30	–
LS-SVM [3]	0.32	0.23
Poly-LFR [19]	0.35	–
Local linear models [21]	1.3	1.1

The blank cells correspond to values not reported in the literature. Bold indicates results obtained by method proposed in this thesis

From the estimated model, we can observe that the predominant non-linearity in the model is cubic in nature (all non-linear model terms are of order 3, except one non-linear term which is quadratic). Indeed, the non-linearity in the true system, as described in [22], is cubic in nature. On the other hand, in [13], the authors first estimate a linear model, and use the resulting estimate to initialize the non-linear identification procedure. Furthermore, the authors resort to a trial-and-error-based approach to select an appropriate structure for the state-space model to be identified. For the Silverbox system, a trial-and-error-based approach for model selection is feasible since it can be well described by a 2nd order state-space. However, for models with higher state dimensions, this is not a feasible approach.

7.4.2 Coupled Drives

The coupled drives dataset was proposed as a benchmark dataset for non-linear system identification in [23]. The setup consists of two electric motors driving a flexible belt around a pulley. The setup is illustrated in Fig. 7.15. The input signal in the dataset is the sum of voltages applied across the two motors, and the output is the pulley velocity as measured by a pulse-counting sensor. The main modelling challenges are (i) short datasets of 500 samples each, and (ii) the sensor is insensitive to the sign of the velocity, resulting in an absolute value non-linearity. The following datasets were used

- Estimation dataset $\mathcal{D}_{\text{est}}$—input $u11$ and output $z11$ (see [23]), is used to estimate model parameters and compute the multi-objective fitness measures in the upper-level optimization problem.
- Validation dataset $\mathcal{D}_{\text{test}}$—input $u12$ and output $z12$ (see [23]), is used to evaluate the final Pareto-set returned by the proposed algorithm.

For this experiment, we use TAG G_{AT} to define the model set since it contains the $\sin$, $\cos$ andabs functions, which are appropriate for the given case study. Just like

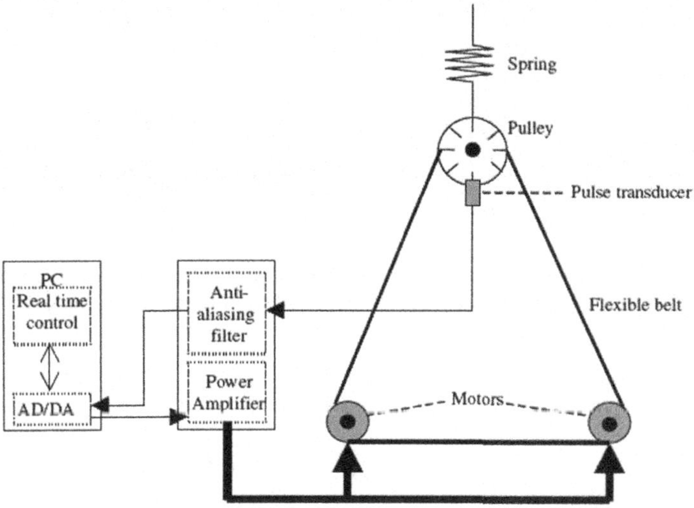

Fig. 7.15 The coupled electric drives setup [23]

in all other previous identification experiments in this chapter, we use crossover rate $p_c = 1$ and $p_m = 0.8$.

The results obtained are depicted in Fig. 7.16. The 2-D projections of the Pareto-fronts of the two experiments are plotted in Fig. 7.16c–f illustrate the estimated Pareto-front evaluated on test dataset $\mathcal{D}_{\text{test}}$. In Fig. 7.16c, d, we can observe that non-linear models including sin, cos and abs perform better than models with simpler dynamics. However, Fig. 7.16e, f reveal that several non-linear models do not generalize well to the test dataset. On the other hand, polynomial non-linear models achieve comparable performance without having the hard absolute-value non-linearity. We know form Weierstrass theorem that polynomials can be used to approximate a continuous non-linear function arbitrarily well, which explains why polynomial models can also perform well in the modelling task.

The one-step-ahead-prediction and simulation response of models that achieved the best prediction and simulation performance on $\mathcal{D}_{\text{test}}$, respectively, is plotted in Fig. 7.16a.

In Table 7.9, the results achieved by the proposed method are compared with the results reported in [2], where the authors used an approach based on Differential Evolution (DE) and Neural Networks (NN) to model the coupled drives system. It can be observed that the proposed method achieves comparable results in terms of prediction error, and marginally better results in terms of simulation error.

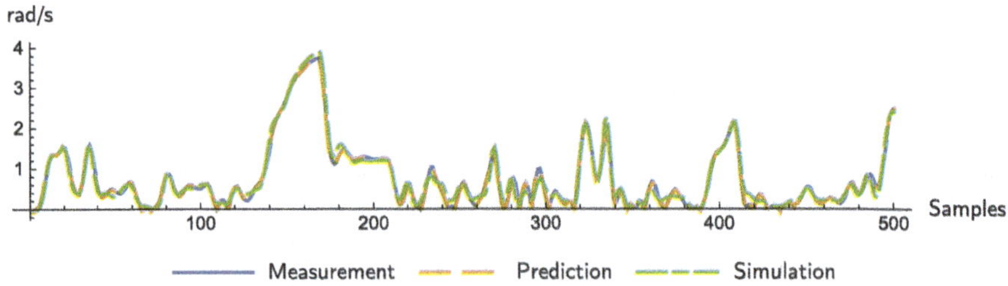

(a) Measured output compared with the prediction and simulation responses computed from models achieving the best prediction and simulation performance on test data $\mathcal{D}_{\text{test}}$.

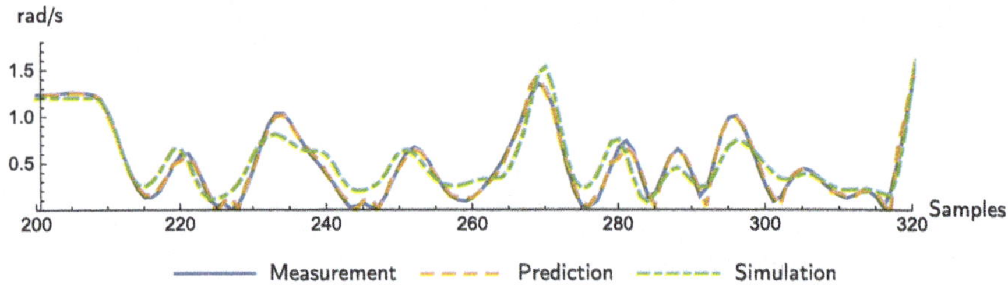

(b) Zoomed-in measured output compared with the prediction and simulation responses computed from models achieving the best prediction and simulation performance on test data $\mathcal{D}_{\text{test}}$.

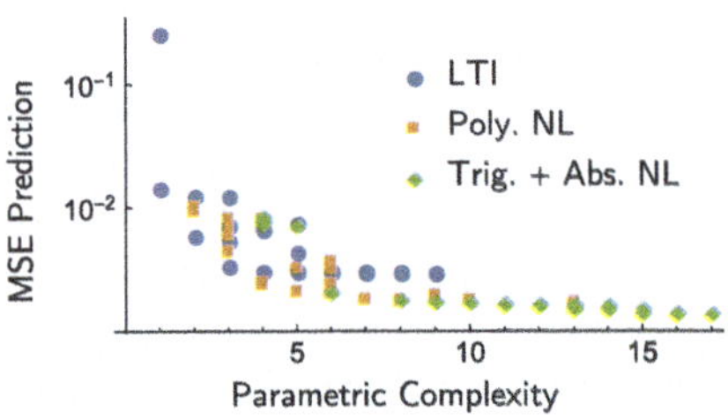

(c) Pareto front evaluated on data $\mathcal{D}_{\text{est}}$ - prediction vs parametric complexity.

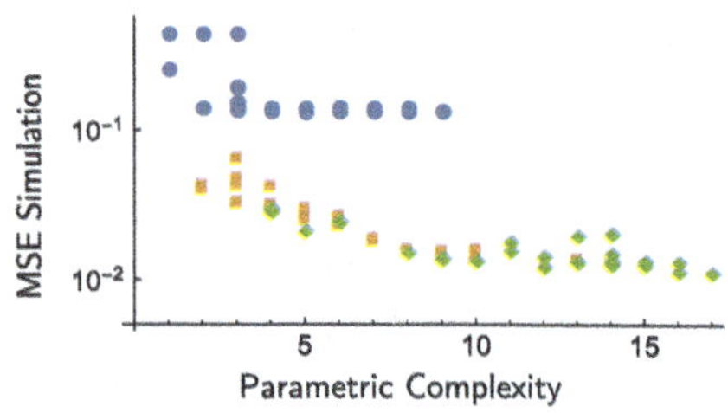

(d) Pareto front evaluated on data $\mathcal{D}_{\text{est}}$ - simulation vs parametric complexity.

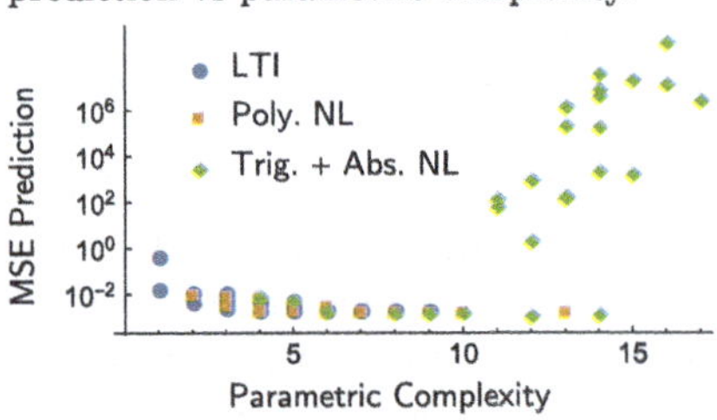

(e) Pareto front evaluated on data $\mathcal{D}_{\text{test}}$ - prediction vs parametric complexity.

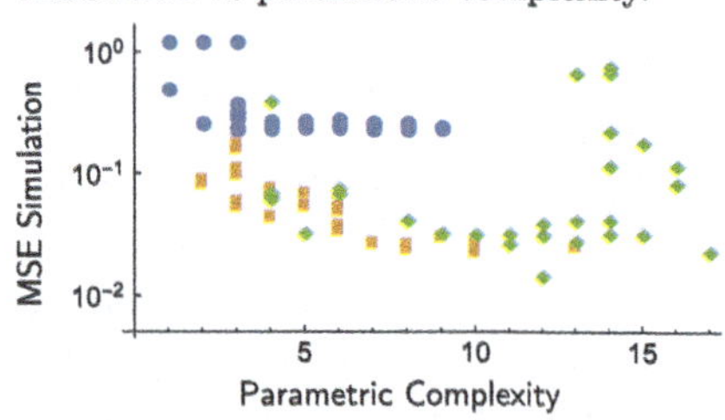

(f) Pareto front evaluated on data $\mathcal{D}_{\text{test}}$ - simulation vs parametric complexity.

Fig. 7.16 Numerical results for the coupled drives benchmark

Table 7.9 Comparison of performance measures for the coupled drives case study

Method	RMS_p	RMS_s
Proposed method	$\mathbf{3.73 \times 10^{-2}}$	**0.12**
GA + DE [2]	4.0×10^{-2}	0.18
LOLIMOT [1]	–	0.30

Bold indicates results obtained by method proposed in this thesis

7.4.3 Wiener-Hammerstein System

The Wiener-Hammerstein benchmark was proposed in [15]. The data is generated from a non-linear electronic circuit which has a Wiener-Hammerstein structure. The Wiener-Hammerstein structure consists of a static non-linearity sandwiched between two LTI systems as shown in Fig. 7.17. The electronic circuit implements the two LTI systems as 3rd order Chebyshev and inverse Chebyshev filters, and the static non-linearity f is based on a diode. More details can be found in [15].

The excitation signal used to generate the benchmark datasets is a filtered Gaussian noise signal. The benchmark dataset is divided into three parts:

- Estimation dataset $\mathcal{D}_{\text{est}}$ consists of the first 70,000 samples of the measured data,
- Validation dataset $\mathcal{D}_{\text{val}}$ consists of the next 30,000 samples of the measured data, and
- Test dataset $\mathcal{D}_{\text{test}}$ consists of the final 88,000 samples of the measured data.

The proposed identification algorithm is used to estimate models for the given measured data. TAG G_N is used to specify the model set, and crossover and mutation rates are chosen to be 1 and 0.8, respectively.

The results derived from the proposed identification method are depicted in Fig. 7.18. The 2-D projections of the Pareto-front evaluated on dataset $\mathcal{D}_{\text{val}}$ and $\mathcal{D}_{\text{test}}$ are plotted in Fig. 7.18c–f, respectively. Figure 7.18c, e demonstrate that, in terms of prediction error, there is no advantage in employing a polynomial IO model over an LTI dynamical model to describe the system. This is not the case for simulation error, as observed in Fig. 7.18d, f, where P-NARMAX models have a significant advantage over LTI models.

The time-domain residuals of the best prediction and simulation model estimated by the proposed identification method are compared with the measured data in Fig. 7.18a. The skewed nature of simulation error, plotted in green, clearly indicates the bias in the simulation response of the model estimate that achieves the best simulation performance. This can also be verified by comparing the results obtained by

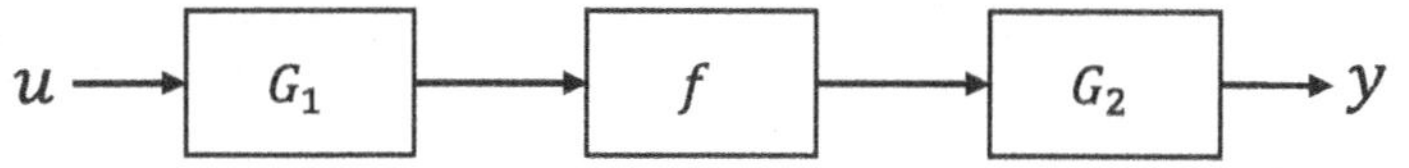

Fig. 7.17 The Wiener-Hammerstein model structure. The static non-linearity f is sandwiched between two LTI filters G_1 and G_2

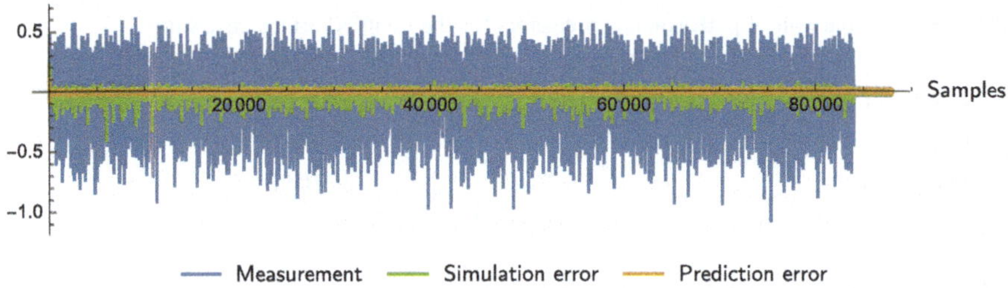

(a) Measured output compared with the prediction and simulation errors computed from models achieving the best prediction and simulation performance on test data $\mathcal{D}_{\text{test}}$.

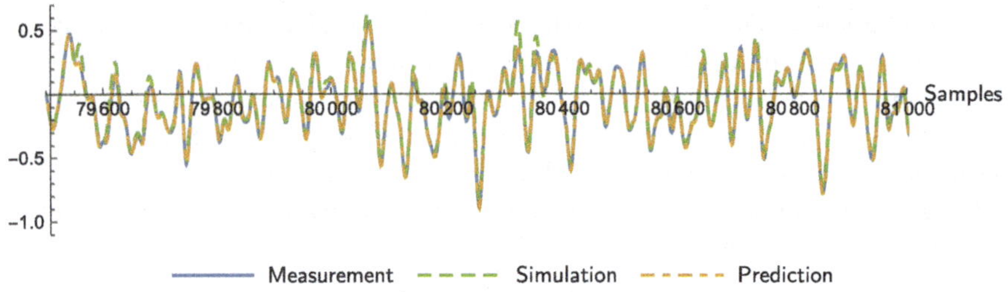

(b) Zoomed-in measured output compared with the prediction and simulation responses computed from models achieving the best prediction and simulation performance on test data $\mathcal{D}_{\text{test}}$.

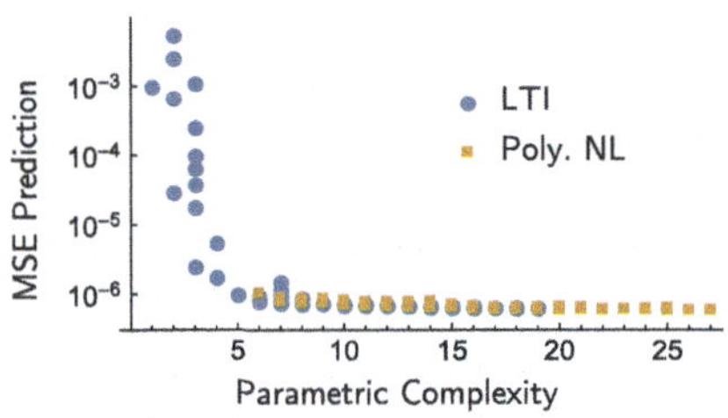

(c) Pareto front evaluated on data $\mathcal{D}_{\text{est}}$ - prediction vs parametric complexity.

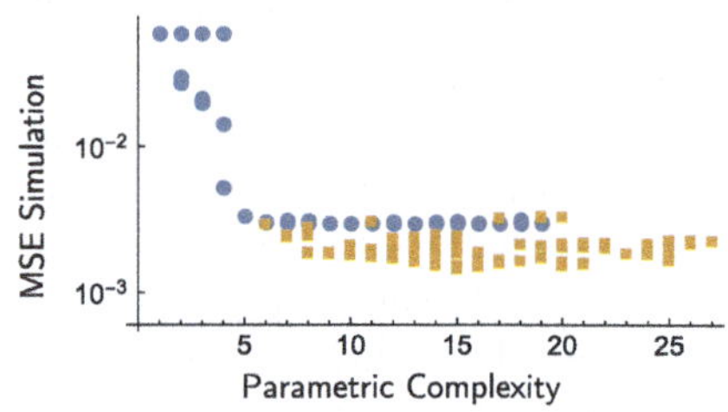

(d) Pareto front evaluated on data $\mathcal{D}_{\text{est}}$ - simulation vs parametric complexity.

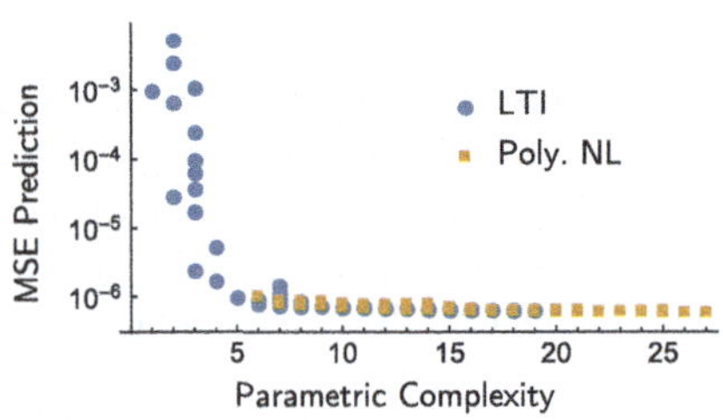

(e) Pareto front evaluated on data $\mathcal{D}_{\text{test}}$ - prediction vs parametric complexity.

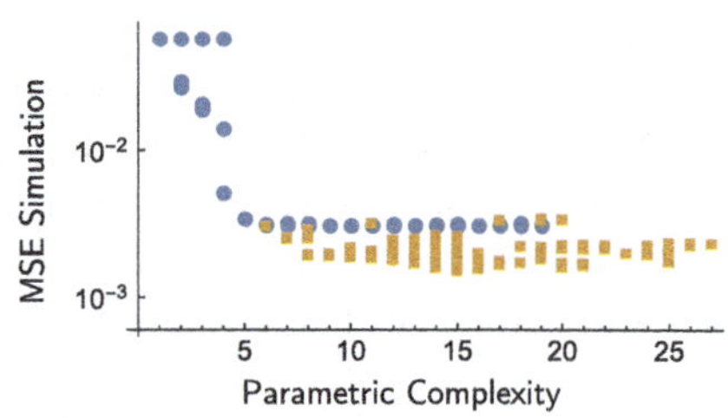

(f) Pareto front evaluated on data $\mathcal{D}_{\text{test}}$ - simulation vs parametric complexity.

Fig. 7.18 Numerical results for the Wiener-Hammerstein benchmark

Table 7.10 Performance of estimated models computed on the Wiener-Hammerstein test data-set and compared with literature

Identification method	Test RMS simulation (mV)	Est. RMS simulation (mV)
Proposed method	**39.0**	**38.0**
Proposed method (optimized for simulation error)	**23.5**	**23.5**
Best Linear Approx. [7]	56.2	–
pNLSS [14]	0.42	0.36
SVM [10]	47.4	47.1
QBLA [16]	**0.279**	**0.270**
Generalized WH model [24]	0.49	0.49
LS-SVM [4]	4.07	4.14
Bio-social evolution [12]	8.55	9.01
Pole-zero splitting, "Advanced method" [17]	0.30	–

The blank cells correspond to values not reported in the literature. Bold indicates results obtained by method proposed in this thesis

the proposed method with that obtained from other non-linear system identification approaches for the same dataset. The comparison is summarized in Table 7.10.

In order to reduce the bias in the estimated simulation models, model parameters of the Pareto-efficient polynomial IO models are optimized for mean-squared simulation error using the method described in Sect. 6.3.2. The mean-squared simulation error achieved by the re-optimized models are compared to the original estimates in Fig. 7.19a, b. By re-optimizing the models for simulation error, the mean-squared error achieved by the best P-NARMAX model was reduced by a factor of 2.6. The improved performance can also be observed in the time-domain response of the models in Fig. 7.19c. The best mean-squared error achieved by the re-optimized models on $\mathcal{D}_{\text{test}}$ is also reported in Table 7.10. However, the best simulation performance achieved, even after optimizing for simulation error, is still about two orders of magnitudes away from the best results reported in the literature. This indicates that there is still much room for improvement. Some suggestions for improvements will be discussed in Sect. 7.5.

7.5 Empirical Perspectives on the Proposed Modelling Approach

In this chapter, we employed the proposed identification framework, developed in Chaps. 4, 5 and 6, to several modelling applications. The modelling applications ranged from simple academic examples to complex real-world physical systems. The results obtained by the proposed method give us insights into the functioning of the algorithm. These insights are discussed below.

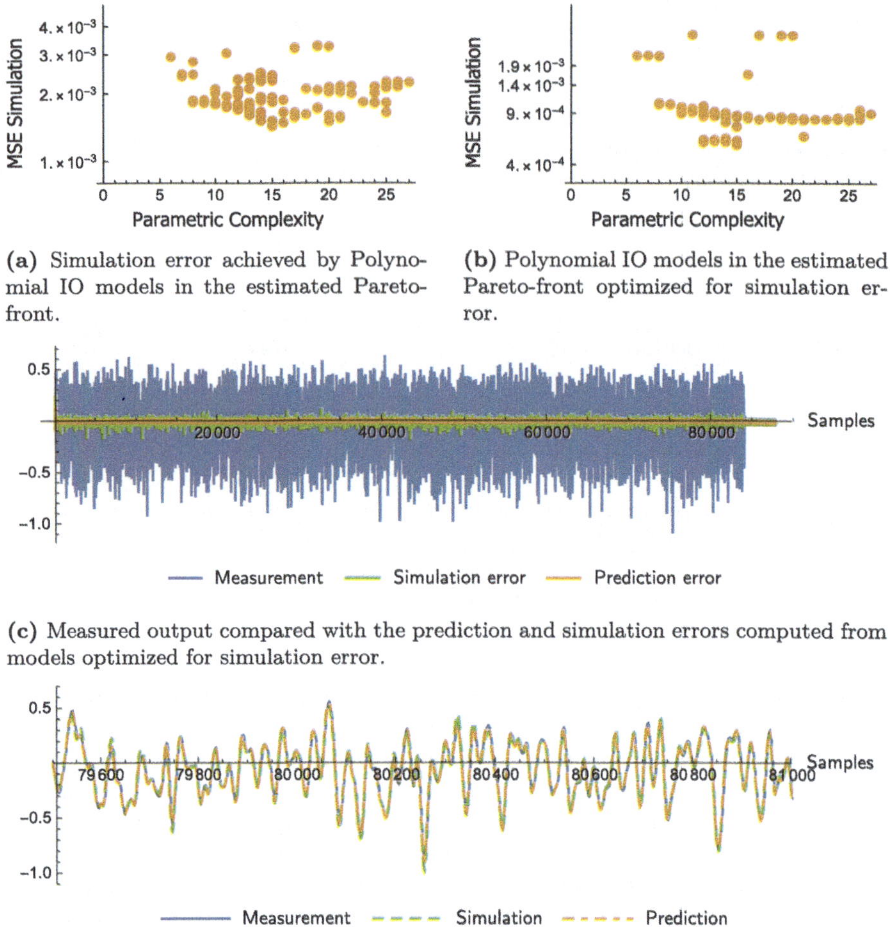

(a) Simulation error achieved by Polynomial IO models in the estimated Pareto-front.

(b) Polynomial IO models in the estimated Pareto-front optimized for simulation error.

(c) Measured output compared with the prediction and simulation errors computed from models optimized for simulation error.

(d) Zoomed-in measured output compared with the prediction and simulation responses computed from models achieving the best prediction and simulation performance on test data $\mathcal{D}_{\text{test}}$.

Fig. 7.19 Results for simulation error optimization of polynomial models in the Pareto-efficient set obtained for the Wiener-Hammerstein benchmark

- The academic example in Sect. 7.2 demonstrates that the proposed method is able to search through complex combinatorial search spaces far more efficiently than random sampling. The proposed method can effectively "learn" from the structure of models that perform comparatively better, which benefits the entire population of models.

- The results obtained for academic example also demonstrate that while many hyper-parameters of the algorithm can be chosen conservatively, the mutation and crossover rate may bear a strong influence on the results obtained. The MC simulations performed for the academic example indicate that medium to high values of crossover and mutation probabilities (i.e., about 0.4–1) yield good convergence properties for the algorithm. On the other hand, poor results are obtained for low values of crossover and mutation rates. Based on these observations, crossover and mutation rates of 1 and 0.8, respectively, are chosen for all other case studies described in this chapter.
- An important aspect of the proposed method is that minimal user-intervention is required to estimate models from search spaces that grow combinatorially. This is illustrated by the fact that the same set of hyper-parameters were used to estimate models for all the case studies described in this chapter, with the exception of (i) the TAG used, and (ii) the maximum tree-depth m_{d}. Hence, the user is not required to pre-specify the structure or complexity of the desired model. Instead, the choice of TAG determines the search space, which may span multiple model classes with varying degrees of model complexity.
- When no prior information is available, a general TAG can be chosen in order to search for models belonging to different model classes. The subsequent estimated Pareto-front illustrates the best models estimated, and their corresponding dynamic class. This makes the model selection problem easier to tackle, since the user is informed of the trade-off being made across the various identification criteria.
- The experimental results presented in Sect. 7.3 illustrate that the modelling approach developed in this dissertation is able to identify model features required to describe the measured data without any prior specification. For the pendulum example, the algorithm was able to identify the non-linear model structure required to accurately describe the flipping behaviour of the pendulum. For the thermal system, the identification method was able to automatically estimate the delay in the system without any prior specification of the delay.
- The benchmark datasets allow us to compare the results obtained from the proposed identification framework with that obtained from other non-linear identification methodologies proposed in the literature. For the silverbox and coupled drives benchmarks, the obtained results were comparable with that obtained from other methodologies. This is noteworthy considering the fact that the algorithm was in no way adapted to the specific benchmark system being identified (except the grammar used), and furthermore, requires very little to no user-intervention. The same is not true for most state-of-the-art identification frameworks that have been applied to the benchmarks examples.
- The weaving mill and Wiener-Hammerstein examples illustrate a shortcoming of the TAG and parameter estimation technique used in this chapter. Both of these case studies feature dynamical systems in which the non-linearity is not directly accessible from measured inputs or outputs, but rather acts on a hidden variable of the system. This results in structural errors in the estimated models, and since model parameters are only optimized for mean-squared prediction error

(for practical reasons), good results are obtained for prediction error only. There are several possible approaches for improving model estimates:

(i) Optimize model coefficients for both prediction and simulation error in each iteration of the algorithm. Of course, this comes at the cost of significantly larger computational time. The simulation error minimization approach discussed in Sect. 6.3.2 was used to optimize the Pareto-efficient models for the Wiener-Hammerstein benchmark, leading to improved simulation performance of the estimate models.
(ii) Use alternate model structures or representations that are more suitable to describe such dynamical systems, e.g., general NARMAX models, SS models or linear fractional representations. These model representations can also be described using TAG (see Sect. 4.4), and hence, it is possible to extend the proposed identification framework beyond the class of IO models. However, the parameter estimation step in the proposed methodology must also be adapted for new model representations.

- The proposed algorithm requires computational time in the order of a few hours. While the computational time required for the proposed method is significantly higher than most other non-linear identification methodologies, this not really a significant cost to the user since no user-interaction is required during the modelling procedure. On the other hand, for modelling complex non-linear systems, expert users may take several hours, days, or more, to develop specialized algorithms for the task at hand. Hence, form the perspective of a user that is not experienced in non-linear system identification methodologies, the proposed identification method can serve as an "AI engineer" that can process measured data to automatically synthesize models, and then report back to the user with the estimated Pareto-front. The computations can also be distributed over a remote computation cluster to reduce computations time or scale up the proposed algorithm.

Appendix 1: Experimental Results and Verification—Simulation Results

Results for Academic Example—Influence of p_c and p_m

In Figs. 7.20, 7.21, 7.22 and 7.23, we illustrate the results obtained for the different values of mutation rate p_m. In Figs. 7.24, 7.25, 7.26 and 7.27, we illustrate the results obtained for the different values of crossover rate p_m. For a discussion on these figures, see Sect. 7.2.

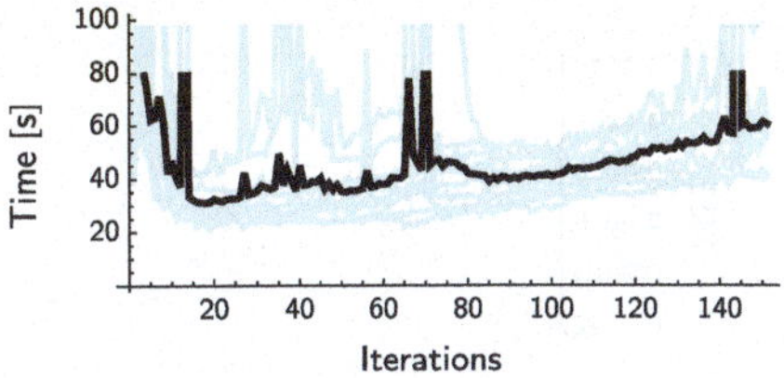

(a) Average time taken per generation for each GP run (light blue) and mean over the 10 runs (black).

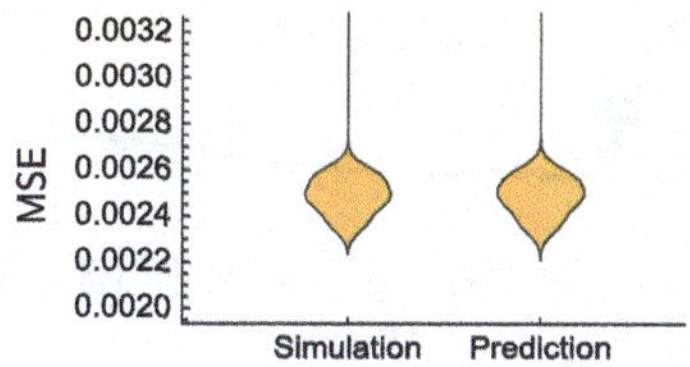

(b) Distribution of the minimum MSE prediction and simulation error evaluated on $\mathcal{D}_{\text{test}}$ over the 10 runs.

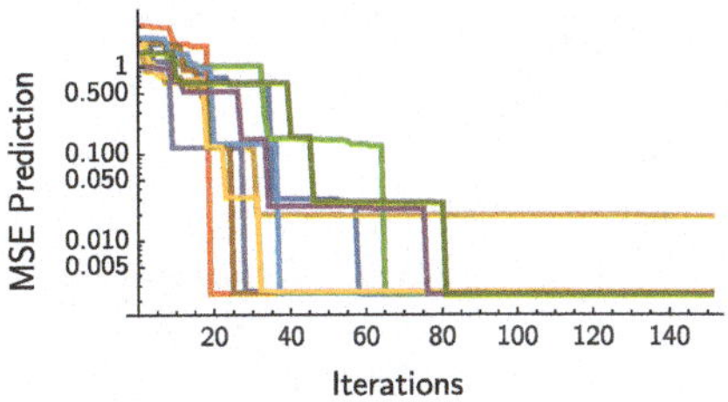

(c) Evolution of minimum prediction error w.r.t GP iterations.

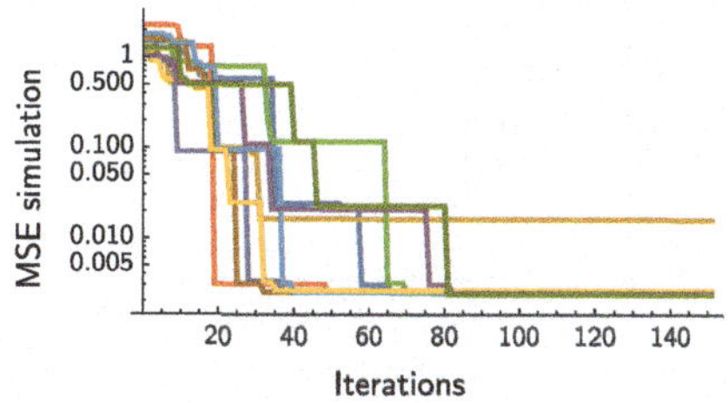

(d) Evolution of minimum simulation error w.r.t GP iterations.

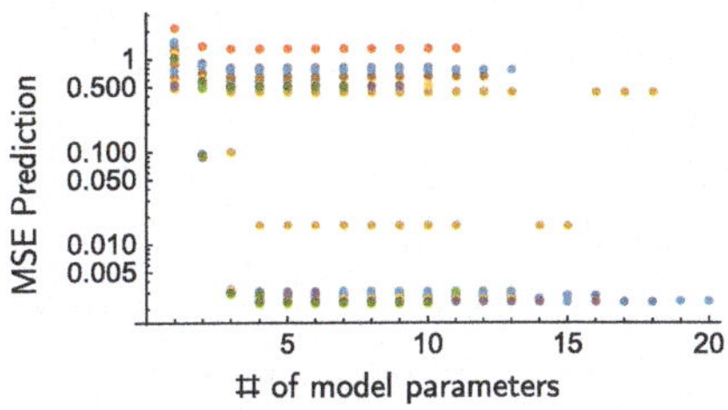

(e) Pareto front - prediction error vs parametric complexity.

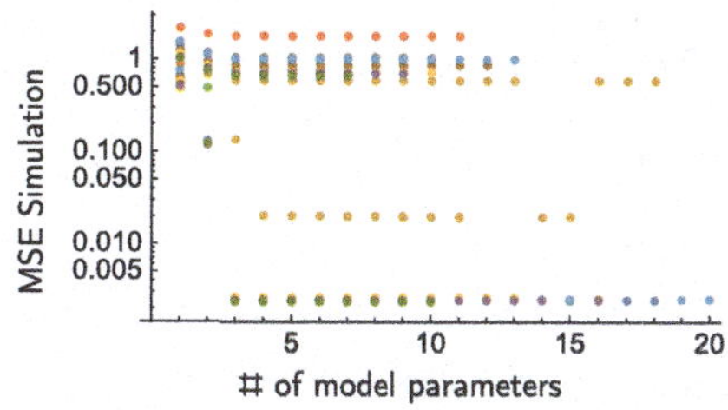

(f) Pareto front - simulation error vs parametric complexity.

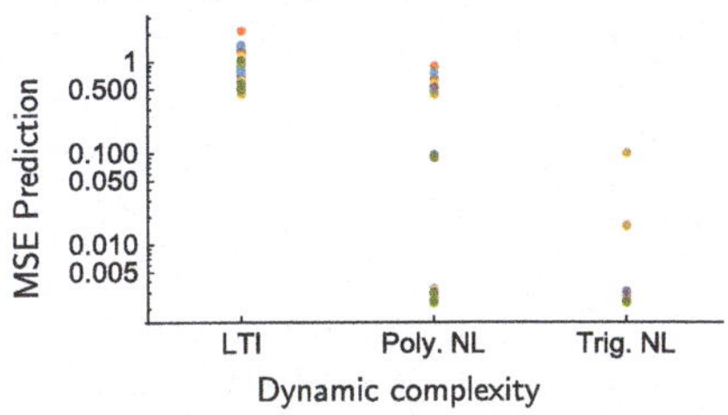

(g) Pareto front - prediction error vs dynamic complexity.

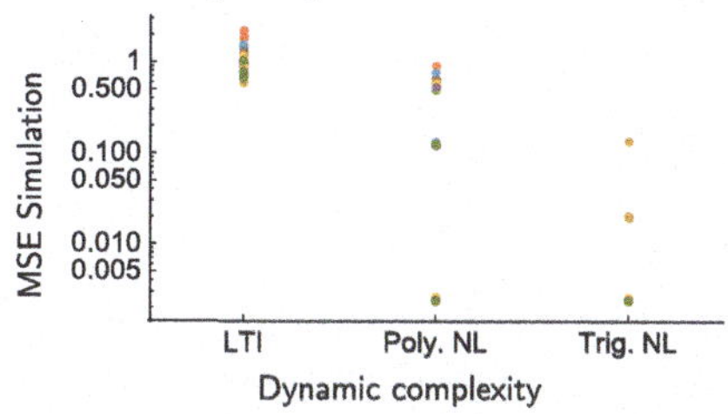

(h) Pareto front - simulation error vs dynamic complexity.

Fig. 7.20 MC simulation results for the academic example with $p_m = 1$ and $p_c = 1$

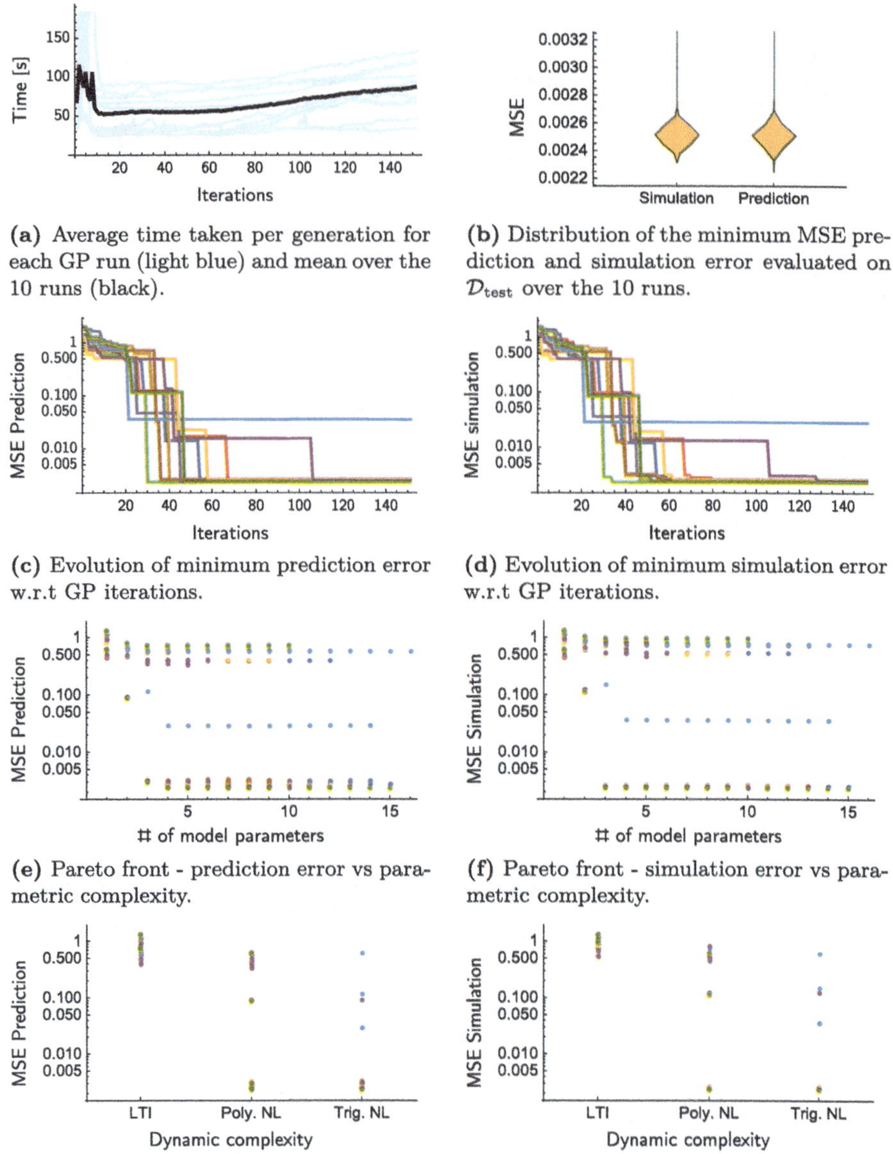

(a) Average time taken per generation for each GP run (light blue) and mean over the 10 runs (black).

(b) Distribution of the minimum MSE prediction and simulation error evaluated on $\mathcal{D}_{\text{test}}$ over the 10 runs.

(c) Evolution of minimum prediction error w.r.t GP iterations.

(d) Evolution of minimum simulation error w.r.t GP iterations.

(e) Pareto front - prediction error vs parametric complexity.

(f) Pareto front - simulation error vs parametric complexity.

(g) Pareto front - prediction error vs dynamic complexity.

(h) Pareto front - simulation error vs dynamic complexity.

Fig. 7.21 MC simulation results for the academic example with $p_m = 0.6$ and $p_c = 1$

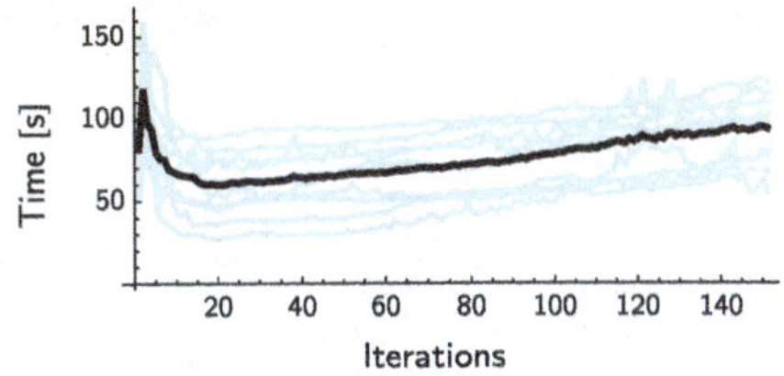

(a) Average time taken per generation for each GP run (light blue) and mean over the 10 runs (black).

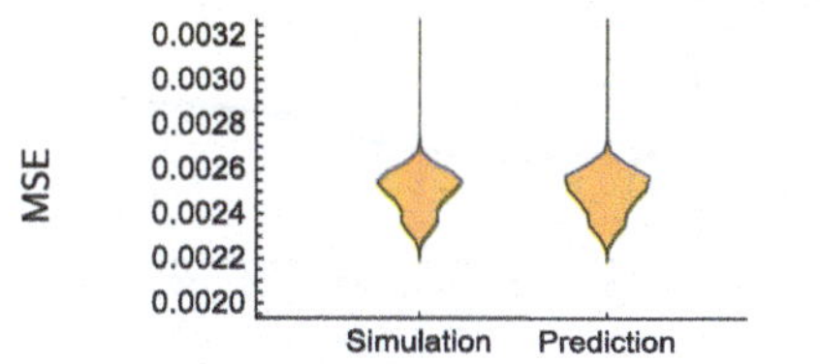

(b) Distribution of the minimum MSE prediction and simulation error evaluated on $\mathcal{D}_{\text{test}}$ over the 10 runs.

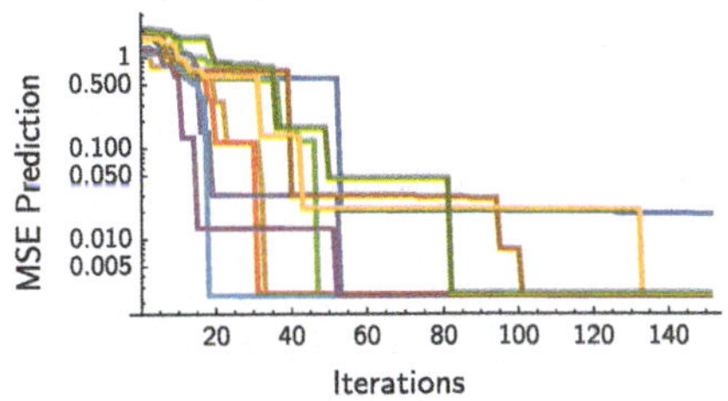

(c) Evolution of minimum prediction error w.r.t GP iterations.

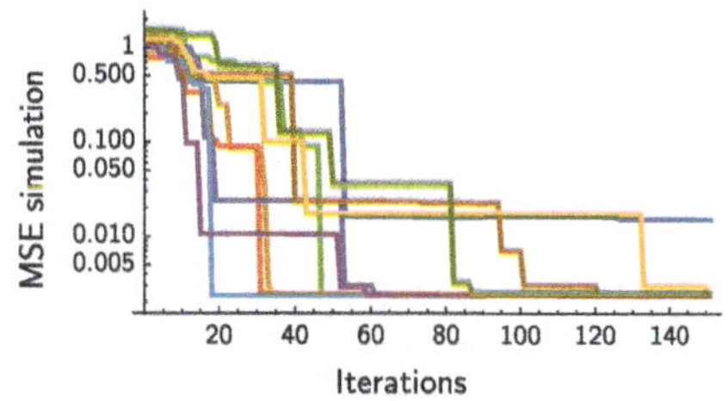

(d) Evolution of minimum simulation error w.r.t GP iterations.

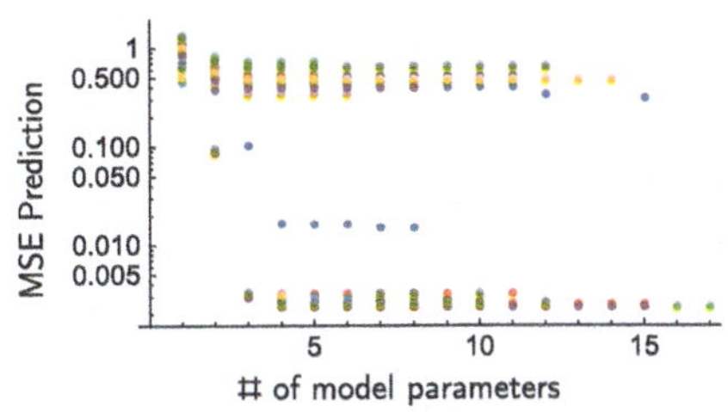

(e) Pareto front - prediction error vs parametric complexity.

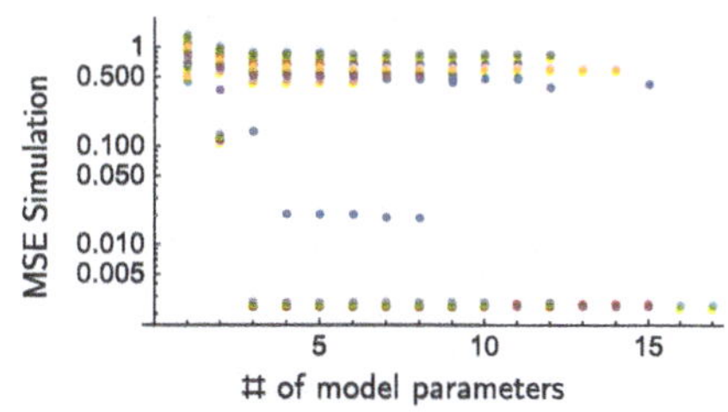

(f) Pareto front - simulation error vs parametric complexity.

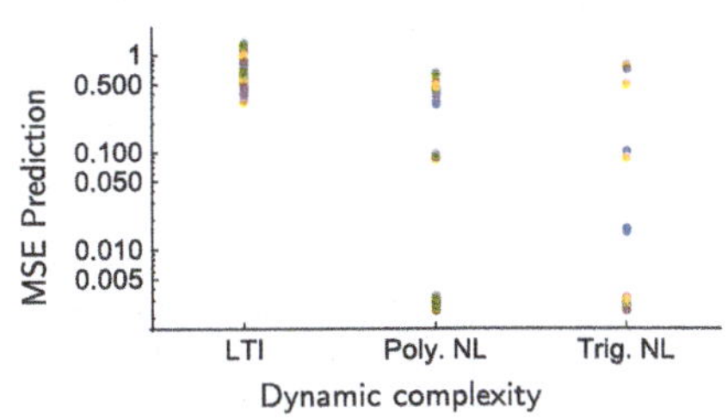

(g) Pareto front - prediction error vs dynamic complexity.

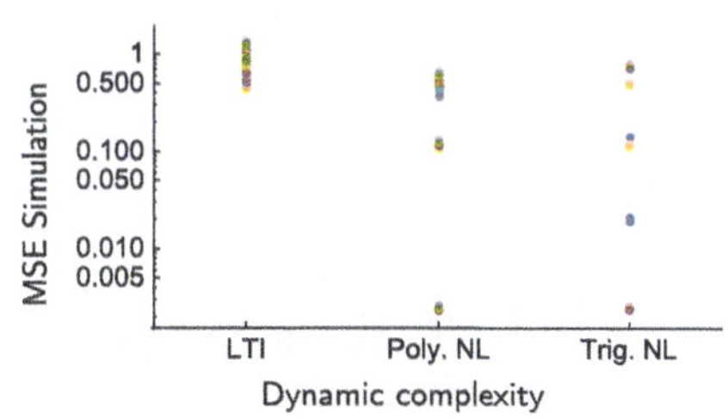

(h) Pareto front - simulation error vs dynamic complexity.

Fig. 7.22 MC simulation results for the academic example with $p_m = 0.4$ and $p_c = 1$

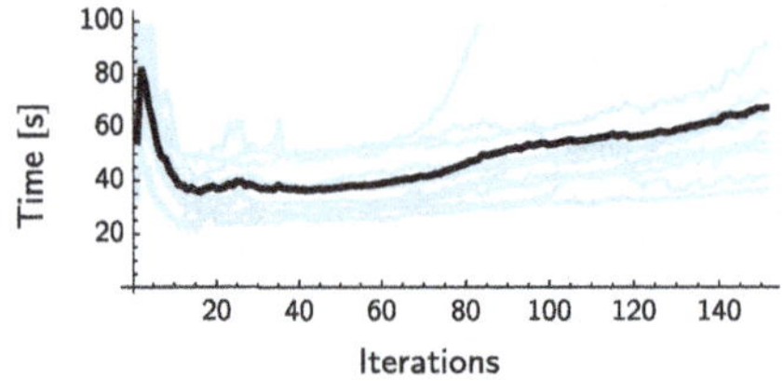

(a) Average time taken per generation for each GP run (light blue) and mean over the 10 runs (black).

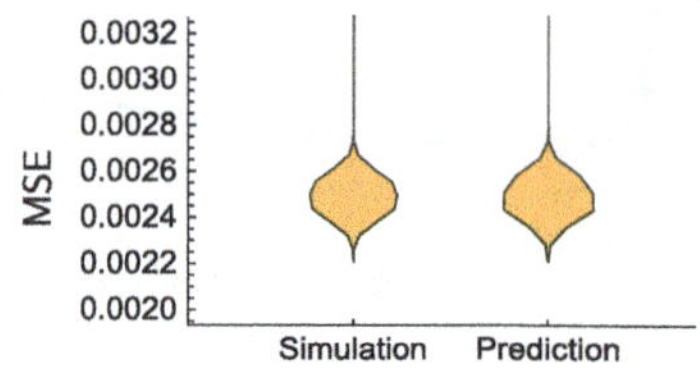

(b) Distribution of the minimum MSE prediction and simulation error evaluated on $\mathcal{D}_{\text{test}}$ over the 10 runs.

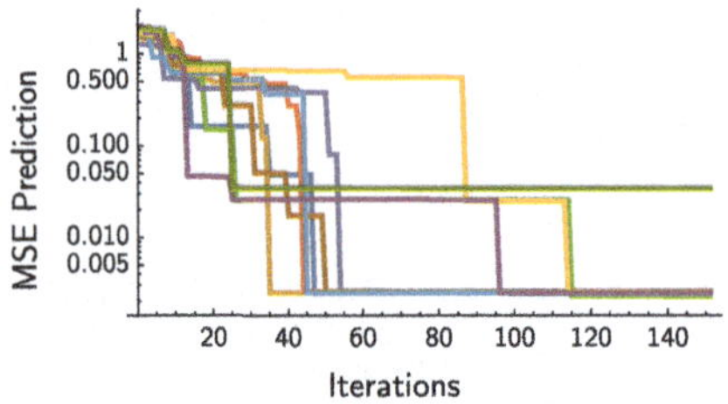

(c) Evolution of minimum prediction error w.r.t GP iterations.

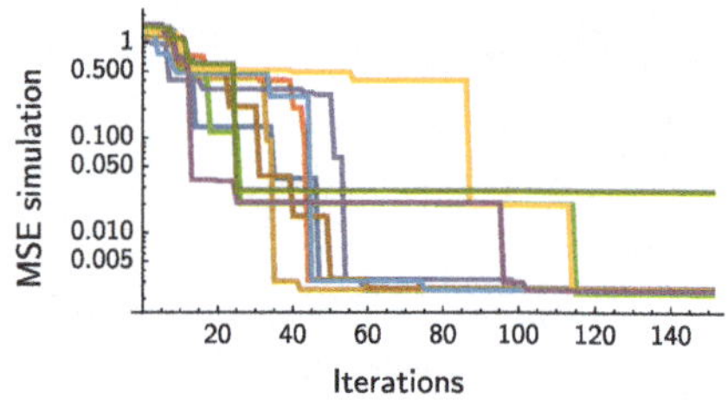

(d) Evolution of minimum simulation error w.r.t GP iterations.

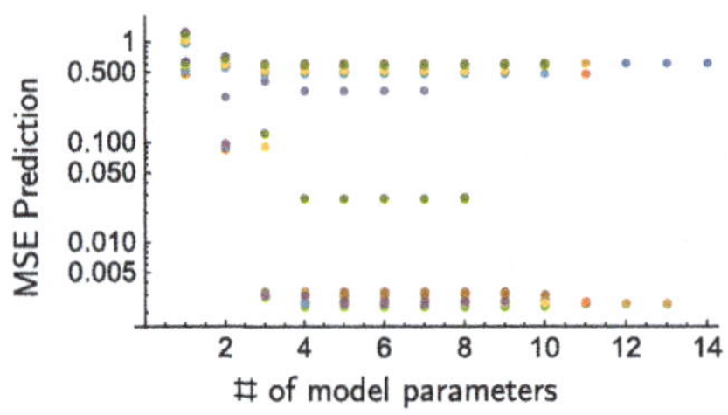

(e) Pareto front - prediction error vs parametric complexity.

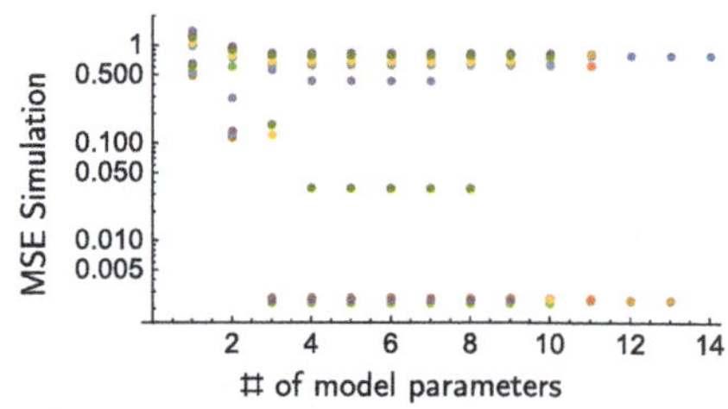

(f) Pareto front - simulation error vs parametric complexity.

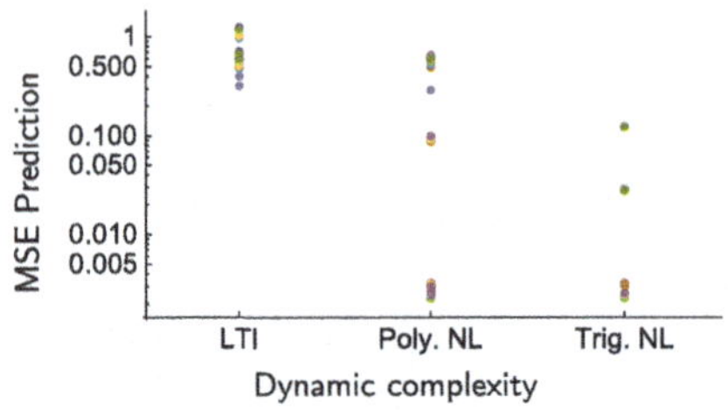

(g) Pareto front - prediction error vs dynamic complexity.

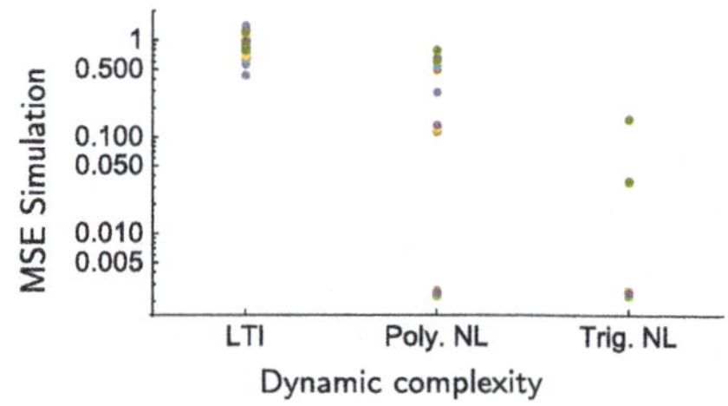

(h) Pareto front - simulation error vs dynamic complexity.

Fig. 7.23 MC simulation results for the academic example with $p_m = 0.2$ and $p_c = 1$

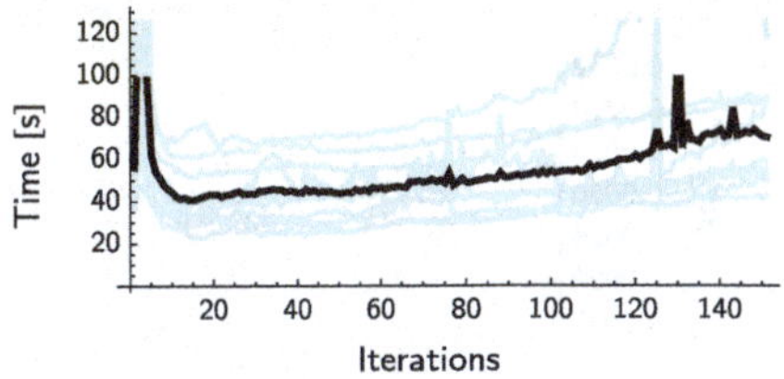

(a) Average time taken per generation for each GP run (light blue) and mean over the 10 runs (black).

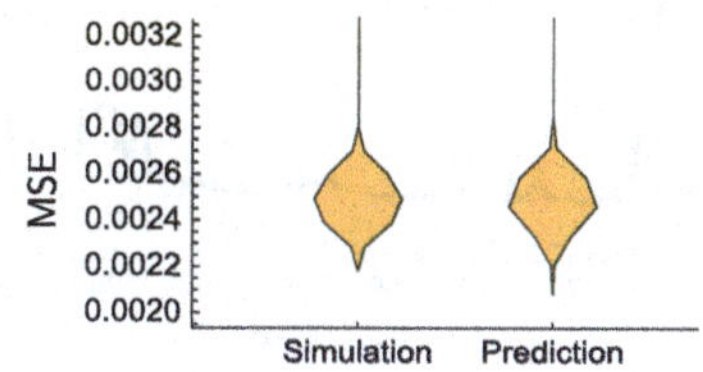

(b) Distribution of the minimum MSE prediction and simulation error evaluated on $\mathcal{D}_{\text{test}}$ over the 10 runs.

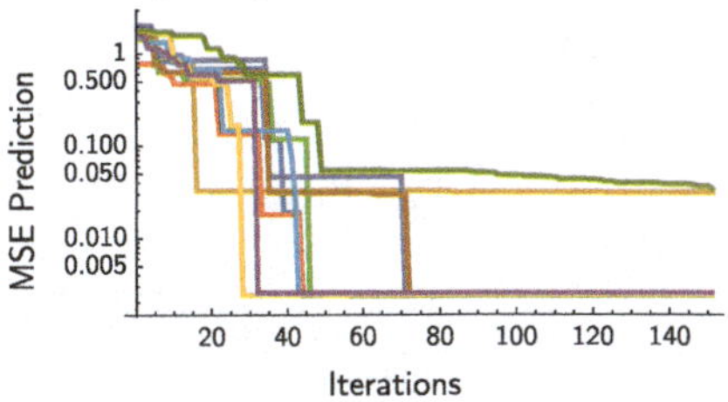

(c) Evolution of minimum prediction error w.r.t GP iterations.

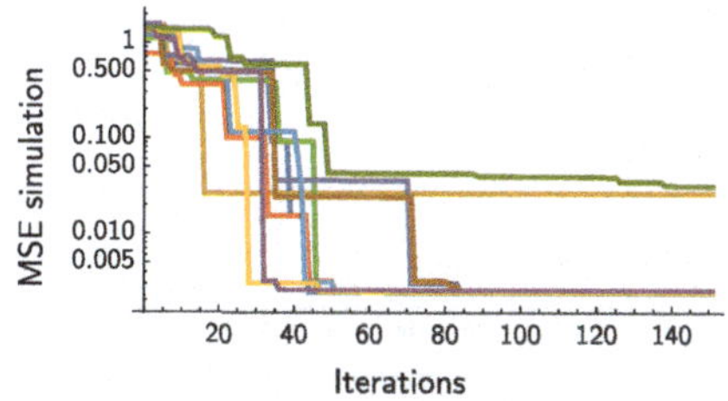

(d) Evolution of minimum simulation error w.r.t GP iterations.

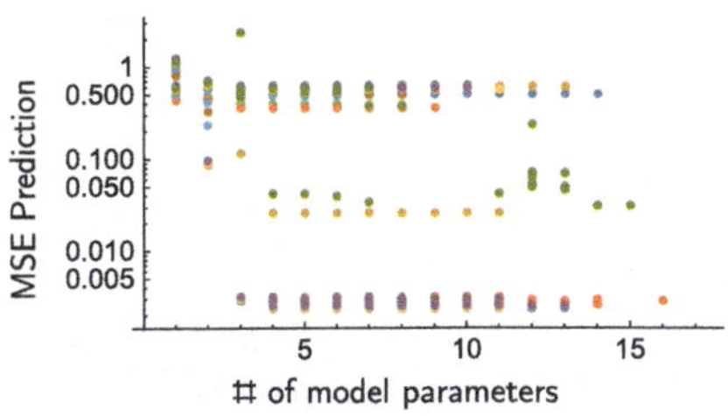

(e) Pareto front - prediction error vs parametric complexity.

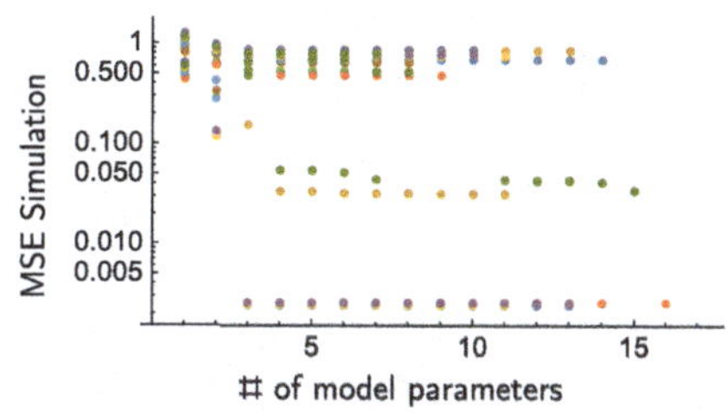

(f) Pareto front - simulation error vs parametric complexity.

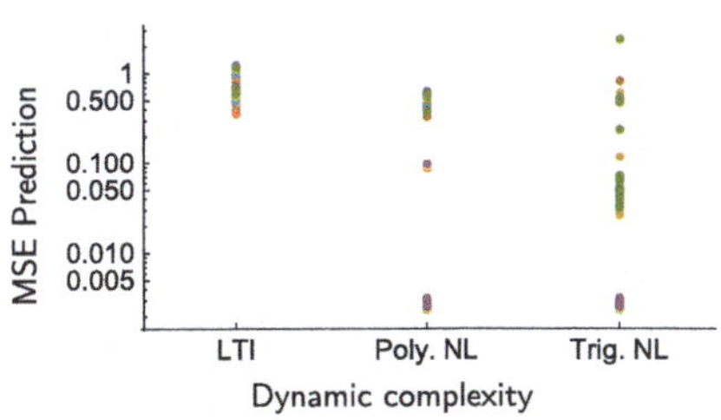

(g) Pareto front - prediction error vs dynamic complexity.

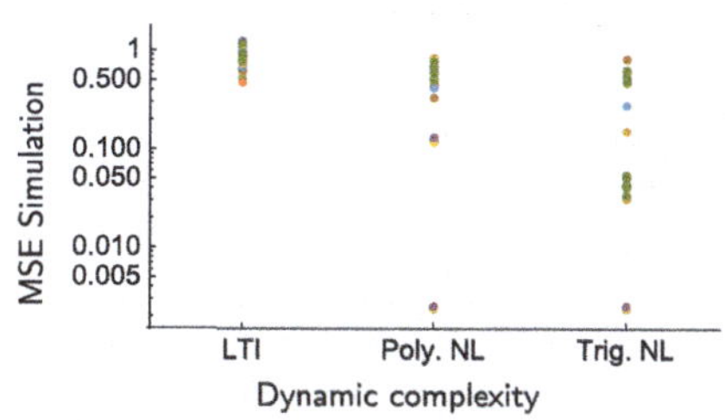

(h) Pareto front - simulation error vs dynamic complexity.

Fig. 7.24 MC simulation results for the academic example with $p_m = 0.8$ and $p_c = 0.8$

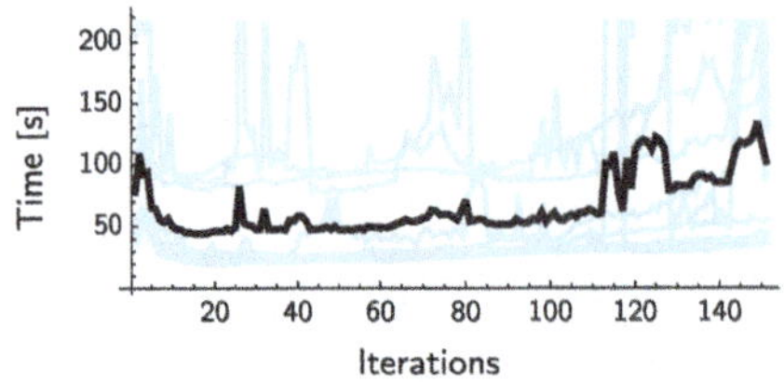

(a) Average time taken per generation for each GP run (light blue) and mean over the 10 runs (black).

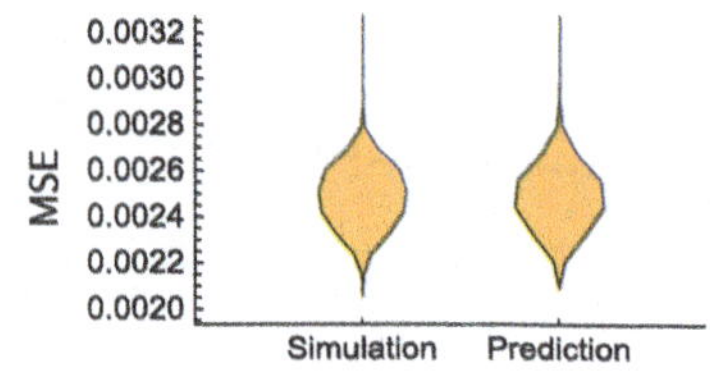

(b) Distribution of the minimum MSE prediction and simulation error evaluated on $\mathcal{D}_{\text{test}}$ over the 10 runs.

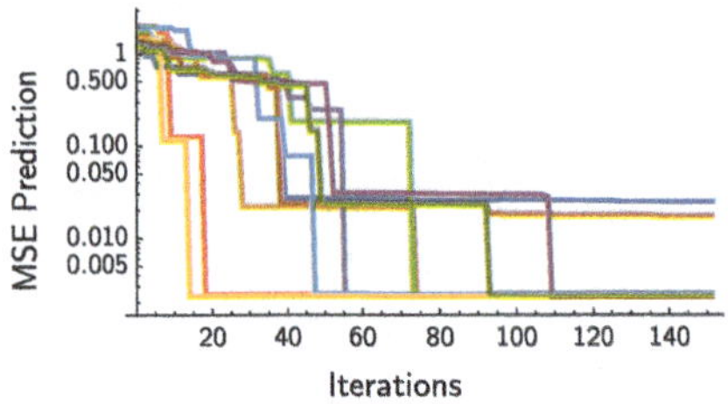

(c) Evolution of minimum prediction error w.r.t GP iterations.

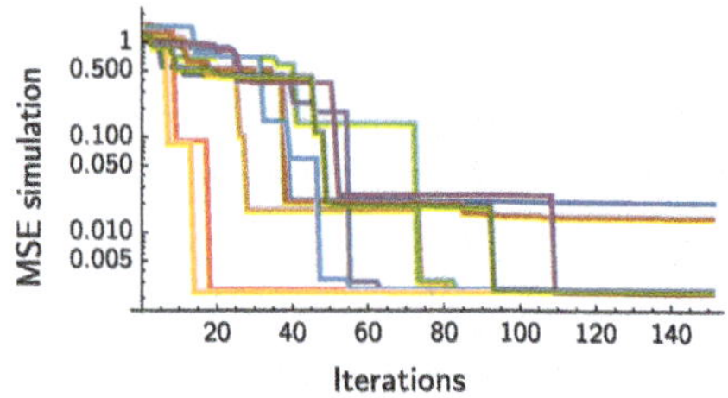

(d) Evolution of minimum simulation error w.r.t GP iterations.

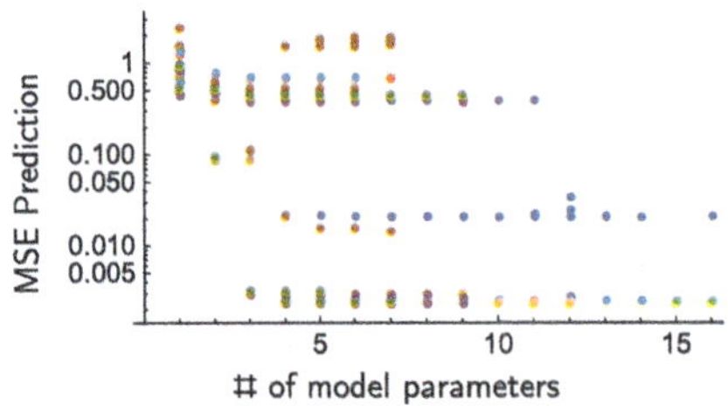

(e) Pareto front - prediction error vs parametric complexity.

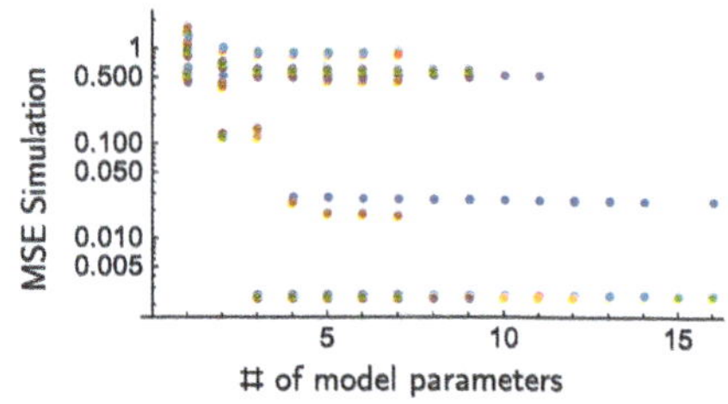

(f) Pareto front - simulation error vs parametric complexity.

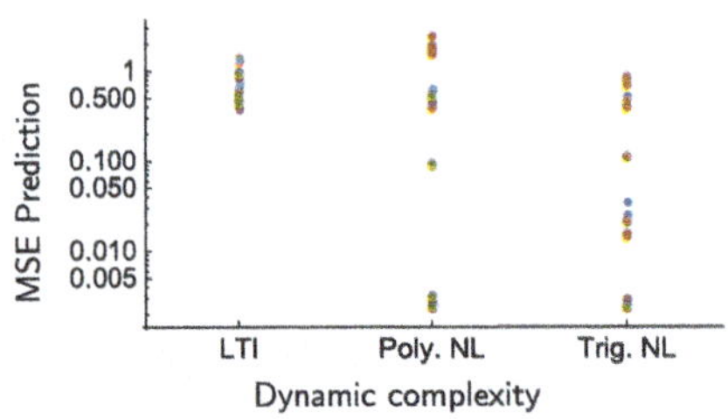

(g) Pareto front - prediction error vs dynamic complexity.

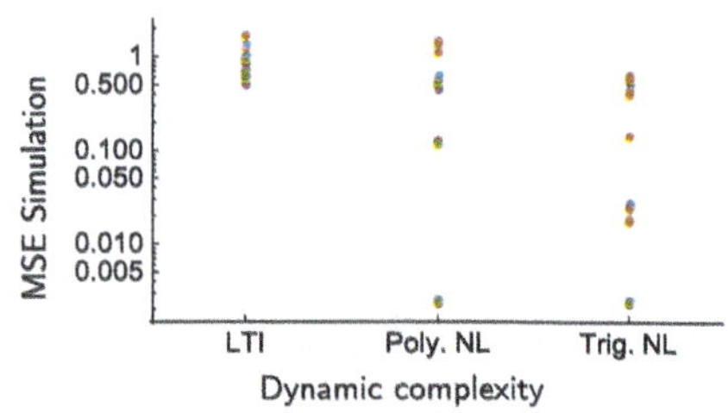

(h) Pareto front - simulation error vs dynamic complexity.

Fig. 7.25 MC simulation results for the academic example with $p_m = 0.8$ and $p_c = 0.6$

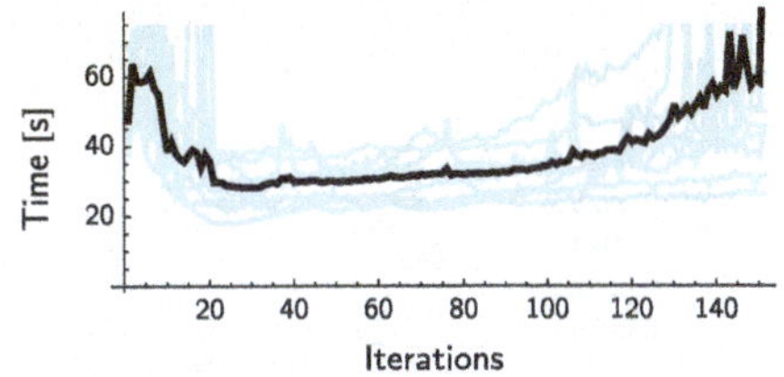

(a) Average time taken per generation for each GP run (light blue) and mean over the 10 runs (black).

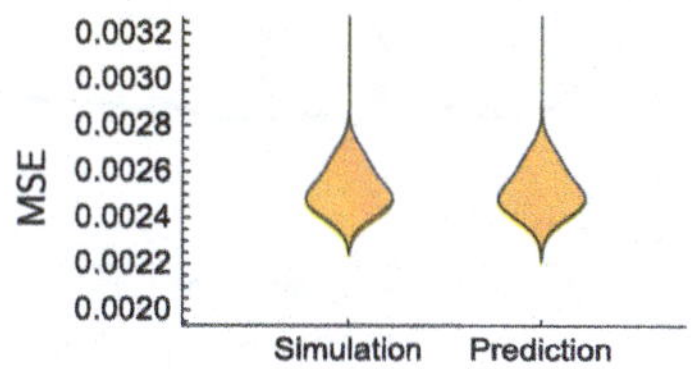

(b) Distribution of the minimum MSE prediction and simulation error evaluated on $\mathcal{D}_{\text{test}}$ over the 10 runs.

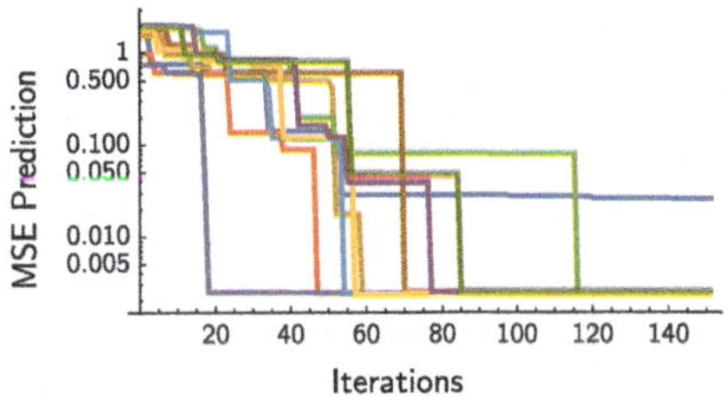

(c) Evolution of minimum prediction error w.r.t GP iterations.

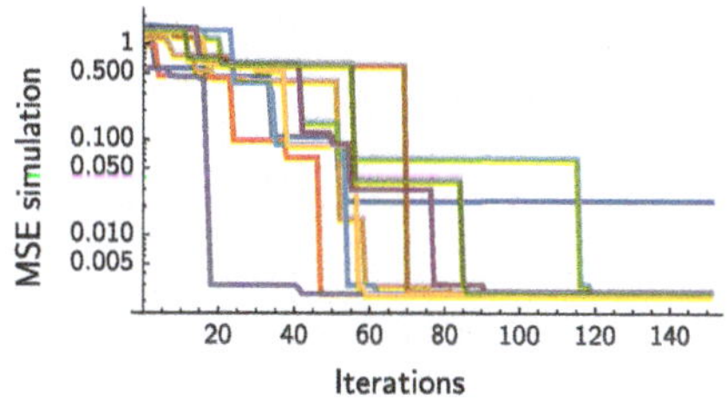

(d) Evolution of minimum simulation error w.r.t GP iterations.

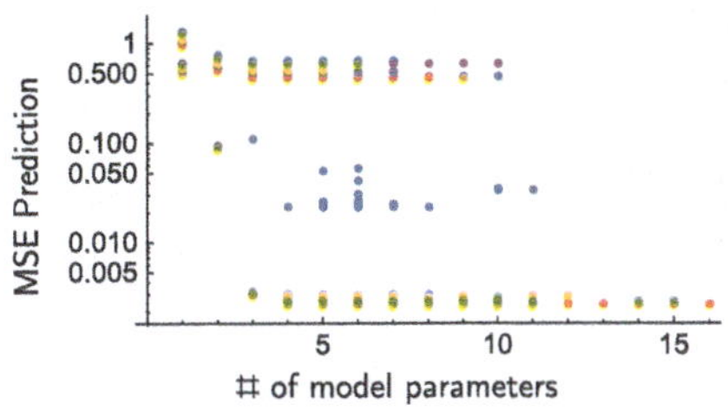

(e) Pareto front - prediction error vs parametric complexity.

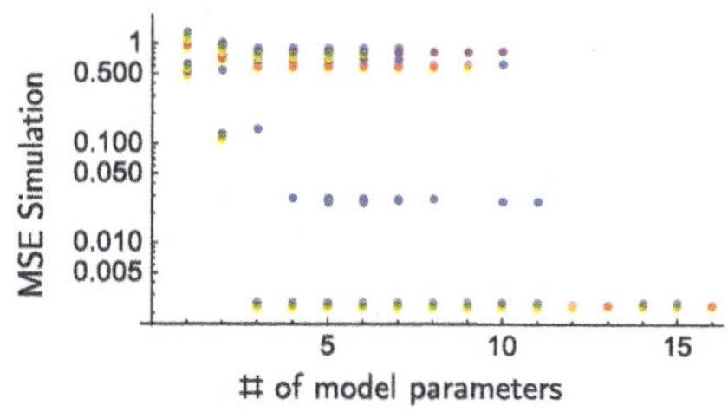

(f) Pareto front - simulation error vs parametric complexity.

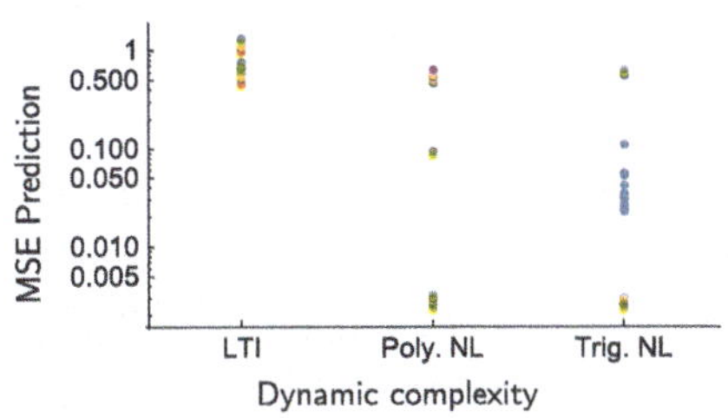

(g) Pareto front - prediction error vs dynamic complexity.

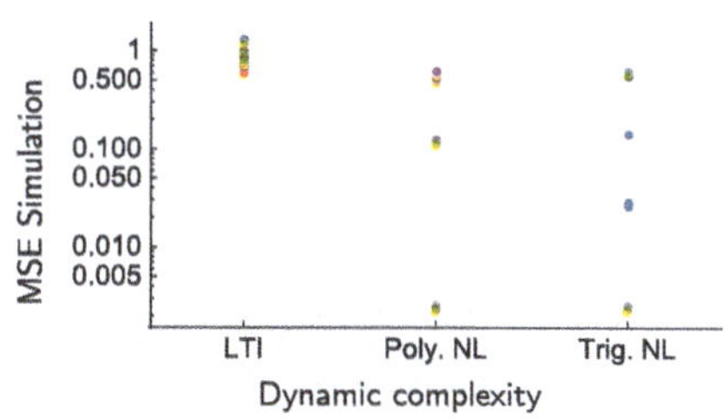

(h) Pareto front - simulation error vs dynamic complexity.

Fig. 7.26 MC simulation results for the academic example with $p_m = 0.8$ and $p_c = 0.4$

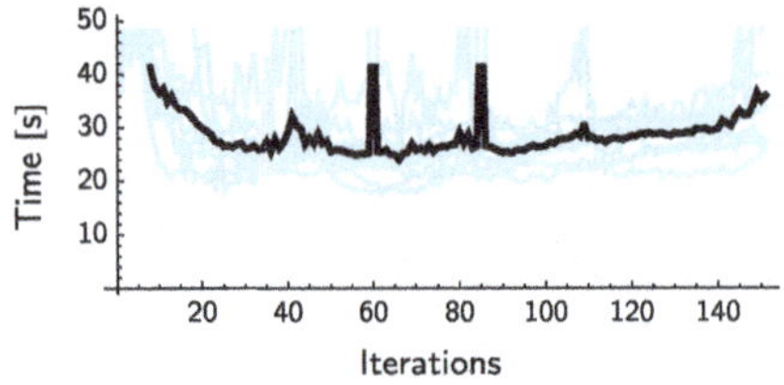

(a) Average time taken per generation for each GP run (light blue) and mean over the 10 runs (black).

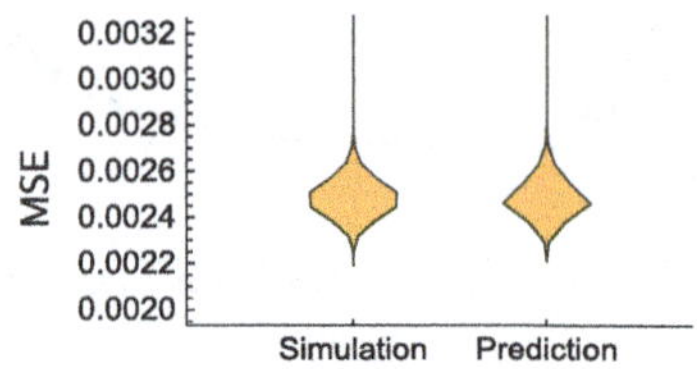

(b) Distribution of the minimum MSE prediction and simulation error evaluated on $\mathcal{D}_{\text{test}}$ over the 10 runs.

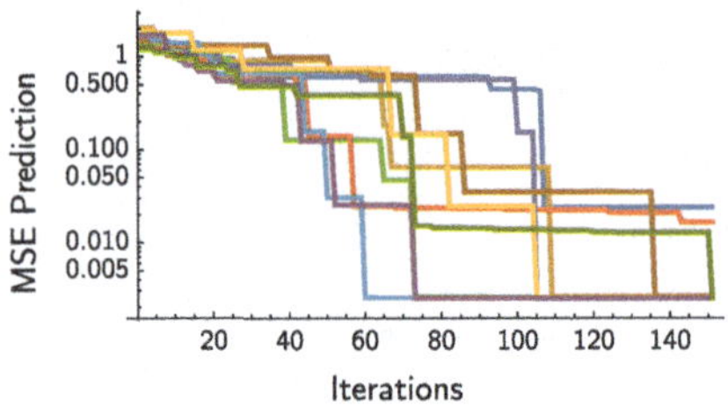

(c) Evolution of minimum prediction error w.r.t GP iterations.

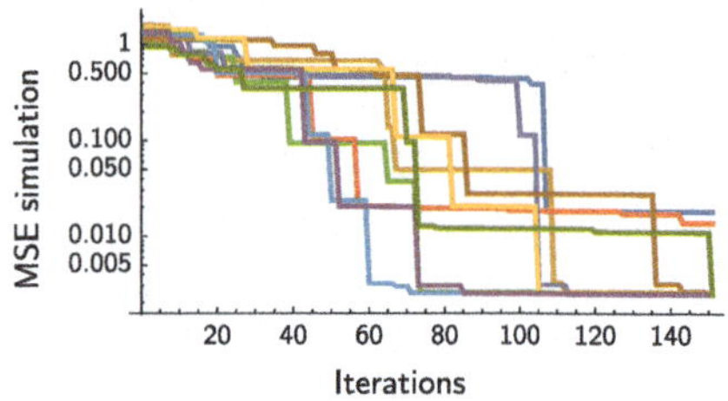

(d) Evolution of minimum simulation error w.r.t GP iterations.

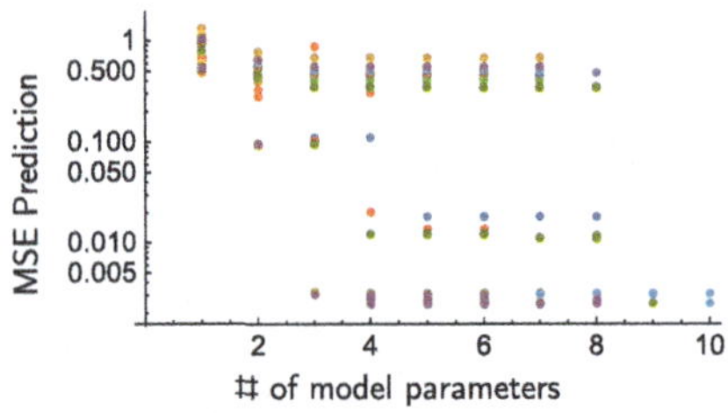

(e) Pareto front - prediction error vs parametric complexity.

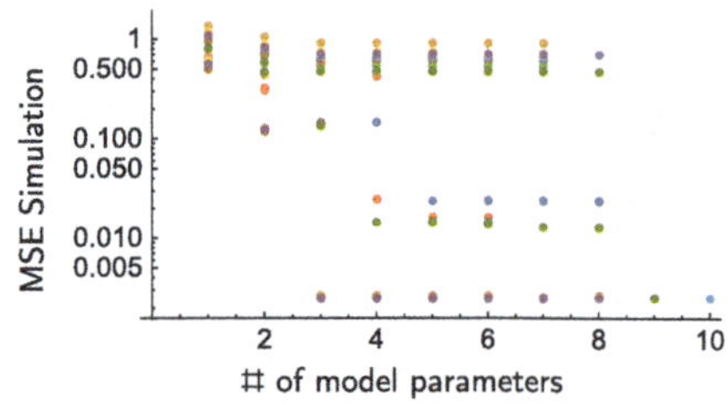

(f) Pareto front - simulation error vs parametric complexity.

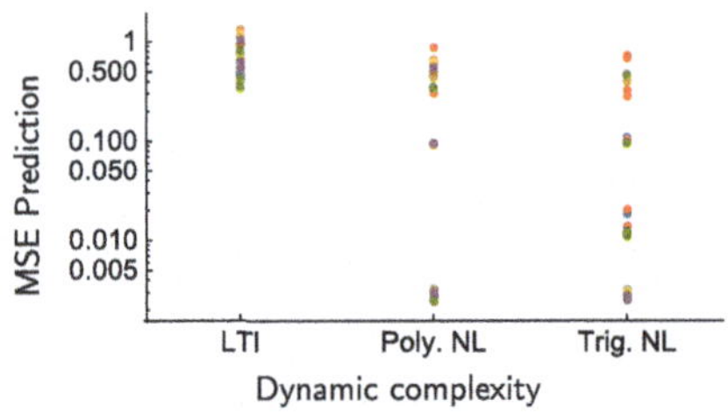

(g) Pareto front - prediction error vs dynamic complexity.

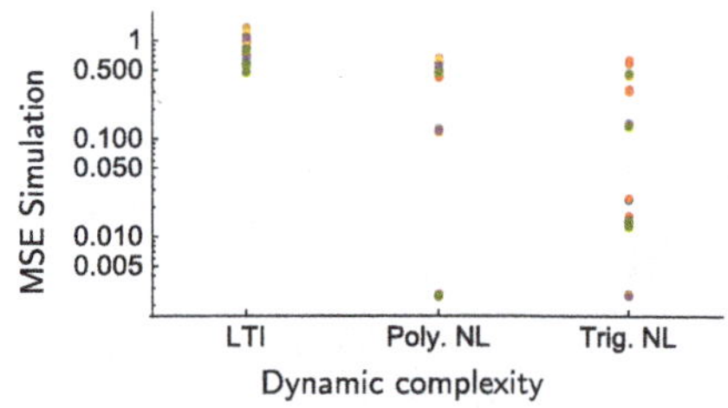

(h) Pareto front - simulation error vs dynamic complexity.

Fig. 7.27 MC simulation results for the academic example with $p_m = 0.8$ and $p_c = 0.2$

References

1. Aleksovski D, Dovžan D, Džeroski S, Kocijan J (2016) A comparison of fuzzy identification methods on benchmark datasets. IFAC Pap OnLine 49(5):31–36
2. Ayala HVH, da Cruz LF, Freire RZ, dos Santos Coelho L (2014) Cascaded free search differential evolution applied to nonlinear system identification based on correlation functions and neural networks. In: 2014 IEEE symposium on computational intelligence in control and automation (CICA). IEEE, pp 1–7
3. Espinoza M, Pelckmans K, Hoegaerts L, Suykens JA, De Moor B (2004) A comparative study of LS-SVM's applied to the silver box identification problem. IFAC Proc 37(13):369–374
4. Falck T, Pelckmans K, Suykens JA, De Moor B (2009) Identification of Wiener-Hammerstein systems using LS-SVMs. IFAC Proc 42(10):820–825
5. Hjalmarsson H, Schoukens J (2004) On direct identification of physical parameters in nonlinear models. IFAC Proc 37(13):375–380
6. Joshi AK, Schabes Y (1997) Tree-adjoining grammars. In: Handbook of formal languages. Springer, pp 69–123
7. Lauwers L, Pintelon R, Schoukens J (2009) Modelling of Wiener-Hammerstein systems via the best linear approximation. IFAC Proc 42(10):1098–1103
8. Ljung L (ed) (1999) System identification: theory for the user, 2nd edn. Prentice Hall PTR. ISBN 0-13-656695-2
9. Ljung L, Zhang Q, Lindskog P, Juditsky A (2004) Modeling a non-linear electric circuit with black box and grey box models. IFAC Proc 37(13):399–404
10. Marconato A, Schoukens J (2009) Identification of Wiener-Hammerstein benchmark data by means of support vector machines. IFAC Proc 42(10):816–819
11. Marconato A, Sjöberg J, Suykens J, Schoukens J (2012) Identification of the silverbox benchmark using nonlinear state-space models. In: IFAC proceedings. 16th IFAC symposium on system identification, vol 16, pp 632–637
12. Naitali A, Giri F (2016) Wiener-Hammerstein system identification—an evolutionary approach. Int J Syst Sci 47(1):45–61
13. Paduart J, Lauwers L, Swevers J, Smolders K, Schoukens J, Pintelon R (2010) Identification of nonlinear systems using polynomial nonlinear state space models. Automatica 46(4):647–656
14. Paduart J, Lauwers L, Pintelon R, Schoukens J (2012) Identification of a Wiener-Hammerstein system using the polynomial nonlinear state space approach. Control Eng Pract 20(11):1133–1139
15. Schoukens J, Ljung L (2009) Wiener-Hammerstein benchmark
16. Schoukens M, Pintelon R, Rolain Y (2014) Identification of Wiener-Hammerstein systems by a nonparametric separation of the best linear approximation. Automatica 50(2):628–634
17. Sjöberg J, Lauwers L, Schoukens J (2012) Identification of Wiener-Hammerstein models: two algorithms based on the best split of a linear model applied to the SYSID'09 benchmark problem. Control Eng Pract 20(11):1119–1125
18. Spector L (2011) Towards practical autoconstructive evolution: self-evolution of problem-solving genetic programming systems. In: Genetic programming theory and practice VIII. Springer, pp 17–33
19. Van Mulders A, Schoukens J, Vanbeylen L (2013) Identification of systems with localised nonlinearity: from state-space to block-structured models. Automatica 49(5):1392–1396
20. Vaseur C (2017) Model based control design for a high performant weaving machine: torsional vibration-and backlash compensation using non-linear feedforward control. Technical report
21. Verdult V (2004) Identification of local linear state-space models: the silver-box case study. IFAC Proc 37(13):393–398

22. Wigren T, Schoukens J (2013) Three free data sets for development and benchmarking in nonlinear system identification. In: 2013 European control conference (ECC). IEEE, pp 2933–2938
23. Wigren T, Schoukens M (2017) Coupled electric drives data set and reference models. Technical Report Uppsala Universitet 2017(024)
24. Wills A, Ninness B (2009) Estimation of generalised Hammerstein-Wiener systems. IFAC Proc 42(10):1104–1109

Chapter 8
Conclusions

In the mental disturbance and effort of writing, what sustains you is the certainty that on every page there is something left unsaid.

Cesare Pavese

In this thesis, we developed an identification framework that brings us closer to the goal of automated system identification. In this chapter, we summarize the main conclusions of this thesis, and examine the extent to which the Research Questions formulated in Chap. 1 have been addressed. Finally, we conclude this thesis with some recommendations for future research challenges.

8.1 Summary

The main research objective adopted in this dissertation was to automate the task of system identification from the user's perspective, such that dynamical models, of appropriate structure and complexity, can be automatically synthesized based on measured data and performance criteria given by the user. An identification tool that achieves such a degree of automation would be invaluable to researchers and engineers (that develop model-based tools, applications and analysis techniques) since it would allow them to focus on the utilization scenario of the model rather than on obtaining the model estimate itself.

In order to achieve the research objective of automatic system identification, we developed an identification framework with the following features:

- The proposed identification method automatically explores different model structures and model complexities without any *a priori* specification of model structure from the user. The user must only select the TAG, which can be chosen conservatively.

D. Khandelwal, *Automating Data-Driven Modelling of Dynamical Systems*, Springer Theses, https://doi.org/10.1007/978-3-030-90343-5_8

- The proposed method is based on a multi-objective framework, which makes it possible to incorporate user-specified performance measures, without the need to specify *a priori* trade-off values between the performance measures.
- Ultimately, the user obtains a set of estimated Pareto-efficient models. This enables the user to select a model *a posteriori*, based on the information contained in the Pareto-front.

The proposed identification procedure does not rely on the expertise of the user to obtain parametric model estimates from measured data, thereby achieving the desired level of autonomy in the identification procedure. Nonetheless, if desired, the user may impose restriction on the model set, or select performance measures that are suitable for the use-case of the model.

In the following, we discuss the key ideas developed and proposed in this thesis in order to achieve the research goal of automated system identification. The discussion is rooted in the context of the three Research Questions posed in Chap. 1.

Research Question 1

In order to ease the burden of model structure and complexity selection on the user, a key challenge was to develop a method that could function across multiple classes of models, without specification from the user. In order to address this research challenge, we developed a framework based on GP and TAG. GP, introduced in Chap. 3, forms the computational basis which allows the proposed identification method to search for Pareto-optimal model structures. GP achieves this by employing a "generate and evaluate" policy for optimization. However, it is challenging to alter the search space explored by GP and to restrict the search space to well-posed or desired model structures. To make GP efficient in exploring model structure, we proposed to use TAG to shape the search space, i.e., the model set. TAG allows the search for model structure to be restricted to "grammatically correct" model structures. The proposed TAG-based model set concept also enables us to incorporate prior knowledge and user preference in modelling of the system. For example, for the silverbox case-study (in Chap. 7), it was possible to restrict the grammar to include only polynomial non-linearities (based on prior information of the system), while for the weaving machine, where no prior information was available, the grammar included other non-linearities such as trigonometric functions.

We combined the two concepts of TAG and GP in Chap. 6, to develop an identification framework that functions across different model structures and complexities automatically. Hence, we can conclude that, within the scope of SISO dynamical systems (without hybrid dynamics) and for IO model representations, we have successfully addressed Research Question 1 in this thesis. The proposed framework is fairly general, and can be extended in the following ways:

- In terms of dynamics of the system, TAG can be used to specify model structures of, for example, hybrid dynamical systems, and GP can be used to explore the resulting model set.
- In terms of model representations, in principle, the proposed concept can be extended to other parametric model representations such as SS models or LFR

models. In Sect. 4.4, we proposed a TAG for parametric SS models for MIMO systems, providing the first steps for automated identification of non-linear SS models.

Research Question 2

The next key research problem was to incorporate user-specified identification criteria relevant to the utilization scenario of the model. We successfully addressed this research question by adopting a multi-objective framework based on Pareto-dominance relations (see Chap. 5). This makes it possible to define the notion of "optimality" in a multi-objective setting without the need to explicitly select or estimate an optimal trade-off between performance measures. In this setting, the optimization of models for multiple performance measures yields a set of Pareto-optimal model estimates. The Pareto-front estimate is an excellent way of proposing trade-offs between the specified identification criteria to the engineer, enabling him or her to make an informed choice *a posteriori*. This feature distinguishes the proposed identification framework from most existing identification methodologies existing in the literature.

In Chap. 5, we also proposed the use of four performance measures—one-step-ahead prediction error, simulation error, parametric complexity and dynamic complexity. These performance measures capture various aspects of the quality of model estimates, giving more information to the user at the model-selection step.

In Chap. 6, the multi-objective framework was combined with the TAG concept to formalize automated system identification as a bi-level multi-criteria optimization problem. Such a formulation makes it possible to incorporate user-specified performance measures without significant changes in the identification methodology. This addresses Research Question 2. To achieve acceleration in computation time for the proposed identification criteria, we made use of specialized estimation techniques (based on LS, gradient descent). In the case of more complex objectives, the proposed method can fall back on global non-linear optimization techniques can such as PSO and DE.

Research Question 3

Having formulated the research goal of automated system identification as a multi-criteria bi-level optimization problem, the final research challenge was to develop an identification algorithm that solves the proposed optimization problem. As mentioned earlier, a modified GP algorithm was developed to address this problem. Our contributions to the standard GP algorithm include:

- Modifications to the crossover and mutation operator to ensure compatibility with TAG. The proposed changes were inspired by Hoai and McKay [5].
- Modifications to the selection operator to ensure compatibility with the multi-objective framework. These changes were based on non-dominated sorting proposed by Deb et al. [2], however, other alternatives also exist.
- Incorporating suitable parameter estimation step in each iteration of the GP algorithm. This step addresses the lower-level problem of the proposed bi-level optimization problem.

In order to make the algorithm computationally efficient, we make use of local optimization techniques that are specially designed for the identification criteria proposed in Chap. 5 at the expense of generality of the algorithm. For arbitrary identification criteria chosen by the user, the proposed identification method can rely on more general numerical optimization techniques can be used, such as PSO or CMAES. This comes at the cost of added computational complexity. Alternatively, suitable modifications must be made to the parameters estimation techniques for the chosen identification criteria.

The performance of the proposed algorithm was evaluated on a number of identification problems in Chap. 7. We found that, with minimal to no changes in the settings of the proposed algorithm, the proposed identification approach was able to automatically infer models for a wide range of dynamical systems. For the pendulum, silverbox and thermal case-studies, the proposed identification estimated model structures that (partially) coincided with the true model structure (obtained from first principles). Based on the information in the estimated Pareto-front, the proposed method was able to reveal the appropriate model order (parametric complexity) and dynamic properties of the system. For example, for the thermal setup, the proposed identification method revealed that an LTI model was sufficient to describe the dynamics of the system. Furthermore, for two of the three benchmark examples considered, the results obtained were comparable to that obtained by other identification methodologies that were specifically designed or tuned for the identification task at hand. The results obtained in Chap. 7 provide some empirical evidence to suggest that the proposed GP-based algorithm can indeed optimize for models in the generalized model set specified by a TAG and for multiple performance measures.

8.2 Main Contributions

In this section, we highlight some of our main contributions to the general system identification literature.

- In Chap. 4, we developed a TAG-based notion of a model set. The proposed notion is more general than that typically used in existing identification frameworks such as PPEM. TAG can be used to specify a model set that spans across different model structures and complexity. TAG can also be used for a variety of parametric model representations. We developed a TAG for both IO and SS model representations.
- In Chap. 5, we demonstrated, through simulation examples, that one-step-ahead prediction error alone should not be used to judge the quality of a model estimate in an "automatic system identification" setting. This is due to the fact that the prediction error model estimate no longer corresponds to maximum likelihood estimate, since the underlying "system in the model set" assumption is routinely violated in the proposed identification framework.
- In Chap. 5, we presented a computationally-efficient technique to compute an (approximate) simulation response of a P-NARMAX model. The proposed tech-

nique takes into account the higher-order moments of the noise distribution, which would otherwise lead to bias in the simulation response of the model.

- In Chap. 6 we formulated the automatic system identification problem as a multi-criteria bi-level optimization problem with the help of TAG and Pareto-dominance ordering. This is a vital contribution as bi-level optimization problems have been studied in many research communities, opening the door to the possibility of employing other optimization techniques developed for this class of problems.
- In Chap. 6, we proposed an GP-based algorithm to solve the bi-level optimization problem, and analysed its asymptotic properties and computational complexity. We proved that, for certain ranges of crossover and mutation rates, and under suitable assumptions on the measured dataset and parameter estimation techniques, the Pareto-efficient models estimated by the proposed identification algorithm asymptotically converges to a subset of the true Pareto-efficient set of models.
- In Chap. 7, we studied the performance of the proposed identification method on a number of real and simulation examples. For case-studies where first principles knowledge was available, we demonstrated that the proposed identification method was able to infer the correct model structure from data. We also demonstrated the value of the estimated Pareto-front in terms of depicting the available trade-offs that can be made by the user. This greatly simplifies the problem of model selection for the user.

8.3 Recommendations

In this thesis, we made a number of choices to restrict the scope of research (see Sect. 1.4), for example, we worked primarily with IO discrete-time representations of dynamical systems. Furthermore, for the Wiener-Hammerstein benchmark and the weaving mill setup in Chap. 7, we reasoned that the restriction of scope adopted in this thesis may cause the poor performance of the proposed methodology. In the following, we recommend some possible extensions and pending research challenges that can be explored as a continuation of the research direction adopted in this thesis.

- In order to develop the proposed framework, we restricted the scope to IO model representations of SISO time-invariant dynamical systems without discrete jumps. However, the basic concepts used to develop the proposed identification framework are fairly general: TAG can be used to describe the structure of various classes and representations of parametric dynamical models, Pareto-dominance-based ordering can be used to incorporate multiple performance measures, and the proposed GP-based algorithm can be used to the solve the resulting bi-level optimization problem for automated system identification. These concepts can be extended to other classes of dynamical systems (e.g., LTV or hybrid systems) and other model representations.
- The results obtained for the Wiener-Hammerstein benchmark and the weaving machine suggest that, in some cases, other model representations may provide

a more suitable structure to describe the given system. As discussed earlier, the proposed TAG-based model set specifications can be extended to other model representations such as SS. The ability to estimate models in different representations would greatly enhance the scope of the proposed identification methodology.

- In some cases, it is more desirable to estimate compact models (e.g., models with fewer parameters or more structured models) that are easier to interpret for a user. In order to estimate compact models of dynamical systems, the concept of TAG can also be used to generate more structured model representations, such as decoupled SS models [3], decoupled NARX models [7] or LFR models [8]. The structure in these models may provide more insight into the dynamics of the true system.
- In Chap. 5, we proposed to use Pareto-dominance-based ordering of solutions to incorporate multiple performance measures. As the number of objectives increase, the probability of a candidate solution to be Pareto-dominant increases, and hence, that ratio of non-dominated solutions with respect to dominated solutions in the population increases. As a result, for large number of objectives functions (typically more than 4), the ability of MOEAs to explore the search space deteriorates. In order to address this issue, several approaches have been proposed in the EC literature, see Ishibuchi [6] for a review. For example, an ordering that is stricter than the Pareto-dominance-based ordering can be utilized. These ideas should be incorporated in the proposed identification framework to deal with with larger number of objectives.
- In Chap. 6, we analysed the asymptotic behaviour of the proposed identification algorithm using a Markov-chain-based analysis. While the asymptotic analysis reveals the importance of certain aspects of the identification algorithm, it does not shed light on the non-asymptotic behaviour of the proposed method. The empirical analysis presented in Chap. 7 illustrates the non-asymptotic behaviour of the system for a simple academic example. However, in order to obtain a complete understanding of the proposed algorithm, the non-asymptotic efficiency of the proposed algorithm should be studied.
- In this thesis, we restricted our focus to systems in open-loop configuration. In a closed-loop setting, the noise process is no longer independent of the input excitation. This introduces a bias in the parameter estimation algorithms and the simulation response computation technique proposed in this thesis. To deal with correlation of noise and input excitation, we need to adopt more advanced parameter estimation techniques such as the instrumental variables approach.
- Once the proposed methodology is extended for the task of closed-loop identification, we can also look to address the challenge of control-oriented identification. This challenge is a combination of the problems of (i) experiment design that is aligned with the control objective, and (ii) identification of *approximate* models for which the model errors minimally degrade the control objective (see Gevers [4]). The proposed identification framework may provide a novel approach to solving these problems due to (i) the estimation of Pareto-efficient models (rather than a single "optimal" model), which provides a new way to explore approximate models, and (ii) the ability to incorporate multiple-objectives, including measures for

the closed-loop performance of the estimated model. More research is required to explore these ideas.

- The proposed identification approach requires significant amount of computational time, especially for large dataset and complex models. A lot of ideas have been proposed in the literature to ease the computational complexity of such algorithms. One promising idea is to use data sparingly and adaptively, such that more data is used only when it brings about significant improvement in the parameter estimate (e.g., see Bongard and Lipson [1]). Another direction to explore is to make use of semantic-based variation operators to make the evolutionary search more efficient. For a survey of such methods, see Vanneschi et al. [9].
- Finally, in terms of practical developments, the identification framework developed in this thesis can be implemented as an accessible toolbox in which the user can simply select the desired model representation (e.g., IO or SS representation) and performance measures from a list of available options. The choice of model representation and performance measures would implicitly determine suitable parameter estimation tools that can be incorporated in the GP-based optimization algorithm. Such a toolbox would require the user to simply provide measured data and preferences. The capabilities of the proposed toolbox (i.e. available options for the user to choose from) can be expanded as the concept of TAG-based model set specification is extended to other model classes and representations.

References

1. Bongard J, Lipson H (2007) Automated reverse engineering of nonlinear dynamical systems. Proc Nat Acad Sci 104(24):9943–9948
2. Deb K, Pratap A, Agarwal S, Meyarivan T (2002) A fast and elitist multiobjective genetic algorithm: NSGA-II. IEEE Trans Evolution Comput 6(2):182–197
3. Decuyper J, Dreesen P, Schoukens J, Runacres MC, Tiels K (2019) Decoupling multivariate polynomials for nonlinear state-space models. IEEE Control Syst Lett
4. Gevers M (2005) Identification for control: from the early achievements to the revival of experiment design. Eur J Control 11:1–18
5. Hoai NX, McKay R (2001) A framework for tree adjunct grammar guided genetic programming. In: Proceedings of the Post-graduate ADFA Conference on Computer Science (PACCS'01), pp 93–99
6. Ishibuchi H, Tsukamoto N, Nojima Y (2008) Evolutionary many-objective optimization: a short review. In: IEEE Congress on Evolutionary Computation (IEEE World Congress on Computational Intelligence). IEEE, pp 2419–2426
7. Karami K, Westwick D, Schoukens J (2019) Identification of decoupled polynomial narx model using simulation error minimization. In: 2019 American Control Conference (ACC). IEEE, pp 4362–4367
8. Vanbeylen L (2013) Nonlinear IFR block-oriented model: potential benefits and improved, user-friendly identification method. IEEE Trans Instrum Measure 62(12):3374–3383
9. Vanneschi L, Castelli M, Silva S (2014) A survey of semantic methods in genetic programming. Genetic Program Evolvable Mach 15(2):195–214

About the Author

He received his Bachelors degree in Electrical and Electronic Engineering from VIT University in Vellore, India in 2013. He pursued his graduate studies in the Department of Electrical Engineering at Eindhoven University of Technology. He received his Masters degree in 2015. The title of his Master's thesis was "Model-based Fault Diagnosis for Inkjet Print-heads".

In January 2016, he started working as a Ph.D. candidate in Control Systems group in the Department of Electrical Engineering at Eindhoven University of Technology, under the supervision of Dr. Roland Tóth. He worked on the project titled "Control And Data-driven modelling Using SYmbolic regression (CADUSY)" supported by the NWO grant 13852. The focus of his research is on automating the process of data-driven modelling of dynamical systems using Evolutionary Algorithms (EA). Since 2018, he has also been co-supervised by Dr. Maarten Schoukens.

In July 2020, he started working at VDL ETG Technology and Development in Eindhoven, The Netherlands, focusing on the development of predictive maintenance and digital twin technologies for high-end semiconductor equipments.

D. Khandelwal, *Automating Data-Driven Modelling of Dynamical Systems*, Springer Theses, https://doi.org/10.1007/978-3-030-90343-5

The manufacturer's authorised representative in the EU is Springer Nature Customer Service Centre GmbH, Europaplatz 3, 69115 Heidelberg, Germany. If you have any concerns regarding our products, please contact ProductSafety@springernature.com

Printed and bound by CPI Group (UK) Ltd, Croydon, CR0 4YY
30/09/2026
02225332-0005